Telecommunications and Energy in Systemic Transformation

Springer
Berlin
Heidelberg
New York
Barcelona
Budapest
Hong Kong
London
Milan
Paris
Santa Clara
Singapore
Tokyo

Paul J. J. Welfens · George Yarrow (Eds.)

Telecommunications and Energy in Systemic Transformation

International Dynamics, Deregulation and Adjustment in Network Industries

With 39 Figures

Springer

Prof. Dr. Paul J. J. Welfens
Universität of Potsdam und
Europäisches Institut für
Internationale Wirtschaftsbeziehungen (EIIW)
August-Bebel-Straße 89
14482 Potsdam, Germany

Prof. George Yarrow
Oxford University
Regulatory Policy Research Centre
Hertford College
31–33 Westgate
Oxford OX1 1NZ, England

Cataloging-in-Publication Data applied for

Die Deutsche Bibliothek - CIP-Einheitsaufnahme

Telecommunications and energy in systemic transformation : international dynamics, deregulation and adjustment in network industries ; with 40 tables / Paul J. J. Welfens ; George Yarrow (ed.). - Berlin ; Heidelberg ; New York ; Barcelona ; Budapest ; Hong Kong ; London ; Milan ; Paris ; Santa Clara ; Singapore ; Tokyo : Springer, 1997

ISBN-13: 978-3-642-64441-2   e-ISBN-13: 978-3-642-60519-2

DOI: 10.1007/978-3-642-60519-2

NE: Welfens, Paul J. J. [Hrsg.]

ISBN-13: 978-3-642-64441-2 Springer-Verlag Berlin Heidelberg New York

Softcover reprint of the hardcover 1st edition 1997

SPIN 10518746 42/2202 - 5 4 3 2 1 0 - Printed on acid-free paper

Introduction 1
Paul J.J. Welfens and George Yarrow

A. Telecommunications in Western Europe: Liberalization, Technological Dynamics and Regulatory Developments 9
Paul J.J. Welfens and Cornelius Graack

1. Introduction 9
2. Liberalization and Market Expansion in Telecommunications 12
   2.1 Global Forces in Telecoms Liberalization 19
   2.2 Privatization and Deregulation in Western Europe 22
   2.3 Politico-economic Deregulation Pressures 26
3. Technological Dynamics 30
   3.1 Digitization 31
   3.2 Integrated Services Digital Network 33
   3.3 Fibre Optics, Fibre to the Home and Optical Networks 35
   3.4 Mobile Communications 38
4. Regulatory Developments 40
   4.1 Regulatory Developments on the EC Level 41
   4.2 National Regulatory Frameworks: Developments and Experiences 46
      4.2.1 Telecommunications Equipment 47
      4.2.2 Value-added Services 49
      4.2.3 Infrastructure 52
5. Prospects and Consequences for Central and Eastern Europe 72

Appendix 78

B. Telecommunications in Systemic Transformation: Theoretical Issues and Policy Options 85
Paul J.J. Welfens

1. Introduction 85
2. Points of Departure in Eastern Europe 90
   2.1 Structure of the Telecoms Industry in an East-West Perspective 94
   2.2 Telecoms Industry as a Strategic Industry for Systemic Transition 97

3. Theoretical Aspects of the Telecoms Industry 99
3.1 Some Problems of Uniform Subscriber Pricing 99
3.2 Competition, Natural Monopoly and Economies of Scope 102
3.3 External Effects of Telecoms Network Expansion 109
3.4 Oligopolistic Telecoms Industries 111
3.5 Regulation of the Telecoms Industry 113
4. Policy Options for a Modern Telecoms Policy 119
4.1 Privatization of the Telecoms Industry 119
4.2 How to Accelerate Telephone Penetration 122
4.3 Special Aspects of Mobile Telecommunications 123
4.4 Innovative Use of Telecoms 125
4.5 Capital Market Aspects 125
4.6 Social Problems 126
4.7 Principles of Efficient Regulatory Policy 127
4.8 Perspectives for Poland 130
Appendix 138

C. Modernization of Telecommunications in the Former GDR 143
Stephen Economides

1. Former GDR Telecommunications System 143
2. German Unification and Telecoms Modernization 146

D. Recent Evolution of Telecommunications in the Region of Central Europe 153
Hanna Kontkiewicz-Chachulska

1. Introduction 153
2. Telecommunication Administration 155
3. Legislative Framework 156
4. Development Strategy 161
5. Network Development 163
6. Private Local Operators 169
7. Privatization of National Operators 178
8. Tariffs 188
9. Cellular Telephony 190
10. Conclusions 192

E. Modernizing Telecommunications in Central and Eastern Europe: A Business Perspective **197**
Marec Béla Steffens

1. Structural Reasons for the Current Eastern Demand for Telecommunications 197
2. Pumping Fresh Money into Old Structures is Not Enough 199
3. How to Attract Investors' Interest 201
4. Special Networks (Overlay, Mobile, Wireless Local Loop) 203
5. Regulators' Possibilities and Contradictions 205
6. Siemens's Role as Supplier and Local Manufacturer 207

F. Telecommunications Reform in the United States: Promises and Pitfalls **213**
Marius Schwartz

1. Need for Pro-competitive Regulatory Reform, and the 1996 Telecommunications Act 213
   1.1 Reassessing Government Intervention in So-called Natural Monopolies 214
   1.2 Challenges to Regulatory Reform 216
2. Increasing the Reliance on Competition and Markets 219
   2.1 Relaxing Legal Entry Barriers 219
   2.2 Liberalizing Restrictions on Foreign Investment 221
   2.3 Making Better Use of the Radio Spectrum 222
3. Role for Government: Media Cross-ownership, Cable Rate Decontrol, Telephone-cable Consolidations 225
   3.1 Media Cross-ownership 225
   3.2 Cable Rate Decontrol 226
   3.3 Cable-telephone Consolidations 226
4. Role for Government: Telephone Competition 228
   4.1 Background 228
      4.1.1 Local and Long-distance Service 228
      4.1.2 Why the 1984 AT&T Breakup: Regulatory Evasion via Favoring Unregulated Affiliates Distorts Competition and is Inefficient 230
      4.1.3 Unbundling Regulated Bottleneck Monopolies from Potentially Competitive Segments: Alternative Approaches 233
   4.2 Bell Companies' Entry into Long-distance Service 236
      4.2.1 Arguments for Bell's Entry: Costs of Structural Separation 237
      4.2.2 Arguments against Bell's Entry: Preventing Access Discrimination 239

4.2.3 Competitive Safeguards in the 1996 Act 241
4.3 Promoting Local Competition 244
4.3.1 Local Telephone Networks and Types of Entrants 244
4.3.2 Benefits of Resale or Partial-facilities Entry 246
4.3.3 Requirements of the 1996 Act: Interconnection and Unbundling 247
5. Reducing Cross-subsidies and Revamping Universal Service 249
5.1 Cross-subsidies, Common Costs, and Economic Efficiency 251
5.2 Distortions in Current System and Tension with Competition 253
5.3 Reforming Universal Service 255
5.4 The 1996 Act 258
6. Conclusion 260

G. Regulation and Tariff Policies in the Energy and Telecommunications Sectors in the Transition Countries: The Case of the Czech Republic 271
Zdenêk Hrubý

1. Introduction 271
2. Features of the Energy Sector 271
3. Price Regulation and Tariffs in the Energy Sector 274
4. Telecommunications Sector Features 279
5. Regulation and Tariffs in the Telecommunications Sector 282
6. Situation and Approaches in CEC and EU 286
7. Conclusions 288

H. The Process of Systemic Transformation and Reforms in the Energy Sector: The Regulatory Issues of the Reforms 295
Andrzej Szablewski

1. Introduction 295
2. Systemic Transformation and Reforms in the Energy Industry 295
3. The Model of an Independent Regulation 298
3.1 The Concept of Reforms in the Energy Sector 299
3.2. The Justification for Independent Regulation 300
4. Controversies Surrounding the Model of Independent Regulation 302
5. Regulatory Authority and Antimonopoly Office 304

I. Restructuring and Regulatory Reform in the Polish Energy Sector: an Assessment **307**
George Yarrow

1. Introduction 307
2. The Energy Sector in Poland 311
3. Policy during the Initial Phase of Economic Transition 314
4. Industrial Restructuring 317
   4.1 Restructuring in Gas 319
   4.2 Restructuring in Electricity 321
   4.3 Conclusions on Structural Decisions 324
5. Regulation 326
   5.1 The Problem of Delegation 326
   5.2 The Trade-offs 327
   5.3 The Consequences of Delay 329
6. Conclusions 330

J. Energy Law Developments in the European Union and Poland **335**
Anita Ronne

1. Introduction 335
2. Energy Policy Developments 335
   2.1 European Union 335
   2.2 Poland 337
3. New Legislation 338
   3.1 European Union 338
      3.1.1 Primary EU Law 338
      3.1.2 Secondary Legislation 340
   3.2 Poland 343
4. Conclusion 347

K. Institutional Choice or Muddling Through - Problems of Transformation in the Electricity Supply Industry in Russia **351**
Petra Opitz and Wolfgang Pfaffenberger

1. Institutional Choice in the Phase of Transformation 351
2. Models for Network-oriented Energy Sectors in the West 353
   2.1 Special Characteristics of the ESI 353
   2.2 Elements of Competition 354
   2.3 Transformation in Different Political Systems 358
      2.3.1 Germany 358

2.3.2 USA 361
2.3.3 European Union 362
2.3.4 Great Britain 364
3. The ESI in Russia 365
3.1 Basic Energy Situation 365
3.2 State of the ESI 368
3.3 State of the Institutional Reform of the Sector 371
4. Analysis of the Reform Process 376
4.1 General Problems of the Present Reform 376
4.2 Regional Interests - Siberia 383
4.3 The Concept of Privatisation in the Discussion 385
5. Cooperation with Western Europe 388

L. Co-operation in Energy Policies: European Union (EU) - Eastern Europe **393**
Peter Palinkas

1. Introduction 393
2. East-West Co-operation in the Electricity and Gas Sector 395
2.1 The Electricity Sector 395
2.2 The Gas Sector 404
2.2.1 Natural Gas from the Former Soviet Union for Western Europe 404
2.2.2 The Outlook: Natural Gas from the CIS in the Context of Meeting Future Requirements 405
2.2.3 New Areas of Co-operation 406
2.2.4 Stable Framework Conditions as a Prerequisite for East-West-Co-operation 407
2.2.5 Examples of Co-operation in the Gas Sector 408
3. Transeuropean Networks (TEN) 411
3.1 Short Overview 411
3.2 Gas Pipeline between Russia and the EU 411
3.3 The East-West High Power Transmission System as an Essential Part of the Baltic Ring 412
3.3.1 Introduction 412
3.3.2 Rough Investment Evaluation for the Baltic Ring 413
3.3.3 East-West High Power Transmission System 413
3.3.4 Conclusions 415
4. Summary 418

M. Regulation and Systemic Transformation **423**
Piotr Jasinski

1. Introduction 423
2. The Process of Systemic Transformation and its Constitutive Characteristics 423
3. Regulation in Market Economies 425
   3.1 The Rule of Law and the Law of Contract as Regulation: an Institutional Approach 426
   3.2 Regulation and Competition 428
4. Regulation in the Strict Sense of the Word 430
   4.1 Regulation as an Institution of Market Economies 430
      4.1.1 Regulation as an Actually Existing Institution in Market Economies 431
      4.1.2 Why Regulation? Regulation and Economic Efficiency 433
   4.2 Regulation of Public Utilities 434
      4.2.1 Public Ownership as Regulation 434
      4.2.2 Regulation and Private Sector 437
         4.2.2.1 Regulation of Private Public Utilities (Network Industries) 437
         4.2.2.2 Regulation and Privatisation 440
            4.2.2.2.1 Procompetitive Restructuring 441
            4.2.2.2.2 Regulatory Agencies and Procompetitive Regulation 445
5. Regulation on the Way Back towards Capitalism 449
   5.1 The Rule of Law and Competition 450
   5.2 Public Utilities and Systemic Transformation 455
      5.2.1 Regulation within the Public Sector: Institutional Reform and Restructuring 456
      5.2.2 Regulation and Privatisation 459
6. Regulation as an Element of the System Being Created: Some Normative Remarks and Suggestions 464

List of Tables 477
List of Figures 479
List of Contributors 481
Index 483

Introduction

*Paul J.J. Welfens and George Yarrow**

Systemic transition in Central and Eastern Europe has raised many new challenges for economic policy, including problems of encouraging adjustments on the supply sides of rapidly evolving markets. The supply-side effects of public policy are of particular significance in respect of the development of infrastructure investment. In part this is because industries such as telecommunications and energy are, for a variety of reasons, liable to be particularly prone to government regulation. In part it is because of the size and significance of the sectors in modern market economies. Both industries are capital intensive, and the expansion and modernization of infrastructures can have strong effects on, and can be strongly affected by, capital market conditions. Since both information and energy are important inputs into many other industries, the innovativeness and international competitiveness of large parts of the economy can be influenced by the performance of the telecoms and energy sectors, which in turn tends to be heavily affected by the government's regulatory stance. Investment in infrastructure, whether to expand systems or to upgrade their quality, has the effect of raising the marginal productivity of private capital in other sectors. Consequently, regulatory reforms and changes in industrial structures in telecoms and energy can have wide ranging effects on aggregate investment in an economy, and hence on its rate of growth. Issues of privatization, deregulation, and regulatory reform are, therefore, of strategic importance for countries engaged in systemic transition.

---

* This book is one of the products of an ACE-funded project on utilities' policies in systemic transformation. We are extremely grateful to the EC Commission for financial support and to Julia Gibert, Cornelius Graack, Piotr Jasinski, Cathryn Ross and Alison Sinclair for editorial assistance.

In the telecoms sector, excess demand for services coupled with poor quality and poor technical reliability was the typical position in socialist Central and Eastern Europe. With the switch to a market economy (and the much greater interchanges of information that this implies) communications have become more crucial for the well-being of households and for the economic performance of firms, industries, and state bureaucracies. In the relatively poor former CMEA countries the key question now is: how can telecommunications networks be modernized quickly and at least cost? And in tackling this question an essential first step is to recognise that new technologies in both the mobile and fixed-link segments serve to undermine the industry's natural monopoly characteristics.

Indeed, in the European Union and in many other parts of the world, the liberalization and deregulation of telecoms networks now ranks among the top priorities of policy, not least because the significance of telecoms competition for international competitiveness has been widely recognised. For example, the supply of telecoms equipment is a high technology industry whose development depends substantially on the competitive environment faced by network operators, and, as has already been noted, performance in telecoms will significantly affect the performances of all those industries that make heavy use of communications services. By 1998 the state monopolies of EU Member States will have come to an end, and in the UK and Sweden competition is already well advanced, as it is in the USA, Australia, and New Zealand.

It is now increasingly accepted that the introduction of competition into the telecoms sector can be expected to bring major economic benefits to the countries of Western Europe (see Chapter A: WELFENS AND GRAACK). In Eastern Europe the legacy of socialist systems is a severe burden but the potential benefits of bold liberalization moves are consequently even more dramatic. Recent theoretical analysis concerning positive externalities in the expansion phases of networks indicates the desirability of very heavy investment in telecoms in the transition economies. In particular, the optimum level of provision of telecoms services can be expected to be much greater than the short-run, profit-maximization strategy of a monopolistic network operator would suggest (B: WELFENS).

Investment has indeed been massive in the former-GDR, where the rapid elevation of Eastern Germany to telecoms penetration rates comparable with those of Western Germany was a top political priority (C: ECONOMIDES). But, whereas clear policy objectives were established early in the former-GDR, this first stage of the policy process has been more protracted in Poland. Complex policy options have been extensively debated but progress has been slower than originally hoped, not least because of changing political priorities. It is also the case that a number of desirable features of a reformed regulatory regime -- such as transparency, accountability, and flexibility -- appear to be difficult to reconcile with the regulatory structures that currently exist in Poland and other former-CMEA countries. These political and institutional factors add greatly to the technical problems that are inherent in any attempt to move to a market structure that is characterised by competition among several operators. Chapter D (KONTKIEWICZ-CHACHULSKA) discusses developments in privatization, competition and the institutional setting in Poland, and compares the Polish experience with that of other Visegrad countries. The delays in developing appropriate regulatory strategies have in turn served to hold back investment, since it is clear that international businesses are not attracted by the proposition that they should pump large amounts of resources into telecoms sectors that are still governed by the old structures (E: STEFFENS).

The developments in the dynamic US market are described and analyzed in chapter F (SCHWARTZ). Since 1984 the US telecoms market has been liberalized, while market volume has expanded. The traditional distinction between monopolistic local telephony and the long distance telecoms is disappearing. There are reasons to expect that the new US deregulation will further intensify competition in the telecoms market. With more players and newly contested markets the complexity of private contracts among the telecoms operators is increasing: whether stable solutions - e.g. for interconnection problems - can be established is an open question.

The policy problems arising in the energy sectors of the transition economies have both similarities with and differences from those encountered in telecoms. For example, both sectors have been characterised by extensive cross-subsidisation: price structures have borne little relationship to cost structures. In consequence, tariff reform can be expected to feature as a high priority on policy agendas in the transition stage in both sectors (G: HRUBY).

Like telecoms, the energy sectors of transition economies have large investment requirements, although the reasons for the size of the demand are somewhat different. In particular, there is no pressing call for a major expansion in the infrastructure networks and in output: for example, energy is currently used in highly inefficient ways and the potential for improvements (which, among other ways, can be stimulated by tariff reform) means that a considerable increase in GDPs could be sustained without increasing energy output. Rather, the demand for new investment arises from the desirability of modernising much of the existing capital stock so as to (i) improve productive efficiency and (ii) reduce the negative environmental impact of current technologies.

At the moment it is also the case that technological progress is much slower in the energy industries than in telecoms. This does not mean, however, that there are few potential gains from introducing competition: a crucially important characteristic of the energy industries is that a large fraction of value added is accounted for by activities that are <u>not</u> natural monopolies. For example, the production and supply (or trading) of gas and electric power -- as distinct from the transmission and distribution of gas and electric power -- are potentially competitive activities. The future performance of these industries will depend, therefore, on a mix of competitive and regulatory influences, although since the development of competition will itself depend upon the conduct of public policy, it is regulation that will be the decisive factor in the transition stage.

A number of CEE countries are now wrestling with the awkward problems of industrial restructuring and regulatory reform in the energy industries. In Poland, for example, initial radical plans to restructure the electricity supply industry and the regulatory system have encountered powerful political obstacles and have, as a consequence, been both delayed and watered down (H: SZABLEWSKI). A contributory factor to this outcome may have been an excessive concern with the issue of the optimal market structure for electricity generation and the insufficient priority given to regulatory reform. Ironically, following the collapse of socialism in Poland there was already a lesser degree of vertical and horizontal integration of businesses than in most Member States of the EU, so that market structure need not have posed a significant obstacle to rapid liberalization (I: YARROW).

Whether the initial loss of momentum toward reform in electricity (and to a lesser extent in gas) can be recovered in Poland will now depend heavily upon energy policy developments in the EU, which Poland is seeking to join at an early date (J: RONNE). In contrast, the objective of harmonising policies with the EU is not a significant influence on reform processes further to the east, including in Russia where energy problems occur on a massive scale as a result of factors such as the large indigenous fuel reserves and a substantial reliance on power from nuclear plants of doubtful reliability. In principle, Russia could opt for any one of a number of current European and non-European (e.g. American) policy approaches for its energy sector, although whether political constraints will allow the development of a clear-cut strategy is still an open question (K: OPITZ AND PFAFFENBERGER). With so many changes going on in Eastern Europe on the one hand and, on the other hand new industrial dynamics in the EU single market problems of East-West cooperation in Energy are quite important (L: PALINKAS). The European Energy Charter envisages new opportunities for trade in energy in Europe and the modernization of the East European energy sector. Liberalization of EU energy markets is, however, crucial to the success of this important field of political cooperation in Europe.

More generally, since both sectors are likely to be subjected to industry-specific regulation for the foreseeable future, energy and telecoms are subject to common influences deriving from the regulatory process. Given the importance of political forces in regulation, the policy process will inevitably show some variations from one country to another, although even here there are significant common factors. One is the desire of many Central and Eastern European (CEE) countries to join the European Union, which means that the EU's policies frequently serve as model to which transition economies seek to converge. A second is the fact -- which distinguishes the transition economies from others -- that the CEE countries are all undergoing a broadly similar economic and political systemic transformation. This gives rise to a particular set of regulatory issues and problems (M: JASINSKI).

The modernization of the telecommunications sector will be crucial for economic opening up, the expansion of the services sector and for the move towards the information society. Given Eastern Europe's lag in the modernization drive of the 1980s bold and rather radical moves are required for successful catching-up with OECD countries. Fast liberalization of mobile and fixed-link telephony - including acceptance of foreign investors - should rank high on the agenda in post-socialist countries. Clearly, specific transformation problems, e.g. those associated with privatization and the creation and function of capital markets, should be taken into account.

The papers in this volume represent the first results of a continuing international research project and are therefore still preliminary in nature. Nevertheless, we are confident that some of the conclusions are robust in the sense that they lend themselves to generalization across CEE countries. Among these, the following merit some emphasis:

- The introduction of competition into network industries has a crucially important role to play in facilitating rapid and efficient network modernisation.
- In liberalized regimes there will be increasingly strong interactions among the different network industries, for example by cross-entry from one sector to another (such as electricity companies entering the telecoms market by routing fibre optic cables along their existing transmission and distribution facilities).
- Moves toward liberalization and deregulation can be expected to encounter strong (political) resistance from entrenched interests who believe they will lose out as a consequence.
- The relatively slow progress in achieving liberalization of network industries in most EU Member States is a factor which itself tends to delay reforms in CEE countries, at a significant cost to the latter.
- The benefits of transparent regulation by a regulatory body with a substantial degree of political independence are large but, since they tend to be realised only in the longer term, they tend to be undervalued by political decision makers with relatively short time horizons.

These last three points suggest that, notwithstanding the accumulating evidence of the benefits of competition in network industries in those countries where extensive liberalization has already occurred, regulatory reform and the promotion of competition will remain a formidable challenge in the transition economies of Central and Eastern Europe.

A. Telecommunications in Western Europe: Liberalization, Technological Dynamics and Regulatory Developments

Paul J.J. Welfens and Cornelius Graack

1. Introduction

Telecommunications is a dynamic growth-field in OECD countries. In the European Union (EU) the share of the telecoms sector in GNP is expected to increase from 3% in 1988 to about 6% in 2000. The telecoms sector represents some ECU 285 bill. worldwide, of which 84 bill. are accounted for by markets in the EU. The equipment market amounted to ECU 82 bill. worldwide and 26 bill. in the Community. Annual growth rates of 6% in the telecoms service sector and 4% in the equipment sector are expected until the year 2000 (EC COMMISSION, 1993a). Modern telecommunication allows the increase of decentralization of firms, facilitates cooperation across space and could help to reduce traffic and energy use via telework, which links households and specialized local centers to company headquarters. Moreover, technological progress could be accelerated by easier scientific communication - a perspective which has stimulated the vision of a global information highway system in the US and the establishment of trans-European information networks in the EU. In 1993 the EC Commission estimated that some ECU 150 bill. of investments are needed over a decade for building trans-European telecoms networks; the first stage between 1994-99 will require ECU 67 bill. The integration of nationally fragmented telecoms markets in the EU thus requires huge investments, which can best be financed by private capital; they call for a national deregulation of the telecoms sector as well as a European framework program with certain standards and uncertainty-reducing rules of the game. Bold visions of the EU and national ambitions to improve international competitiveness of firms via better information networks and improved communication ultimately raise the question of demonopolization and deregulation in the telecoms industry.

Telecommunication consists of infrastructure (line network and exchange switches), the direct use of telephones for voice communications and the exchange or transmission of data and video. Moreover, there is an increasing significance of value-added services offered by specialized providers using the telecoms network:

e.g. for banking services, TV program emission, or the exchange of research results within a network of university or industry R&D labs. The telecoms network is important to all households and industries. The development of a modern telecoms sector is a crucial ingredient for international competitiveness and growth, particularly in the industrial sector. In principle, competing private networks could provide customer-tailored networks with access to differentiated services, leading to different prices paid by various industrial demanders. A telephone is considered a meritorial good for private households; governments in Western Europe have long favored state-owned telecoms networks with uniform prices. Uniform subscriber charges and uniform user prices for domestic phone calls were a marked contrast to discriminatory prices for international phone calls. Extra profits from international phone calls and supra-normal profits from peak-time phone calls allowed PTOs to subsidize local phone calls and to grant cheap (uniform) subscriber rates to people in the countryside. These uniform pricing policies under a regulated telecoms monopoly are also a marked contrast to almost all developments in normal markets, where there is a broad variety of services and products sold at divergent prices. Greater product diversity allows greater satisfaction of diverse consumer preferences. Price differentiation sometimes creates bigger overall markets, thus improving welfare. Appropriate regulation could therefore help to build an efficient and dynamic telecoms industry.

Until the end of the 1970s, the telecommunications sector consisted of highly regulated and protected national markets, with state owned monopolies in most cases. Since the beginning of the 1980s, however, a trend towards the liberalization of telecommunications markets can be observed, starting with the divestiture of AT&T and followed by developments in the United Kingdom and Japan. Today, liberalization has become a worldwide phenomenon.[1]

In the European Union, regulatory decisions are made on two different levels: on the European Community level and on the national level of EC member states. In connection with the creation of the single European market, a growing influence of EC legislation on national telecommunications decisions and regulatory framework can be observed. Until the first half of the 1980s, telecommunications policy and regulatory decisions were exclusively made on national levels; European coordination or cooperation rarely existed, except as technical interconnection rules. Since then, harmonization and liberalization of the European telecommunications market has been promoted by EC legislation, partly against the opposition of national governments. The necessity of reforms due to international and European developments, technological changes and national requirements, however, has led to an agreement concerning the creation of an open European telecommunications market. Today, progress in liberalization in certain fields like telecommunications equipment, value-added services and transmission of data is considerable. The last step towards full liberalization in telecommunications is expected to be made in 1998, when voice telephony services will formally be exposed to competition. Until then, a completely new regulatory framework on the European and national level has to be created since competition requires other regulatory means than the regulation of a monopoly.

A single network provider still exists in most European countries, with some considerable exceptions like the UK and Sweden. As a result of politico-social goals, those monopolists are obliged to offer uniform services at average prices. This causes price distortion and massive cross subsidization, which cause welfare losses in comparison to market-oriented pricing. The latter is the expected result of competition; therefore, the introduction of a more competitive environment is required.

If networks are exposed to competition, a period of transition in regulatory bodies may be necessary to create a new regulatory framework including interconnection and service obligations rules; the incumbent network provider will use this period to rebalance prices.

The developments of the Central and Eastern European telecommunications markets must be viewed in the context of European developments. In those ex-CMEA countries intending to become members of the European Union, the new regulatory framework for telecommunications has not only to anticipate present and future technological changes, but should also adapt to EC legislation.

Liberalization and market expansion in telecommunications are described in Section 2. Since technological changes play a major role in the discussion of the liberalization of the telecommunications markets, a review of technological developments is given in Section 3. Section 4 is devoted to regulatory developments. The first part is a description of those EC regulatory policies, which directly affected the regulatory frameworks of EC member states. The development of telecommunications markets on a national level is hence the subject of the second part of Section 4. Perspectives on and consequences for Central and Eastern European countries resulting from European developments and experiences (with special regard for the situation of Polish telecommunications markets) are discussed in Section 5.

2. Liberalization and Market Expansion in Telecommunications

The telecoms business in the OECD rapidly expanded because of technological advances and the growth of the computer industry, the break-up of the US telecoms giant AT&T in 1982 (which accelerated competition in North America), the liberalization of telecoms in OECD countries and the growth of multinational companies, which typically require extensive intra-company communication across countries.

The telecoms industries are very important for economic development and prosperity. Some basic data for global and markets of the EU (European Union/EC) is of special interest (EC COMMISSION, 1993b): The world market for information services reached some ECU 500 bill. in 1991, of which telecoms services amounted to ECU 330 bill. and software and computing services ECU 170 bill. If equipment for information services is included, the global market for information amounted to between ECU 900 and 1000 bill. at the beginning of the 1990s. More than 50% of gross domestic product is in one way or another influenced by the use and availability of information technologies. In electronics, Europe's trade balance with the rest of the world recorded a deficit of ECU 30 bill.; the US and Japan are the two main supplier countries for the EU. The important EU-US relationship in telecoms services is characterized by a structural surplus of some ECU 300 mill. which is a considerable amount since US suppliers operate in EU markets through local subsidiaries and associated firms. National traffic amounted to more than 90% of total traffic in the EU in the early 1990s, leaving some 10% for international, of which intra-EU traffic is only one part. Comparing this to the US, where more than 1/3 of overall telecoms services revenues were from long distance traffic, the EU traffic is still dominantly regional. While deregulation has opened up mobile communication services and value-added services for competition, most EU countries still have a national main operator accounting for roughly 95% of total revenues. Intra-Community services were impaired by high tariffs and technical problems; the share of network operator's receipts from intra-Community traffic only reached between 3% and 14% (HENRY, 1993, p.60). Given the rising potential of intra-Community traffic, network operators face few adverse effects from deregulation in an environment of high growth and still limited competition.

The EC's liberalization of the telecoms sector started with the Green Paper on the Development of the Common Market for Telecommunications Services and Equipment (June 1987). It aims (1) at achieving integrated European networks with easy connections between suppliers and users of telecoms services throughout the Community, and (2) at establishing broad varieties of services produced at least costs, which should be particularly valuable in the context of the single EC market. National public networks are, however, expected to continue providing universal service. Another major Commission directive (May 1988) opened up markets for terminal equipment by 1993. The single market competition in the equipment industry will hence increase, while the size of the single market should facilitate the exploitation of static and dynamic economies of scale as well as standardization. Two further Commission Directives (June 1990, September 1990) established (1) an EC internal market for telecoms services through the implementation of open network provision and (2) opened up infrastructure procurement to all EC suppliers. The open network provision basically requires standardized interfaces with which service providers can connect to networks. Access to networks should be cost-related and under non-discriminatory terms. In Spring 1994 the EU Commission published a Green Paper on mobile telecommunications.

An EC telecoms standardization body was created in 1987, which can make proposals for the Community. These measures were seen as elements of EC competition policies, while the aim of improving the competitiveness of European equipment producers in outside markets was also realized. Despite information-related Community R&D programs such as ESPRIT (1984) and RACE (1985) - with prominent indirect participation of US and Japanese firms - the US and Japan improved their respective global market positions. The share of EC telecom industries in global exports fell by 22 percentage points in the 1980s, while the US and Japan increased their respective shares by 5 and 11 points. Germany lost its leading export position and with a share of 9% (compared to 16% in 1980) ranked only third behind the US and Japan (SCHNÖRING, 1992, p. 22).

The information industry consists of the software industry, service providers, network operators and the hardware industry. While the US (AT&T) and Canada (Northern Telecom) are vertically integrated industries with high barriers to entry, European industries are less integrated and relatively open to outside investors. Western Europe has particular national relations for network operators, which are traditionally state-dominated and state-owned PTOs. Highly aggregated data for the US and Canada suggest that overall network operation is characterized by economies of scale. However, a strong critique by KNIEPS (1993) emphasizes the problem of data overaggregation and of some theoretical problems with monopoly operators; for adequate regulation policies the relevant subset with sub-additive costs has to be identified, and there are major differences in local network operation and in the operation of the long distance trunk network. Economies of density impair the functioning of competition at the local level; marginal costs fall as subscribers in the same neighborhood all use the same local distribution system causing sub-additivity and natural monopoly to be relevant. Through the progress of cellular communication and satellite technology, however, corporate or groups of residential users find it cheaper to by-pass the local network. Although the trunk network might show some economies of scale for a given demand in a country, there is no need to not allow the lease of lines and resale of transmission capacity. Leasing lines allows the building of group-specific networks (e.g., for banks or universities), which could then apply fast and customer-tailored communication protocols at increased transmission speed. Moreover, the resale of lines helps optimally to exploit a given line capacity, but indirectly also creates competition for the network operator. Mobile network operators can lease lines from the main network operator. Here, a strategic question concerning the interconnection charge to the terrestrial network arises. Non-discrimination requires interconnection charges to be equal for all domestic and foreign companies. The structure of the network operation will generally require some interconnection of the main terrestrial network and mobile network elements (Fig. A1). If the main network operator has a mobile communications subsidiary, interconnection charges for other mobile service operators might be raised artificially, while interconnection points could be chosen in a discriminatory way with respect to the private mobile competition. Moreover, the routing of telephone calls could be done in such a way as to impair the quality of the private competitor. If competition in the telecoms network business is to be efficient, there must be generally applied quality standards, interconnection charges and access rules for all mobile network partners.

Fig. A1: Structure of Telecoms Network

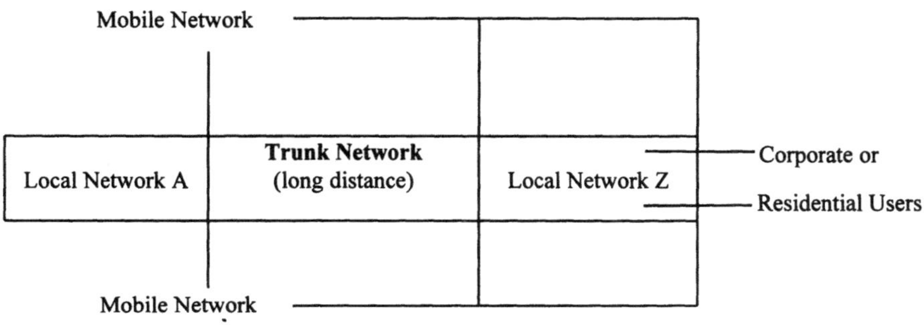

Exploiting static economies of scale under monopoly means the move from A to B on a given marginal costs curve. If - as can be assumed - technological progress is slower in a monopoly situation than under competition, the costs savings by the monopolistic move from A to B are not necessarily the optimal resource allocation. With a faster downward shift of the marginal costs curve society will reap the benefit from the move from A to C on the new marginal cost curve k'_1; static economies of scale are not optimally exploited here, but dynamic gains from competition could be the more important effect. Moreover, a monopoly network operator could weaken the competitiveness of the equipment industry so that the benefits from exploiting economies of scale are achieved at the costs of excessive hardware prices. In a true natural monopoly industry competition would duplicate investment and inflate call charges, however, the potential costs saving from static scale economies under monopoly could easily be offset by X-inefficiencies and non-optimal use of network capacity.

Fig. A2a: Network Operation - Role of Demand Shifts over Time

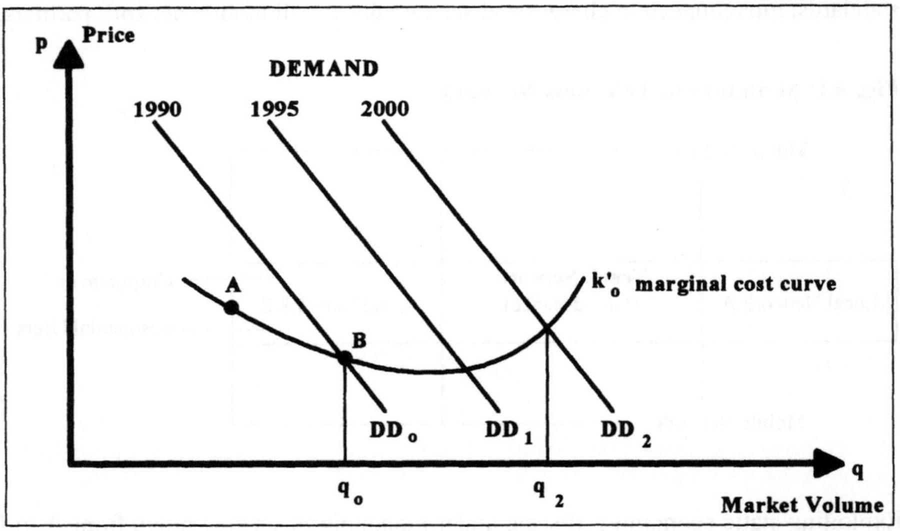

Fig. A2b: Shift of Cost Curve under Competition vs. Static Scale Economies

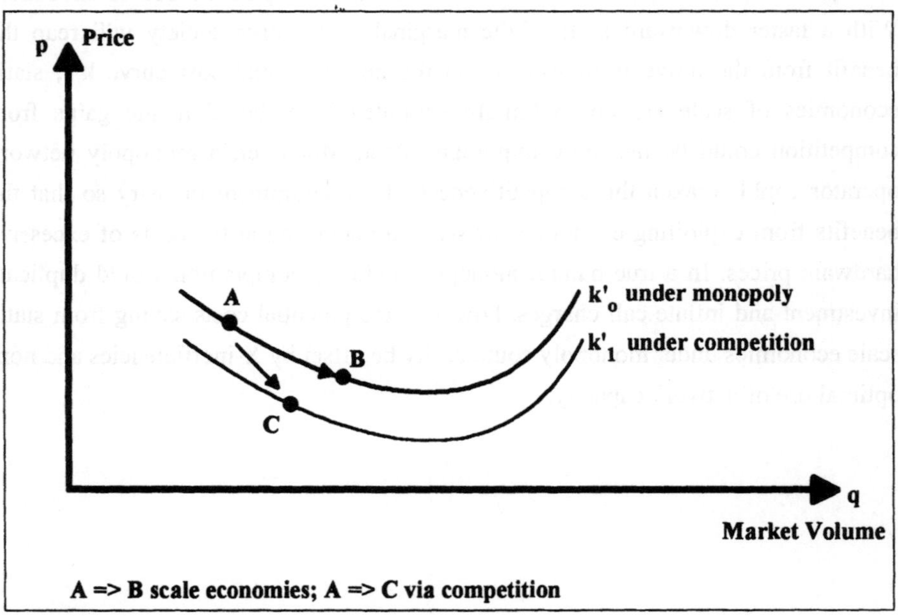

The argument of a monopoly arguing that two cables connecting city A and B are a waste of resources associated with competition is distracting attention from the problem that one cable whose capacity utilization rate is much under 50% is an equal waste of resources - and even worse, competition would offer more diversified, cheaper and more customer-oriented network access and more useful services. The real issue is network access and technical compatibility of network operation, and this could indeed be achieved by a pure network company which would have to offer open access to potential network users. In Western Europe for historical reasons state telecoms network operators are also offering (universal) services, but compared to the US and Canada there was no vertical integration with the equipment industry; a partial exception in this respect was Italy where government influence through state holdings was important until the early 1990s.

Fig. A3: Prices under Monopoly, Regulation and Competition

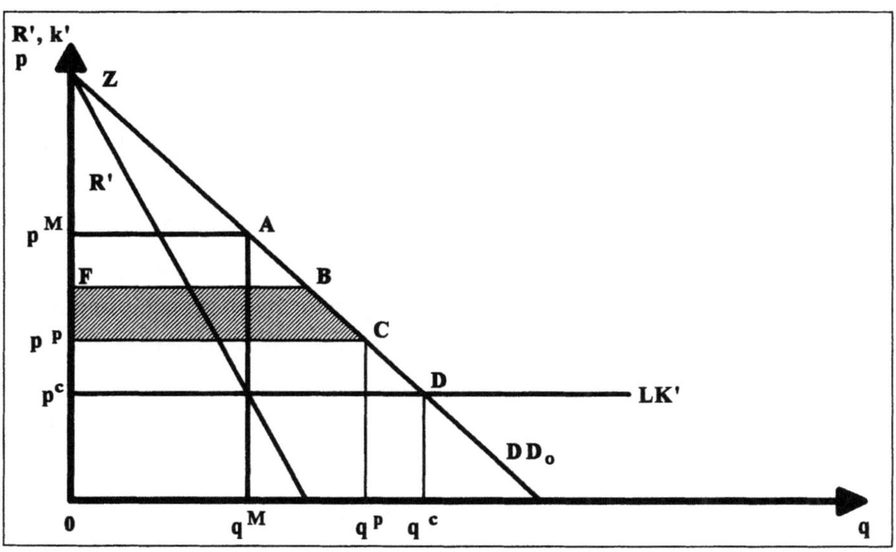

One important question concerns the role of demand shifts over time. Falling marginal costs could occur in the beginning of network operation or at low per capita costs. With positive network externalities, which make the telecoms service more valuable for the existing user group, each additional subscriber causes the demand curve to shift to the right over time. Increasing per capita income will reinforce this process so that trunk operation will hardly be characterized by falling marginal costs over the long term. In the long run the demand curve (DD_2) could intersect a positively sloped marginal costs curve.

If local networks are subject to regulation because of economies of scale (or scope), one major problem arises: The regulator does not know the true long run marginal cost curve which for simplicity's sake is assumed to be horizontal (Lk'). With a given demand curve (DD_o) a single network operator without regulation would impose the monopoly price p^M, which is in contrast to the price under competition p^C. The monopoly price means that consumer rent falls from the triangle ZDP^C to the triangle ZAP^M. Regulation causes the imposition of direct or indirect price regulation. Let us assume that this political price is p^P - it lies above the price under competition. Compared to competition, there is still a loss of consumer rent equivalent to CDP^Cp^P. Regulation itself will cause some costs, and if we assume that the costs of regulation (regulatory agency, compliance costs) are equivalent to the area $FBCP^P$, the recovering of consumer rent from regulation is only the area $ABFP^M$. Following the Chicago critique, one could assume that even this consumer gain is largely eroded as the monopolist pursues rent-seeking, i.e. invests part of its extra profits in lobbying efforts for favorable political regulation. However, regulation could have been imposed because of market power and major distortions in the telecoms sector; in this case, regulation might be part of a second-best policy approach. But optimal regulation could aim at nurturing competition.

The telecoms sector is a high-technology industry of strategic importance for economic growth in industrialized countries. Both OECD countries as well as many newly industrializing countries have launched new initiatives to expand and modernize the telecoms industry in the early 1990s. In Europe the telecoms industry is undergoing a radical transformation as most countries of the European Union consider privatization of the national PTO a viable medium-term option. Moreover, the single market project in the EU is stimulating the emergence of bigger EU firms facing more intense competition. Flexible organization of firms across a wider space and better exploitation of communication flows are thus particularly valuable.

At the start of the EC single market, national telecoms markets in Europe were nationally fragmented and - except for the U.K. - characterized by a monopoly network operator (disregarding mobile cellular networks). It is interesting to analyze which different strategies have been chosen in selected EU countries. Moreover, questions arise as to which options for liberalizing telecoms markets for the benefit of business and households generally exist, and what lessons the short liberalization history of Western Europe can offer.

2.1 Global Forces in Telecoms Liberalization

Intra-company networks for international data transmission have become a rapidly expanding and profitable field of the telecoms sector, and new international consortia for serving Multinational Companies' demand emerged in the late 1980s and the early 1990s. Foreign direct investment grew twice as fast as international trade in the 1980s. Moreover, multinational companies (MNCs) often deal intensively with each other and increasingly form strategic alliances to raise communication needs. Thus the need for greater company networks as well as for inter-MNC networks has increased.

Big multinational companies typically expect services to be offered to all foreign offices of the multinational company, causing the need for a global presence - or at least for the coverage of Europe, North America and Japan. Multinational companies expect a seamless network, which creates the fundamental problem of international network interconnection. In April 1994, thirty leading European MNCs signed a contract for a private network to be operated by a consortium consisting of BT and an alliance of AT&T and Unisource. Big MNCs in Europe will thus enjoy an effective choice of carrier for a large part of their voice traffic, even before the 1998 EU deadline for the end of PTOs voice monopoly. Demand for communication, however, is not only growing in the multinationals' business community. The international media industry is creating additional demand as pay-per-view systems and pay-on-demand systems allow personalized and individually charged cross-border services. The movie industry, dominated by the US, as well as some press conglomerates perceive new growth opportunities depending on cheap international communication networks.

The US long distance operator MCI joined forces with BT, which took a 20% share in the US company. Deutsche Telecom and France Télécom forged another alliance (after the German state monopoly's attempt to create an alliance with the Japanese NTT and the British BT faltered in 1991). Public telecommunications operators from EFTA countries formed the alliance Unisource, which comprises the PTT Telecom Netherlands, Telia of Sweden and the Swiss PTT Telecom; Unisource will also include the Spanish Telefonica as of 1994. Another perspective partner of Unisource is AT&T. Deutsche Telecom and France Télécom envisaged a global alliance with AT&T, but the EC Commission signalled that such an alliance would face difficulties from the competition policy side. The number three in US long distance communication, Sprint, agreed to sell 20% to France Télécom and Deutsche Telekom for $ 4.2 bill. AT&T protested that the deal should be stopped unless France and Germany opened up their markets more quickly. Alliances between state-owned and private companies are particularly difficult, because two divergent corporate cultures have to cooperate efficiently. Commercially-oriented flexible private firms typically find it difficult to cooperate with slowly reacting state-owned companies.

Tab. A1: The World's Leading Telecommunication Operators (Ranked by 1992 Revenue)

	Revenue ($ bill.)	% change 1991-92	Lines in millions	% change 1991-92
NTT (Japan)	51.4	1.7	57	2.7
AT&T (US)	39.6	2.0	n.a.	n.a.
DBP Telecom	34.6	14.3	35	5.6
BT**	23.4	0.7	26	1.9
France Télécom	23.2	5.9	30	3.4
SIP	17.5	10.8	24	2.8
Bell South	15.2	5.2	19	3.4
Nynex	13.2	(0.9)	16	1.9
GTE	12.6	0.6	17	3.6
Bell Atlantic	12.1	2.5	18	2.4
TOTAL	242.6	7.1	242	3.2

* AT&T does not offer local exchange line services
** indicates year ending March 1993

Source: Financial Times (1994a)

Hardware Market

The market for telecoms hardware is an international oligopoly in which AT&T, Alcatel, Ericsson, Siemens and Northern Telecom, as well as some Japanese companies are the leading players. Falling computer prices and the increasing need to interconnect national networks while making better use of existing transmission capacity, however, are gradually increasing the role of software.[2]

Close relationships between national equipment firms and their respective national PTO were typical for Europe in the 1980s; this is unlikely to continue as pressure from the EC single market and privatization of major European PTOs occurs in the 1990s. Moreover, the single market requires an EU-wide opening-up of government procurement, causing an expected rise in the internationalization of the equipment business. Central and Eastern European countries wanting to join the EU in the future should take these facts into account.

The liberalization of goverment procurement implies intensified competition in the EU telecoms equipment market, which will cause a fall in the price and intensified innovation. Rising concentration in the equipment industry is to be expected as well. Firms facing falling prices in domestic markets will aim at higher foreign market shares, which seem necessary to recover high R&D costs (plus other fixed costs) and better to exploit static and dynamic economies of scale. Rising telecommunications and digitization require more powerful computers and chips; prospects for specialized chip producers (IBM, Toshiba, Motorola, Siemens and others) are positive.

Mobile Telecommunications

Cellular telephones and radiopagers have created mobile and personal telecommunications based on wireless transmission. Mobile communications still require the use of an established network. A network for mobile telecommunications can only operate if all major regions are covered by a net of receivers, which are in turn linked to the terrestrial network. Modern digital mobile technologies for wireless communication are GSM, CT2, DECT, CDMA (plus IS-54, PDC), while older analogue systems comprise AMPS (American Mobile Phone System), TACS (Total Access Communication System) and NMT (Nordic Mobile Telephone). Some 40 mill. users of GSM are expected around the year 2000. GSM (Global System for Mobile Communication) and its sub-standard DCS1800 (operating in the 1.8 GHz band width) are expected to set the global standard in the 1990s. In contrast to GSM, which uses Time Division Multiple Access (TDMA) to break up speech channels into multiple time slots, Code Division Multiple Access

(CDMA) more efficiently exploits available frequencies by using a spread-spectrum technology.[3]

Mobile telecommunications is typically offered by specialized service providers using the existing monopoly network, or one of the competing network operators (three long distance operators in the case of the US). In Germany and the U.K., there are three competing networks for mobile communications; several service providers compete for new subscribers, who can often obtain the equipment at subsidized prices. Italy has two cellular phone companies: No. 1 is the state-owned SIP, while in March 1994 a second licence was awarded to Omnitel-Pronto Italia, headed by the Olivetti group. The Omnitel-Pronto Italia network paid a concession fee of ECU 300 mill. for the licence; 1 bill. ECU will be invested by 1997. Huge investment requirements are one reason for the inclusion of banks/financial intermediaries in mobile telecoms consortia. The Omnitel-Pronto Italia consortium reflects the strong international orientation of mobile network operation: Omnitel, which had a share of 70%, is comprised of Olivetti (51%), Bell Atlantic (16.6%), Cellular Communication (14.7%), Swedish Telecom (9.7%) and Lehman Brothers (8%). Pronto Italia is comprised of Pacific Telesis (34%), Mannesmann Eurokom (15%), Banca Roma (15%) and other Italian partners. Italy is an interesting country for mobile telecoms due to its rather uneven population density. After 18 months 40% of the territory will be covered by Omnitel-Pronto Italia, and after 30 months some 80%. One may assume that the more prosperous and densely populated areas will be served first, but it is remarkable that the 80% coverage target should be reached within three years. With a rising territorial coverage, the usefulness of the network for its users as well as demand will gradually increase. Obviously, the higher the number of subscribers, the greater the communication business.

2.2 Privatization and Deregulation in Western Europe

Most Western European public telecommunications operators (PTOs) will no longer be state monopolies by the end of the 1990s. There are four major reasons for this development:

- New technologies will erode the spatial monopoly for voice and data communication, where two global satellite projects, Teledesic (Gates/McCaw) and Iridium (Motorola), are the boldest challenge to national telecom monopolies.
- Government deficits drive privatization in Europe: High state budget deficits in most European countries leave the PTO with insufficient government funding in a period of modernization needs, and favorable prospects for increased tax revenues of a privatized PTO and high expected privatization proceeds reinforce the need quickly to privatize PTOs. The EU telecoms voice market will be liberalized by 1998; Spain, Ireland, Greece and Portugal will enjoy an extra grace period of five years, but Spain does not intend to use it fully since a slow-down in adjustment implies high opportunity costs.
- Liberalization and deregulation of national telecoms markets in some European countries will lead to the establishment of modernized, more advanced and cheaper telecoms networks and services. A modern telecoms industry provides valuable factor inputs for all industries and facilitates flexible and lean organization on a broader scale; it contributes both to international competitiveness of industries and to the locational advantage of a country as perceived by potential foreign investors: "Footdraggers" in the modernization of the telecommunication sector will suffer from declining export shares and diminished investment inflows.
- Seizing global player opportunities requires the formation of international consortia with appropriate management incentives and a flexible workforce. Since PTO employees enjoy the status of a civil servant (as in Germany, Switzerland, Belgium, Poland, etc.), it is quite difficult to implement efficient wage incentives and flexible workplace organization modes; furthermore, a national PTO faces serious limitations when going abroad.

Privatization Stages

The stage prior to privatization is the transformation of the state PTO into a joint-stock company fully or partly owned by the government. Finland, France, Ireland and Sweden will transform their national PTOs into state-owned corporations. The advantage of this organizational modernization is the establishment of legal autonomy and increase in operational independence. Germany, Belgium, Spain and the Netherlands offer further examples in this respect; but these countries have already prepared the privatization of their state-owned network operator.

The change of corporate status brings a minimal degree of entrepreneurial freedom in organizing a more flexible and internationalized company. Foreign and domestic investors can obtain a minority participation, and managers will enjoy greater freedom to pursue a market-oriented accounting and marketing approach. Privatization could be on a full scale or involve retention of a major government equity stake; strategic foreign investors are expected to contribute special know-how, while remaining equities would be floated on the stock exchange. In Belgium, Belgacom has created an alliance with the regional US operator Pactel in order to build up a new digital cellular network and to improve management. A full privatization might come later. Even Greece had prepared - prior to the Papendreou government - plans to sell 35% of the Greek PTO to a strategic partner, which would have guaranteed $ 7 bill. in investment over a decade. The government intended to sell 14% in the form of equities to the general public while retaining a stake of 51%.

PTOs of Spain, Italy, Denmark and Portugal are listed companies. In Spain, the government owns only a minority stake (1/3). Nearly 50% of the Italian telecommunication holding STET, with its six operating and manufacturing companies is privately owned. In Portugal 49% of Marconi - one of three state network operators - is owned by the private sector. The Netherlands will sell a first tranche of share in the telecoms operator KPN in 1994 and a further tranche will follow in 1995/96, reducing government ownership to some 30%. The flotation of BT in the UK and the introduction of competition through Mercury were the most direct path to privatization and deregulation in Europe. The experiment was not fully successful to the extent that the Thatcher government's privatization of BT in effect created a private near-monopoly enjoying a market share of over 90% and high barriers to entry. In Germany the PTO will be privatized by 1997. Competition is emerging in the form of three licensed operators in the cellular phone business: the D1 network, a subsidiary of Deutsche Telekom, the D2 network which is operated by a consortium led by Mannesmann and the E+ network. Deregulation in Western Germany started at a slow pace in the late 1980s: Data transmission, value-added network services, terminal equipment and the margin of the physical network (satellite communications; mobile radio) were liberalized. Moreover, the use and resale of leased lines was increasingly allowed in Germany.

In the US there are three long distance companies which might face additional competition from cable TV companies in the 1990s. New alliances between cable TV networks and long distance telecoms companies as well as regional telecoms companies are emerging in the US, where only local networks and parts of intrastate long distance telecommunications are subject to government regulation.

Fig. A4: Stages of Privatization in Western Europe

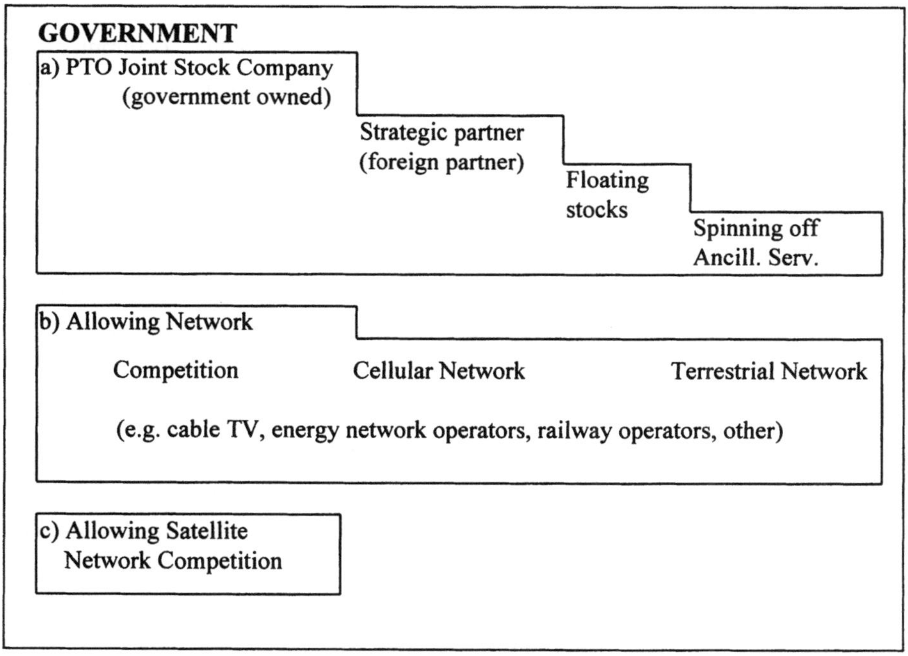

Ancillary business of the PTO/the telecoms network operator could be floated-off. Directory assistance could indeed be required to be a separate business strictly regulated by government to ensure that all potential users have access to directory information. Number portability is essential for effective network competition because it increases the degree of substitutability across networks.

Revenue Aspects in Privatization
The direct and indirect government revenue from privatizing a PTO depends on the regulatory framework applying to the industry. Direct government revenue comes from the floating of PTO stocks; indirect revenues are the capitalized future tax receipts from a profitable telecoms industry and from higher profits from industries using or providing more modern telecoms services. The apparent short-term trade-off between higher revenue from privatization (of a monopoly) and competition will be irrelevant and if the time horizon of policymakers is sufficiently long: Creating competition, thus raising future tax revenues, is the true important issue.

Seamless Network
Network competition is expected to spur innovation and reduce communication costs. The latter might be true for each network, but the real question is whether users can really switch between networks for a given period or, more extremely, for each call or bit transmission. In economic terms it would be useful to allow some flexibility in users' choice to the extent that this helps fully to exploit existing network capacities. One major problem with establishing network competition while maintaining a seamless network is the insurance that open access standards and common technical protocol are applied; government authorities as well as private standardization bodies are required to achieve this. In order to not prevent innovations in operating modes and basic technologies one might, however, tolerate a small number of technological tracks, even if no full compatibility is achieved at a given moment (technological progress might solve this problem in the future).

Directory information should not be allowed to become proprietary knowledge because this information can in effect be used to prevent market access. Government regulation should impose general principles for directory numbers and rules for access to directory information.

2.3 Politico-economic Deregulation Pressures

The typical regulatory policy in both the US and some EC countries had the dual goal of recovering costs and expanding the network. A pricing policy to stimulate network expansion can be justified during the early stages of network development by network externalities: Additional users raise the utility of the network for the stock of users. In this context pricing schemes were developed in which long distance telecoms users subsidized local telecoms users, thereby promoting network expansion. A major pitfall in the course of network expansion was, however, the maintenance of a standard cross-subsidization policy, even after completion of the network (meaning high penetration rates). Given the high penetration rates in the OECD countries in the 1980s, it is clear that positive network externalities were insignificant once a full network had been established (VON WEIZSÄCKER, 1984; MEYER et al., 1980).

Before deregulation in the US and Germany, long-distance users heavily subsidized local telephone networks (FOREMAN-PECK and MÜLLER, 1988; BLANKART and KNIEPS, 1989, p.586). In West Germany (and similarly in the US before 1982) pressure groups before deregulation could be characterized as follows (BLANKART and KNIEPS, 1989):

(1) the national telecoms operator and its unionized personnel benefitted from monopoly rents and thus favored maintaining the *status quo*;
(2) dominant suppliers of the national network operator benefitted from the actual network monopoly regime by privileged supply and cosy relations, resulting in extra profits (and reduced risks with respect to recovering R&D costs);
(3) small users and users in distant locations caused high installation and maintenance costs in the local network relative to the number of profitable long distance calls they were making. These groups benefitted from internal subsidies enforced by the existence of legal entry barriers;
(4) business users with frequent long-distance calls were paying a huge part of the internal subsidy to the other users. The group of business users was clearly against the monopoly *status quo*.

BLANKART and KNIEPS (1989) view the following as the main reason for early US telecoms liberalization, when compared to Germany's slow deregulation: decisions on telecommunication deregulation were centralized at the federal level in Germany, while regulation policies in the US were assigned to both the state and federal level. Groups in favor of deregulation thus had multiple options to undermine the status quo network and service arrangement in the US. Such a politico-economic approach to explain regulation and deregulation seems to be quite relevant for European countries, too. Countries having no major equipment supplier should find it easier to deregulate the telecoms sector than countries with important cosy equipment-network operator links. Countries with politically influential peripheral regions - facing higher telecoms costs immediately after deregulation - will find it more difficult to deregulate the telecoms sector. One may add two additional observations: Foreign equipment suppliers that claim open access to government procurement can destroy the traditional equipment-network link; the EC single market project to open-up government procurement is quite important in this respect. Secondly, countries with a high share of multinational companies can be expected to lobby strongly for telecoms deregulation since a broader multinational business community can easily internationally compare communications costs and threaten to relocate future overall investment outlays in favor of countries with low communication costs.

Competing networks in the US as well as in some EC countries always emerged first in the business community. Exclusive data transmission networks or video networks were created. There is no economic reason why private firms should not be allowed to operate special networks. On the contrary, competition in this field should be allowed and promoted as long as basic technical standards and rules are obeyed. Modern computer technologies which digitize voice, data and video have blurred the demarcations of traditionally segregated networks. This will create intensified competition in the 1980s, provided that government regulations will allow it to unfold. One must not overlook the fact that process innovations in the telecoms sector benefitting the business community will also benefit residential users in the long term.

Another problem consists of the high prices for leased lines in Europe as compared to the US: In 1994, there was a 10-to-1 price differential in leased lines (FINANCIAL TIMES, 1994b, p.2). This hampers the dissemination of information and services in Europe since interconnection charges to the public network are high. This could reflect inefficiency of continental European network operators or monopolistic price setting along with cross subsidization of other PTO services; all this supports the case for network competition.

Conclusions for Network Growth

Two fundamental conclusions to be drawn depend upon the difference in pricing regulatory policies in the two stages of network building. (1) Countries with low penetration rates should only transitorily adopt a pricing policy that promotes network expansion; this implies that conditional regulations are economically efficient. (2) Telecoms deregulation will be easier once a critical minimum of foreign investors has been attracted, and a critical number of domestic multinational companies has been established. It is not surprising to find that major source and host countries of foreign direct investment (the UK, the Netherlands and Sweden) are leaders in telecoms deregulation policies in Europe. (3) Reduced interconnection charges for leased lines as a consequence of network competition could generate much additional demand for telecoms services as prices for such services can be reduced.

Price Regulation in Established Networks

Network competition can start once one network has been completed (or even before). A primary stage of competition concerns third party access (TPA). The incumbent could deny access or quote prohibitively high prices for network access. To avoid this problem one would have to divide the telecoms business into a pure network operator and a supplier of services/final products. If splitting the network operation is not considered, there is a need for price regulation. Economic optimality requires marginal cost pricing to be applied. This, however, is a major problem in an industry with declining marginal costs, since fixed costs could then not be recovered. If the network operator is to stay in business, the government must allow prices above marginal costs to be realized at least during some time or for some services (e.g. during peak load time). However, LAFFONT and TIROLE (1994) have argued that marginal cost pricing is indeed the best regulatory principle. They point out that one should redefine marginal costs (not only extra "wear and tear" on the system). Network operators also in the market for final products will suffer revenue losses to any newcomer in the final product market. Facing a reduced market share due to a newcomer, the network operator loses contributions which should be included as opportunity costs in the interconnection price. This pricing strategy is dubbed "efficient component pricing". The regulated final product price should mimic the outcome of a contestable market as is argued by BAUMOL and SIDAK (1994). This suggests that there is no real need to separate the network from the supply of final products. In New Zealand and in California, efficient component pricing is applied. In reality it will be difficult to impose optimal interconnection prices. Asymmetric information is especially a problem because the telecoms sector is characterized by high technological dynamics.

3. Technological Dynamics

At the beginning of the 1980s, the telecommunications sector could be characterized as a highly regulated and protected market with strong political influences; the data-processing sector, in contrast, has developed in a relatively liberalized market influenced by international and national competition. During the last decade, technological changes (especially digitization) led to a merger of the telecommunications and data-processing sector, two sectors with a totally different regulatory framework, making the construction of a new regulatory framework necessary. The question, of whether to open up the telecommunications markets or to regulate the market for data-processing was answered by a more or less rapid liberalization of parts in the telecommunications sector in several European and non-European countries, especially in the United States, the United Kingdom and Japan.

The construction of a (new) regulatory framework without the anticipation of present and future technological changes will hamper the development of an efficient and adaptable telecommunications sector. During previous years the following technological inventions and innovations had a strong impact on the telecommunications sector (UNGERER and COSTELLO, 1990, p.37):

- micro-electronics, in particular the manufacture of complex circuits on a single integrated-circuit chip;
- digitization of networks and switches;
- introduction of stored-programme-control (SPC) exchanges;
- new transmission techniques: optical fibre, improvement in microwave and satellites and the revolution in communications and the audio-visual sector resulting from them;
- new software technologies for methods and tools.

Some detailed information on the technological developments is given in the following paragraphs.

3.1 Digitization

Before the era of digitization, voice telephony had been transmitted by analogue techniques. Signals in the form of acoustic waves were transformed into electromagnetic waves and transmitted through networks and switches to the receiver, where the electromagnetic waves were re-transformed into acoustic waves. Analogue techniques have three disadvantages in comparison to digital techniques:

- need of more transmission capacity
- worse transmission quality
- difficulties in using analogue networks for data transmission.

Digitization enables the encoding, transformation and transmission of any information as bits, using signals consisting of "0" and "1". Transmission in digitized form leads to higher quality, easier handling in both multiplexing and switching, transmission of both voice and data in the same form and to more efficient use of existing infrastructure capacity. For the latter, two explanations can be given. First, digitization leads to a merger of existing infrastructure which before could only be used for the transmission of voice telephony or data transmission. Second, digitized lines enable the transmission of up to 32 telephone calls (and even more in the future) at the same time, due to a kind of time sharing of a single transmission line. It is not necessary to transform the whole phone call into signals; only fractions are encoded, which are sufficient for the transmission of the whole phone call. The time span between the transmission of the fractions of a phone call can be used for the transmission of fractions of other phone calls.

The degree of telephone network digitization of European Public Telecommunications Operators in 1990 and 1992 (1993/94 respectively) is described in Tab. A2. Obviously, there are strong differences between the national PTOs, from 100% digitization at Mercury to 12.0% subscriber line digitization in Germany; the special situation in Germany is due to unification and must be taken into consideration.

Tab. A2: Degree of Digitization of the Telephone Network of European Public Telecommunications Operators

	subscriber lines digital in percent		trunk lines digital in percent	
	1990	1992	1990	1993/94
U.K. (Mercury)	100.0	100.0	100.0	100.0
Ireland	55.0	68.0	100.0	100.0
Netherlands	33.0	86.0	83.0	100.0
Germany	12.0	12.0	80.0	70.0*
Sweden	56.0	60.0	75.0	n.a.
France	70.0	83.2	74.0	87.6
Denmark	29.3	39.0	70.0	100.0
U.K. (BT)	45.0	64.0	n.a.	70.0
EC Average	34.9	65.3	n.a.	n.a.
Spain	28.4	36.4	59.6	86.0
Portugal	30.0	54.0	50.0	63.0
Belgium	37.2	48.0	n.a.	95.0
Italy	33.0	48.0	n.a.	48.4
Luxembourg	31.0	70.0	n.a.	n.a.
Greece	15.0	15.0	25.0	n.a.

* including East Germany, data for 1992

Source: OECD (1993), ITU (1994), PTOs annual reports, Mercer (1994), own calculations.

Reasons for the PTO's choice of a digitization strategy include the following: higher quality services, lower costs of maintenance and, in the longer term, a lower unit cost per line. There are still other advantages which should be mentioned: digital exchanges occupy less floor space and require less manual intervention by operators (OECD, 1993, p.47).

As Tab. A2 shows, digitization rates of trunk lines on the national level vary from 100% in Ireland and the UK (Mercury) to 25% in Greece. The main reason for digitizing networks is to avoid capacity bottlenecks. Digitization rates of trunk lines are generally higher than those of subscriber lines in the local loop as trunk lines are used more frequently. Above-EC-average digitization rates are found especially in countries with a modern national economy such as the Netherlands, Germany, Sweden, France, Denmark and the UK. Another important reason for digitization seems to be relevant for the UK and Sweden: competition requires telecommunications services at low prices. To be competitive, Mercury has built up a 100% digital network to offer higher quality at lower prices than BT.[4] The data for 1992 shows a strong increase in digitization, with Germany as an exception due to unification.

The main benefit of digitization is that a PTO can carry out many different types of services (voice, text, data or image) over the same physical network, leading to technical convergence between telecommunications, computing and television. Therefore, digitization is the basis for the Integrated Services Digital Network (ISDN).

3.2 Integrated Services Digital Network

The ISDN is defined by the CCITT[5] as follows: "The main feature of the ISDN concept is the support of a wide range of voice and non-voice applications in the same network. A key element of service integration for an ISDN is the provision of a range of services using a limited set of connection types and multi-purpose user interface arrangements. ISDN supports a variety of applications including both switching and non-switching connections. Switched connections in an ISDN include both circuit-switched and packet-switched connections and their combinations."

For the introduction of ISDN, a 100% digitized network from one user to the other is necessary. The incremental costs of a full-scale digital network via existing copper cable are between 20% and 30% of the total investment (UNGERER and COSTELLO, 1990, p.49). The total number necessary for a country like Germany can be calculated as follows. The average installation costs per subscriber line are estimated to be at $ 1500 according to international organizations like the OECD and the World Bank (BERLAGE and SCHNÖRING, 1992, p.7). In 1992, the amount of mainlines in Germany was approximately 32.1 mill. This leads to a total required investment of $ 48.15 bill. Therefore, the total amount necessary for a 100% digitized network will be in the range of $ 9.63 to $ 16.05 bill.

Tab. A3: Main Lines and Telecommunications Investment for Network Expansion and Network Modernization in EC Countries (1980-90)

	Main lines in 1990 ('000)	Pene-tration rate[4] (1990)	Cummulative investment[2] 1980-1990 (US$m)	Network expansion in %	Network moderni-zation in %[3]
Germany	29 981	47.41	102 280.78	13.85	86.15
U.K.[1]	25 404	44.25	43 680.06	26.47	73.53
France	28 085	49.78	65 073.28	28.09	71.91
Sweden	5 849	68.33	12 198.63	12.65	87.35
Denmark	2 911	56.63	5 597.20	18.36	81.64
Luxembourg	184	48.17	332.53	23.46	76.54
Netherlands	6 940	46.42	9 761.26	31.47	68.53
EC	139 678	41.14	347 580.16	29.20	70.80
Italy	22 350	38.77	67 220.86	20.83	79.17
Belgium	3 913	39.26	7 557.48	28.78	71.22
Ireland	983	28.06	3 380.05	22.19	77.81
Spain	12 603	32.35	34 521.96	23.35	76.65
Portugal	2 379	24.13	4 322.20	48.22	51.78
Greece	3 945	38.94	3 852.50	65.36	34.64

1 Includes estimates for Mercury and Kingston Telecommunications as well as BT.
2 Constant 1990 values and exchange rates, corrected for gross fixed capital formation (GFCF) inflation.
3 Network modernization is assumed to be equal to total investment minus network expansion investment.
4 Penetration rate per 100 inhabitants in 1990.

Source: OECD (1993), p.10 and p.45, own calculations.

Between 1980 and 1990, the cumulative telecommunications investment in Germany amounted to $ 102.280 bill. at constant 1990 values and exchange rates, leading to an average amount of approximately $ 10 bill. per year. The bigger part of the investment (86.5%) has gone towards modernization, notably the change from analogue to digital systems, and the introduction of new services. The data for the other EC countries are given in Tab. A3.

National PTOs obviously pursued different strategies. In countries with an above average EC penetration rate, network modernization was the prime target. In countries with below average penetration rates a uniform strategy cannot be found.

Countries like Italy, Belgium, Ireland and Spain have fostered network modernization to fulfill requirements of the business sector to offer higher quality and reduce call charges. According to UMBERTO DE JULIO, vice president of the networks division of SIP, the Italian state operator, current network investment alone would reduce cost of calls by 20% within five years - irrespective of staff productivity (FINANCIAL TIMES, 1994b, p.2). Portugal and Greece pursued a strategy of network expansion, mainly to achieve the politico-social goal of higher penetration rates.

ISDN offers a wide range of advantages for the support of the future telematics market: a multitude of services via a single subscriber access, simultaneous operation of two channels via one "basic access", and easy connection of digital PABXs[6] via ISDN "primary access" facilities at 2 Mbit/s, which is equivalent to 30 simultaneous channels. In Germany, for example, the amount of ISDN basic access increased from 7,615 in 1990 to 138,800 subscribers in 1992, the number of ISDN primary access from 624 in 1990 to 11,300 in 1992 (DBP TELEKOM, 1992 and 1993).

Moreover the quality of the telephone service will be improved and the access of a broad range of new services will be enabled through ISDN. The extent of new services via ISDN is not yet fully measurable. According to an EC Recommendation, ISDN in Europe will offer high quality telephony with additional services like call waiting, calling line identification, etc.; high speed data communication at 64 Kbit/s, with special attention paid to the connection of personal computers, and a range of other advanced services like facsimile at 64 kbit/s and videotext at 64 Kbit/s will also be offered (UNGERER and COSTELLO, 1990, p.50). The most enormous potential of future services through digitization and ISDN has, however, not been mentioned yet: the value-added services sector.

3.3 Fibre Optics, Fibre to the Home and Optical Networks

In comparison to copper or coaxial cable, glass-fibre cable offers a vast capacity potential. During the last few years the demand for additional capacity and higher, flexible transmission rates has risen constantly. Even if PTOs continue to replace copper wire by glass-fibre cable in the long distance sector, the replacement of copper cable via glass-fibre in the local loop, or fibre to the home (FTTH), is still being contested.

Tab. A4: Development of Optical Fibre Networks of Selected PTOs

	1990	1991	1992	Measure
Belgium	14,000	24,740	25,640	Fibre km
France		2,700	9,000	Fibre km
Germany	30,900	42,100	56,200	Fibre km
Greece			500	Fibre km
Ireland	3,500	7,000		Fibre km
Italy	733	890	1,467	Cable km
Netherlands			10,000	km
Portugal	3,960	6,681	13,251	Fibre km
Spain	15,131	20,290	22,056	km
United Kingdom	1,444,000	2,045,000	2,337,000	km

Source: OECD (1995), National PTOs: annual reports, various issues.

The construction of a second glass-fibre network in addition to the existing copper cable leads to cost duplication under the current capacity demand situation. Nevertheless, a second network can be efficient under certain circumstances. Therefore it is advantageous to distinguish between the business and the private sector.

The up to date situation shows that higher transmission capacities and more flexible transmission rates are required by the business sector. This bottleneck could best be solved by the interconnection of already existing high capacity Local Area Networks (LAN), Wide Area Networks (WAN) and Metropolitan Area Networks (MAN) with national or European wide data highways, consisting of broadband ISDN glass-fibre cable (FROHBERG, 1993, pp.57-60). This idea is demonstrated in Fig. A5.

Fig. A5: Interconnection of LAN, WAN, MAN and ISDN-B

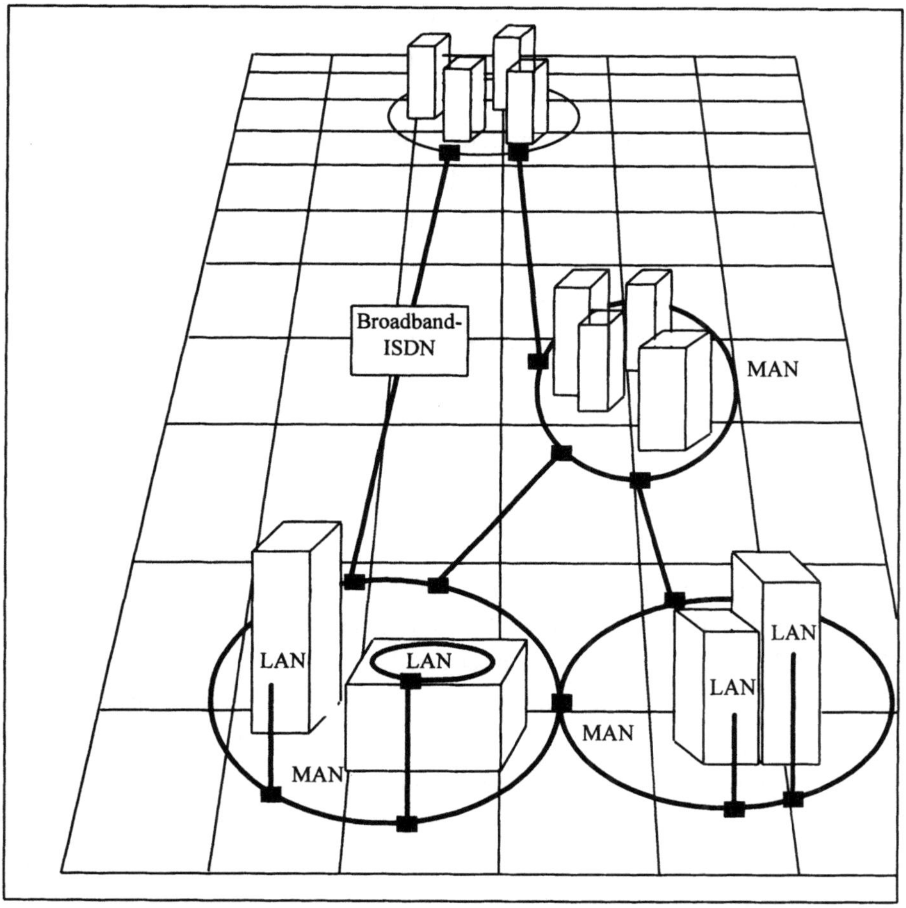

Source: Frohberg (1993), p.59

At present, customer distribution circuits are used on average only about ten minutes in every 24 hours (HARPER, 1991, p.44); the installation of a second network therefore seems to be economically unwise. This holds especially for an additional glass-fibre network. Nevertheless, technical developments and new services have to be taken into consideration (GRUPP and SOETE, 1993). On the one hand, new services like videotelephone, high definition television (HDTV) and pay-per-view television will require broadband networks with transmission rates between 100 Mbit/s and 1 Gbit/s. On the other hand, technical developments will improve the maximum transmission rate of copper and coaxial cable as well as airwave transmission. Today's sophisticated signal processing also enables unshielded twisted-pair telephone wiring to carry 125 Mbit/s over distances of up to 100 meters (VAN AS, 1993, p.10).

The use of optical transmission via optical-fibre networks offers a substantially higher channel bandwidth which can carry almost unlimited amounts of information. Yet the application of optical networks in the near future will be weak due to technical problems concerning optical switching. Moreover, the need for almost unlimited amounts of information in local areas and optical transmission to private homes is incredible, even in the medium run.

3.4 Mobile Communications

With regard to mobile communication, a distinction between land-based and satellite networks has to be made.

Land-based telecommunications is nothing new; the first public mobile network in Germany was introduced in 1958 (Fig. A6). Since that time, the quality, mobility and maximum number of subscribers have considerably increased. Today, three different mobile systems can be observed in Europe: elderly national analogue systems (e.g. NMT, TACS, C-net), GSM (Groupe Spéciale Mobile) networks and DCS-1800 networks (Personal Communications Networks [PCN]). The analogue system is based on analogue technology, which hampers the transmission of data; it allows roaming and handover, but its use is restricted to national borders. Even if the current number of subscribers is still weighty, the system will probably be replaced by more efficient digital mobile networks like GSM and PCN systems. The European GSM network is fully ISDN compatible. Moreover, data and speech services can be used throughout Europe. From a technical point of view, the expected number of PCN subscribers could be four times the expected number of GSM subscribers. PCN telephones are smaller than GSM telephones and are expected to be much cheaper. In addition, due to smaller cellular areas, scarce frequency bands can be used more efficiently (BARTH, 1992, p.97). In 1993 first Personal Communications Network in the world started in the United Kingdom; a second is expected to start in Germany at the beginning of 1994 (DEUTSCHE TELECOM E.V., 1993, p.14).

In the 1980s, the technology and quality and capacity of satellite communications systems have improved a lot. In 1958, the first geosynchronous communications satellite "Early Bird" transmitted up to 300 voice circuits, or one television channel. Today, a single satellite is able to relay several dozens of channels, or up to 80 000 voice channels (UNGERER and COSTELLO, 1990, p.68).

Fig. A6: Development of Existing and Planned Airwave Systems in Germany

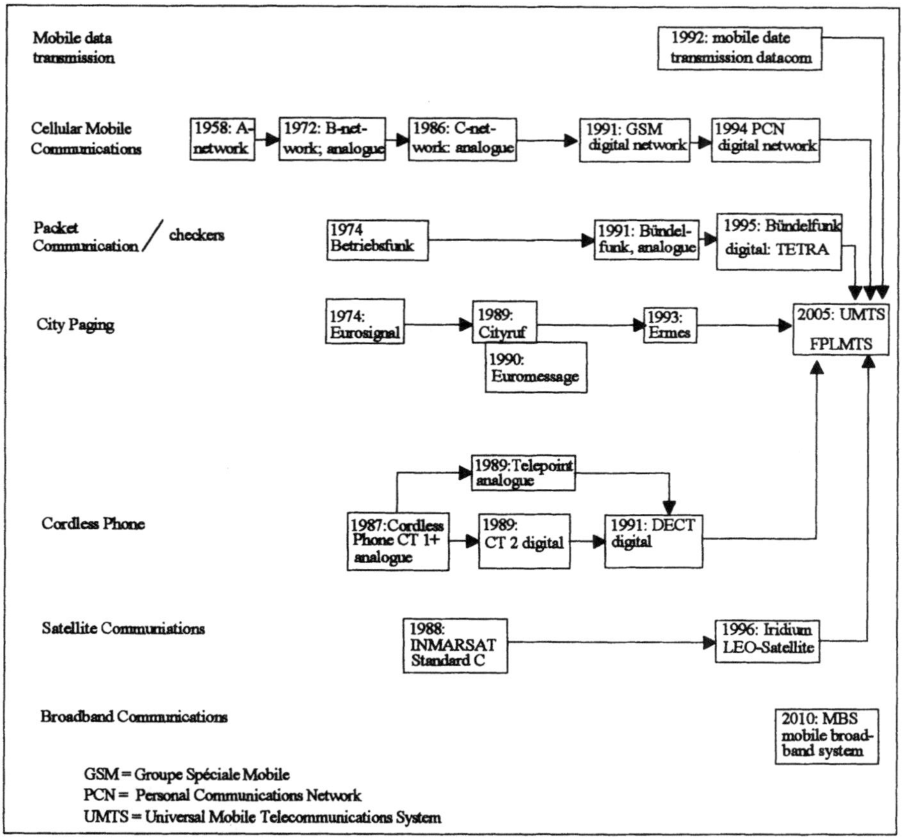

Source: Decker and Walke (1993), p.13.

In the near future, the creation of several low earth orbit satellite systems (LEO-systems) is planned. Those new systems enable the worldwide use of handy, dual mode telephone sets for personal communications purposes. The dual mode telephone sets make an alternative use of land based or satellite based airwave communications possible. Some technical data of selected planned LEO-satellite systems are given in Tab. A5. This clearly indicates that investment costs are tremendous and that only a small number of subscribers can be covered. Moreover, satellite phones and calls could be highly expensive. The first satellite phones are expected to cost around $ 3,000. A call will cost around $ 3 a minute (ECONOMIST, 1993, p.20).

Tab. A5: Planned LEO Satellite Systems

System	Odyssey	Globalstar	Aries	Iridium
Altitude	10.370 km	1.389 km	1.018 km	765 km
Typ	MEO	LEO	LEO	LEO
No. of satellites	12	48	48	66
No. of subscribers	2,3 m	2-5 m	1 m	1,4 m
Costs (billion $)	0,8-1,4	0,8-1,5	0,3-0,5	3,4-4,0

Source: Lutz (1993), p.29; Economist (1993), p.19.

Another project is planned by Calling Communications, a small Californian company, mainly to offer fixed telephone services to 20 mill. subscribers in remote areas. The company will launch 840 tiny LEO satellites, at the cost of some $6.5 bill. (ECONOMIST, 1993, p.19).

4. Regulatory Developments

Regulatory policy encompasses economic conduct like pricing, market entry and market exit. Whether a market should be regulated or not has to be an economic question in the first place. Price regulation has to be introduced if the prices set by the firms are too high due to a lack of competition, and if the entry of new firms is improbable because of natural barriers to entry or political decisions. Artificial barriers to entry built up by a regulatory framework are required if only a few firms, or in the extreme case a single firm, are advantageous from an economic point of view. The latter is an example of a natural monopoly, which is able to provide a market with lower prices, due to a sub-additive cost function, than a host of firms. A sub-additive cost function can be illustrated by inequality (1) for a single product:

$$(1)\ C(x_t) < \sum C_i(x_i)$$

If inequality (1) is fulfilled, the production costs of a monopoly (C) for a certain amount of products (x_t) are lower than the added costs [$\Sigma C_i(x_i)$] of a number of firms (i = 1,...,n) for the same amount of products ($x_t = \Sigma x_i$). In this case a regulatory framework has to be set protecting the monopolist by artificial entry barriers.

A sub-additive cost function for a multi-product firm can be described as follows:

$$(2)\ C(x_t,y_t,z_t) < \Sigma C_i(x_i,y_i,z_i).$$

Inequality (2) shows a cost function where the production costs for the total amount of products x,y and z produced by a single firm are lower than by a number of different firms. The reason for a similar sub-additive cost function can be found in economies of scale and economies of scope. If the inequality (2) is valid, all three markets should be protected by entry barriers.

The telecommunications sector can be subdivided into three different markets: the equipment market, the service market and the network market, or simply the infrastructure. A monopoly for all three markets can only be efficient from an economic point of view, if inequality (2) is fulfilled. Until the beginning of the 1980s, the sub-additive cost function was one of the main arguments for a national monopoly in most European countries, especially as far as the infrastructure was concerned. Due to technological changes and a growing demand for services and transport capacities, this argument was contested more and more. In addition, the successful liberalization of the telecommunications markets in the United States, Japan and the U.K. made it clear that competition is more efficient than a regulated monopoly; therefore, a new regulatory framework has to be constructed (WELFENS and GRAACK, 1996). There are arguments supporting the view that telecoms are no longer a natural monopoly. In Europe, regulatory developments have to be regarded on two different levels: the European Community level and the national level of EC member states.

4.1 Regulatory Developments on the EC Level

Telecommunications and trans-European telecommunications networks have become one of the major challenges for the completion of the internal Community market and a prerequisite for the creation of a single information market as discussed in the 1993 White Paper of the European Commission (EC COMMISSION, 1993a, p.88).

The first move towards the harmonization of European telecommunications markets has been made by the 1984 "Community action programme on telecommunications". The programme included some primary steps towards the creation of a Community market for telecommunications equipment and terminals, promotion of advanced telecommunications services and networks, improved access for less favoured regions of the Community and the coordination of negotiating positions within the international organizations dealing with telecommunications (UNGERER and COSTELLO, 1990, p.135).

The second phase towards harmonization started with the Green Paper on telecommunications, published in 1987. According to the preface, the Commission's intention was to create competitive European markets to face the challenge of rapid changes in telecommunication and computer technology and the growing pressure of non-European markets (EC COMMISSION, 1987, preface). The Green Paper included a range of different purposes (EC COMMISSION, 1987, p.184):

- free (unrestricted) provision of terminal equipment within member states and between members, subject to type approval;
- further liberalization of the service markets, excluding the provision of a limited number of basic services such as voice telephony, which should still depend on national decisions;
- strict requirements regarding standards for the network infrastructure and services provided by the telecommunications administrations or service providers, in order to maintain or create Community-wide inter-operability;
- introduction of open access to national networks to trans-frontier service providers (Open Network Provision - ONP);
- separation of regulatory and operational activities of telecommunications administrations;
- full application of the Community's common commercial policy to telecommunications.

These general goals have been realized through several EC Directives that are binding EC legislation, namely the EC Equipment Directive, the EC Service Directive and the ONP Directive (Appendix A).

The EC Service Directive included the introduction of full competition across the Community for all value-added services, the full liberalization of data communications services, including the simple resale of capacity from 1 January 1993; a longer period of transition was accepted for countries with underdeveloped public data networks until 1st January 1996. National choice was given over the liberalization of voice telephony services and networks.

The ONP Directive was a necessary supplement to the Service Directive and included the interconnection rules for service providers. The ONP Directive was based on the principles of non-discrimination and equal access for European service providers to national networks as well as the transparency of access conditions, subject to objective access criteria. The required tariffs should be cost-oriented, transparent and set according to objective criteria (EC COMMISSION, 1990). ONP also requires the development of certain advanced services (such as direct calling, call forwarding and Community-wide "Green Numbers"), a defined minimum set of telecommunications services available on a Community-wide basis (such as the offering of basic voice telephony) and the non-mandatory application of ONP principles to certain services such as data communications and to ISDN. The provision of such additional services, as well as the mandatory provision of a minimum set of leased line offering, is viewed as a key element in reinforcing the international competitiveness of business in the Community, as well as promoting the development of new services (EC COMMISSION, 1993d, p.4).

Within the liberalization programme it was agreed that further steps towards liberalization should be subject to an intensive review of telecommunications services in 1992. The Commission stated that progress concerning liberalization and harmonization of the European telecommunications markets had already been made. Nevertheless, considerable problems still exist, such as high tariffs for intra-European long-distance calls when compared to national long distance calls.

Tab. A6 shows the price per minute for a 4-minute call in 1991 US$. To demonstrate the differences between the price of a national and an intra-European long distance call, consider the following example: The price of a call from Muenster (Germany) to Paris (France) was $ 0.69 per minute in 1991, while a national long-distance call, for example from Muenster to Munich (which is approximately equivalent to the distance from Muenster to Paris) cost $ 0.43 per minute (without tax), roughly two-thirds of the international price. Through the tariffs of France Télécom, a phone call from Paris to Muenster cost $ 0.67, while a phone call from Paris to Bordeaux cost only $ 0.37 per minute (without tax) in 1991 (FRANCE TELECOM, 1993, p.16). This shows that for national PTOs the law of one price does not hold - no surprise in a monopoly situation.

Unreasonable price differences between national and international long-distance phone calls hamper the dissemination of information on the international and especially the European level. Moreover, high international call prices can be considered as market entry barriers to foreign service providers. The price of a value-added service consists of the price of the service itself plus the call charges. Even if the service itself can be offered at a lower price by the foreign service provider than by the national provider, the total price for the service cum international call charges could be higher. In this case, the difference between international and national call charges can be considered as an import duty for foreign telecommunications services.

Tab. A6: Peak Rate Intra-European Telephone Calls between European Member States, December 1991

	B	DK	F	D	GR	IRL	I	L	NL	P	E	UK
B	--	0.88	0.63	0.63	0.88	0.63	0.88	0.44	0.52	0.88	0.88	0.63
DK	0.51	--	0.51	0.45	0.58	0.51	0.58	0.51	0.51	0.58	0.58	0.51
F	0.67	0.67	--	0.67	0.67	0.67	0.67	0.67	0.85	0.67	0.98	0.67
D	0.69	0.69	0.69	--	0.69	0.69	0.69	0.69	0.69	0.69	0.69	0.69
GR	0.84	0.84	0.84	0.84	--	0.84	0.84	0.84	0.84	0.84	0.84	0.84
IRL	0.98	0.98	0.98	0.98	0.98	--	0.98	0.98	0.98	0.98	0.98	0.77
I	0.91	0.91	0.80	0.80	0.80	1.05	--	0.80	0.91	1.05	0.91	0.91
L	0.43	0.61	0.61	0.61	0.61	0.61	0.61	--	0.43	0.61	0.61	0.61
NL	0.59	0.59	0.59	0.59	0.78	0.78	0.78	0.59	--	0.78	0.78	0.59
P	1.00	1.00	1.00	1.00	1.00	1.00	1.00	1.00	1.00	--	0.95	1.00
E	1.17	1.17	1.17	1.17	1.17	1.17	1.17	1.17	1.17	1.17	--	1.17
UK	0.58	0.58	0.58	0.58	0.58	0.56	0.58	0.58	0.58	0.58	0.58	--

This table is expressed in US$ exchange rates. Figures show the price per minute of a 4 minute call (without tax), in 1991 US$.

Source: OECD (1993), p.55

One of the lowest peak rates was applied in the UK, characterized by competition in network operations. Spain, in contrast, had the highest rate at double the rate of the UK. This indicates inefficiencies or welfare losses (prices exceed marginal costs) in Spain's network operation. The Madrid-London phone call is twice as expensive as the London-Madrid call. This violates the international law of one price.

Concerning further regulatory steps, the Commission took four options into consideration (EC COMMISSION, 1992, p.42):

(1) no further liberalization and maintenance of the status quo;

(2) radical regulatory measures with the aim of solving the problem of different tariffing of intra-European phone calls;

(3) full liberalization of international and national voice telephony;

(4) liberalization of intra-European voice telephony.

The Commission worked closely with the High Level Committee of National Regulatory Authorities established by the Council of Ministers in its Resolution of 17 December 1992 (EC COMMISSION, 1993d, p.3). In June 1993, the EC Council accepted the proposals of the Commission to open up the European markets for voice telephony by the 1st January 1998. Exceptions are being made for EC countries with an underdeveloped or very small telecommunications network like Spain, Portugal, Greece, Ireland and Luxembourg. For those countries, the expected deadline is the 1st January 2000 (Luxembourg) and the 1st January 2003 (for all other countries) (EC COMMISSION, 1993c, p.68).

In addition, the Commission advocates, in the short term (1993-1995), the use of alternative infrastructure such as networks of railway companies or electricity utilities (to a greater extent for private networks or closed user-groups) as well as the supply of liberalized services via cable-television networks (HDTP, 1993, p.11). Those proposals have been rejected by the EC Council. Nevertheless, the Commission still sticks to the idea of the use of alternative infrastructures. It has already been announced that further steps towards liberalization will be taken, if necessary without the Council's approval (DEUTSCHE TELECOM E.V., 1993, p.5).

The liberalization of the telecoms sector requires a regulatory framework solving the problems of universal services. In the Communication from the Commission of 28 April 1993, the political goal of universal service was identified as: "making available a defined minimum service of specified quality to all users at an affordable price" (EC COMMISSION, 1993d, p.4). Community-wide definition of universal service principles have been developed within the framework of the ONP rules. Main elements of the universal service concept are: (1) the basic provision of services, (2) quality of services, (3) tariff principles i.e. cost-orientation of tariffs, combined with tariff flexibility (which allows, for example, specially targeted tariff schemes for low income groups, (4) dispute resolution mechanism, (5) special public service features (for example, operator assistance and emergency services) and (6) certain Community-wide service features.

As a first step, the major elements of universal service in telecommunications whose provision each Member State must ensure, either through binding obligations or by the operation of market forces, have to be defined. The Commission considers that the ONP principles are sufficient at this stage to form the basis of universal service conditions, whilst being open to further evolution on the basis of technological and market developments, as well as in the light of practical experiences (EC COMMISSION, 1993d, pp.7-8).

4.2 National Regulatory Frameworks: Developments and Experiences

Different regulatory structures and procedures are necessary, whose details depend on different market structures. In a period of transformation from a monopolistic to an oligopolistic market structure, the task and the goals of a regulatory body will certainly change.

In a monopolistic market structure, the prime target of regulation is to reduce the market power and the power of setting monopoly prices. Those goals can best be achieved by incentive regulation, like price cap or sliding scale regulation. Due to asymmetrical information between the regulator and the regulated monopoly, regulation will never be as efficient as effective competition. Therefore, competition should be introduced whenever economic reasons like sub-additive cost functions do not require a single provider.

According to technological changes and falling transmission and switching costs, competition between basic network providers seems theoretically possible; this is underlined by experiences of the United States, the UK, Japan, Sweden, Australia, Finland, Canada and New Zealand. Economic reasons for regulatory market entry barriers concerning terminal equipment, mobile communications, satellite or value-added services can hardly ever be found. Nevertheless, a period of liberalization has to be accompanied by a dynamic and appropriate regulatory framework. Here, two questions have to be answered: What are the goals and challenges to a regulatory body, and which is the regulatory structure most suitable for managing those challenges?

4.2.1 Telecommunications Equipment

As far as economic reasons are concerned, a special regulatory framework for telecommunications equipment seems not to be necessary. The admission of equipment should be given due to technical and safety requirements; technical tests should be made by independent laboratories. A temporary limited licence for products, including a prolongation option, is adequate in view of rapid technological changes.

Since the liberalization of the French terminal equipment market in 1987, a liberal licensing procedure was introduced in France. In 1991, 1060 applications for terminal equipment approval have been made, of which 900 have been successful.

Tab. A7: Applications for Terminal Equipment in France 1991

	Applications	France	Other EC-Countries	Others
Switches	100	70%	15%	15%
Data terminals	320	75%	8%	17%
Other terminals	160	75%	8%	17%
Mobile terminals	480	25%	45%	30%

Source: Pospischil (undated), p.15.

The average rate of success concerning other terminals was 85%, with a slight advantage for French terminals (90%) in comparison to other EC countries (83%) and non-EC countries (75%). Even if this data gives the impression of an open market, non-regulatory entry barriers still exist. First, certification procedures are notoriously slow. Second, due to a monopolistic position, France Télécom still maintains discriminatory procurement policies for the purchase of sophisticated network equipment (NATA, 1990, p.56).

The French example shows that the restricted power of the regulatory body concerning non-regulatory market entry barriers can hardly influence the procurement policy of a monopolist by direct rules. Tab. A8 indicates that most segments of the telecommunications equipment market are subject to procurement policies of public telecommunications operators.

Tab. A8: World Distribution of Telecommunications Equipment by Type

	1986 (US$)*	1986 (in %)	1991 (US$)*	1991 (in %)	Annual growth
Switching equipment	28,81	44,3	33,36	40,4	3,0
- Private switching	12,33		12,19		-0,2
- Public switching	16,74		21,17		4,8
Transmission equipment	9,14	14,0	16,07	19,5	11,9
- Microwave links	2,13		2,52		3,4
- Mobile communications	4,75		9,83		15,6
- Space communications	2,26		3,72		10,4
Cables	5,76	8,9	6,43	7,8	2,2
- Metallic cable	4,53		4,13		-1,8
- Optic fibre	5,76		2,31		13,3
Other transmission equipment	12,88	19,8	13,74	16,6	1,3
Terminal equipment	8,48	13,0	13,01	15,7	9,0
- Telephone terminals	4,99		6,67		6,0
- Telex and telegraphs	1,03		0,77		-5,6
- Fax	2,49		5,57		17,4
Total	65,07	100,0	82,35	100,0	4,9

USD* = billions of 1991 fixed value dollars;
annual growth = average annual growth
other equipment: modems, repeaters and amplifiers

Source: OMSYC (1991), own calculations.

Double digit growth rates are observed for communication with mobiles, space communication, optical fibres and fax machines. These are rapidly expanding sectors. The trend towards wireless communications and fixed networks with high capacity is obvious. The increasing number of fax machines and decreasing number of telex and telegraphs indicates that the telex and telegraph service (reserved service to national PTOs) are being replaced by fax services. Fax machines obtaining national approval can easily be connected to the network without or with only few additional costs. Falling prices of fax machines make their use attractive, even to private households.

Competition in public switching equipment, transmission equipment and cable equipment is hardly imaginable in a market dominated by a national monopsony with special preferences to national suppliers. Even if a regulatory body has little direct influence on the procurement policy of a monopsonist, an attitude change can be achieved via monitored procurement, via network competition or via price regulation of the incumbent PTO (RPI-X regulation). These latter two measures are assumed to induce a reduction of revenues of a national public telecommunications operator, which could lead to a more cost oriented and less preferential supply system.

4.2.2 Value-added Services

A detailed definition of a value-added service (VAS) cannot be given, as the borderline between value-added services and basic service is not quite clear. The term value-added services comprises all services except the simple transport of speech or data, therefore implying "intelligent delivery of information to the subscriber" (UNGERER and COSTELLO, 1990, p.53).

The future dimension of new service is not yet imaginable, its impact on the economy and society life not yet predictable. Today, VAS have only a relatively small but growing influence on economic and social life. In 1986, the revenue from VAS amounted to 800 to 900 mill. ECU in Europe, which is slightly more than 1% of total telecommunications service revenues. The projected growth rate of VAS is between 20% and 30% per year; according to forecasts this service sector will accounted for 5% of GNP in 2007 in Germany (VON WEIZSÄCKER, undated, p.7). The still relatively weak significance of VAS in comparison to basic services is demonstrated in Tab. A9.

Public switched telephony network (PSTN) revenues include connection charges, line rental charges and call charges. The PSTN revenues generate about 75 - 85% of total income of national PTOs. In countries with a liberalized telecommunications sector like the UK and Sweden, PSTN revenues are less important than in strongly regulated countries like France, Germany and Italy. There are two reasons for this. First, competition leads to diversification and new value-added services, which increase the share of other revenues. Second, competition forces national PTOs to decrease and rebalance charges, which results in lower PSTN revenues.

Tab. A9: Source of Revenue for Public Telecommunications Operators (1990)

Country	Total revenue (US$ million)	PSTN revenues* (in %)			Other reven. (%)
		Gespräche	Anschluß- und Grundgebühren	gesamt	
Austria	9 292,2	64,69	13,76	78,45	21,55
Belgium	3 216,8	54,09	16,54	70,63	29,37
Denmark	2 581,3	n.a.	n.a.	n.a.	n.a.
Finland	1 968,3	40,71	27,07	67,78	32,22
France	23 163,0	65,00	n.a.	n.a.	n.a.
Germany	34 543,7	50,24	18,57	68,81	31,19
Greece	1 584,8	60,42	21,56	81,98	18,02
Ireland	1 383,9	62,25	20,77	83,02	16,98
Italy	20 207,5	54,07	27,65	81,72	18,28
Luxembourg	231,4	n.a.	n.a.	n.a.	n.a.
Netherlands	5 965,3	44,43	n.a.	n.a.	n.a.
Portugal	2 061,3	58,38	19,30	77,68	22,32
Spain	11 660,2	56,74	24,94	81,68	18,32
Sweden	6 042,8	27,29	18,92	46,31	53,69
UK[1]	26 500,2	52,51	16,86	71,37	28,63

1 British Telecom
*PSTN = Public Switched Telephone Network

Source: OECD (1993), p.22, own calculations

Low connection charges in France, Germany and Greece could be an indicator for subsidization as an incentive to extend the subscriber base. Different rental and call charges indicate a similar strategy if fixed costs (rental charges are costs of ownership) are relatively low in comparison to variable costs (call charges are costs of network use).

The category "Other Revenues" includes leased lines, data communications services (including packet switching), other text services (including electronic mail), videotext, image-based services (including video-conferencing and inter-regional TV programme exchange) and mobile services (including cellular radio, public mobile radio and paging), plus other miscellaneous or experimental services. The share of value-added services like "other text services, videotext and image-based services" seems to be relatively low in comparison to the total revenue of the national PTO. From this point of view, three aspects have to be considered. First, the share of PSTN revenues has already diminished over the last few years, which indicates an increase of diversification of national PTOs. Second, the true figure is probably much higher since many privately-owned operators and service providers are excluded from this table. Third, the EC average of 16.7% is far below the OECD average of 26.9%, as PTOs in countries like Sweden, Finland, Australia and the United States earn more than 30% of their total revenue from non-telephony services (OECD, 1993, p.23).

An example of a successful VAS System provided by a national PTO is the French Minitel system, which was launched in 1985. The Minitel terminal is a simple specialized terminal equipped with screen and keyboard. The terminal enables the subscriber to obtain information from private service providers and to communicate with other terminal holders. At the end of 1992, 6.3 mill. terminals had been installed and over 17 200 services provided over the network (FRANCE TELECOM, 1992 and 1993). Three key factors of this rapid growth should be emphasized. First, the French administration chose an easy-to-use version of videotext with low production costs. Second, a very liberal policy created a booming service environment. Third, the "Kiosque" system simplified payment for services since France Télécom passes payment to the service providers without paperwork or special arrangement (UNGERER and COSTELLO, 1990, p.56). A comparison to the German Bildschirmtext (Btx) system, which was launched in 1983, underlines the success of French Minitel: In 1992 the DBP Telekom counted 340 000 Btx subscribers and approximately 2 800 service providers (DBP TELEKOM, 1993).

A regulatory framework has to anticipate the present and future needs of customers concerning VAS. As demand for new services like video-conferencing, telework, telepublishing, multimedia communications and information services grows, regulatory market entry barriers for service providers should be avoided. A free market entry and exit can best serve the customers demand for special services since niche enterprises are able to provide specialized services according to customers' needs. A licensing system can be ingenious if a national licence for service providers is a prerequisite for the provision of services in other EC countries.

The attractiveness of a country as a place of business for national and foreign companies and multinationals is determined by the country's high level of qualified VAS. Therefore, the non-regulatory environment for the development of the enhanced service sector has to be examined. Access barriers to public networks, high access charges, high costs of data and service transmission and the prohibition of closed networks will certainly hamper the evolution of a qualified service sector.

4.2.3 Infrastructure

Telecommunications have become an essential tool in business and in private life. Therefore, the development of an efficient, technical up-to-date and dynamic telecommunications sector is necessary. The infrastructure, which consists of fixed and mobile networks, is a key element for telecommunications, since it permits the transmission of all kinds of different services.

New Telecommunications Model

In the past, several networks for several purposes have been constructed, since different services needed different transmission modes and transmission rates. Due to technological developments (digitization, ISDN), the mandatory link between special networks and special services no longer exists; this encourages the optimal use of existing (and future) infrastructures, but only if the regulatory framework is appropriate. The following example elucidates this idea:

The present Dutch telecommunications situation is described in Fig. A7b. In the Netherlands, like in most European countries, other network operators besides the Public Telecommunications Operator exist: e.g. authorized cable-television providers, Dutch Railways, Electricity companies, corporate networks and companies responsible for managing the "national" radio spectrum. Due to regulation, the use of those private networks is restricted to special services (like TV broadcasting), or to private use only. Three different sectors can be distinguished: the infrastructure, the network or basic services sector and the value-added service sector. Fig. A7b demonstrates that no competition in the infrastructure or basic service sector is possible due to regulatory restrictions and vertical links. The difference from a possible new telecommunication model, as shown in Fig. A7a, is quite obvious. Technological changes and deregulation make competition possible on all three levels.

Fig. A7: Present Situation and Telecommunications Model

Fig. A7a: TELECOMMUNICATION MODEL

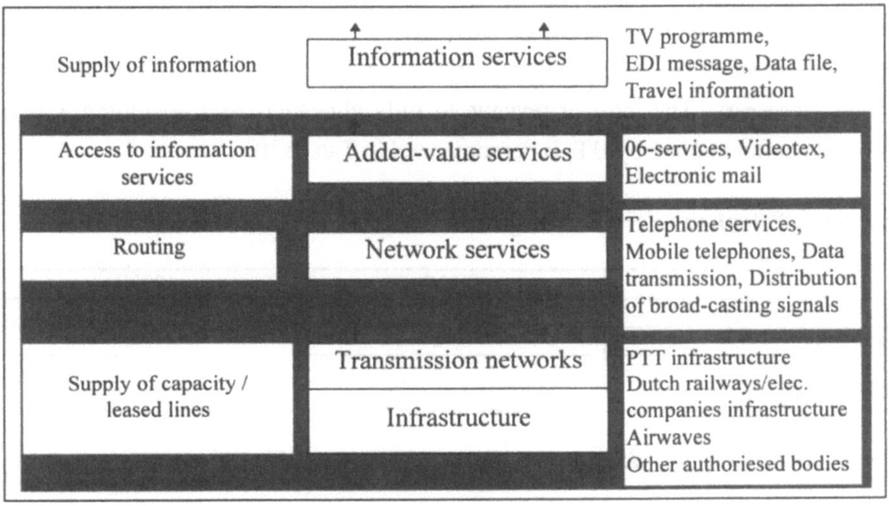

Fig. A7b: PRESENT SITUATION

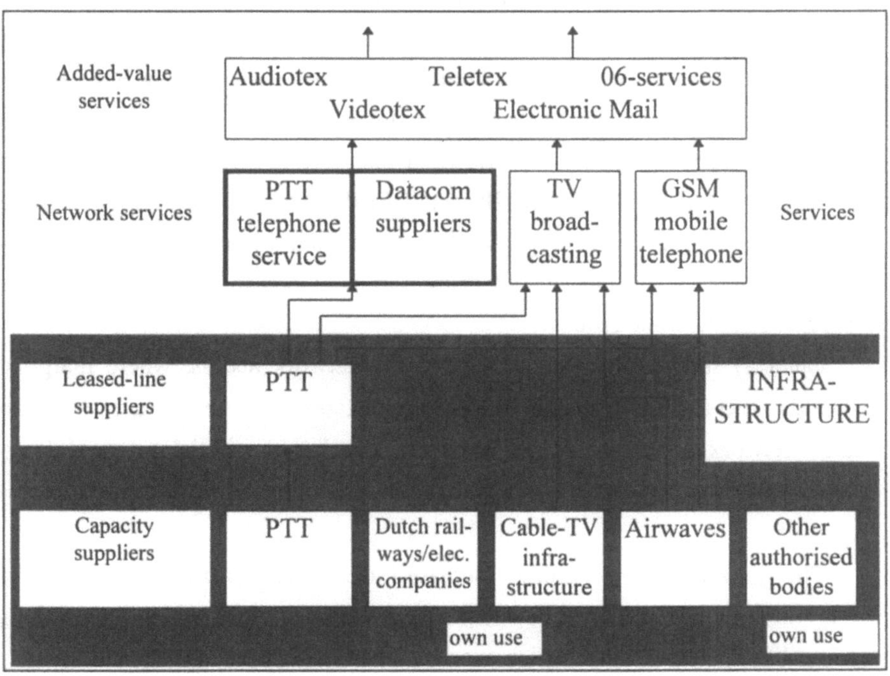

Source: Adapted from HDTP (1993)

Rebalancing of Prices and Welfare Gains

An effective competition leads to prices equal to marginal costs. Prices set by national telecommunications monopolies are often distorted. Due to politico-social goals and the principles of average tariffing, the prices of basis access are well below its incremental costs, making large cross-subsidies from other telephone services necessary. The ratio of revenue to fully allocated costs, including a 17% return on capital applied by BT, is demonstrated in Tab. A10.

Tab. A10: Telephone Service Margins in the UK

	1982/3	1990/1
Exchange line rentals	0,69	0,56
Local calls	0,72	1,46
National calls	2,24	2,00
International Calls	1,23	1,59

Source: Attenborough (1993), p.6.

Although there are inevitable difficulties and uncertainties in allocating elements of common costs, Tab. A10 underlines the above-mentioned price distortions.

If effective competition leads to cost-oriented prices, welfare gains or losses subject to price elasticity occur. In the United States, price elasticity of basic access is near zero (in other words inelastic), while price elasticity of long-distance services varies from about -0.25 to -1.2 (and therefore elastic), depending on the type of services (HAUSMAN, TARDIFF and BELINFANTE, 1993, p.178). If those elasticities are also valid for Europe, welfare gains from a price reduction of long-distance calls exceed welfare losses from a price increase of basic access, as described by Fig. A8a and A8b.

Panel a) shows an inelastic demand for network access, where marginal costs are k'_A. It is assumed in panel b) that long-distance call rates are used to cross-subsidize network access: marginal costs are k'_L, but the initial price of long-distance calls is PL_0. Due to cross-subsidization, the initial price of network access is PA_0.

The welfare loss from a price increase for basic access from PA_0 to PA_1 is equal to the triangle A'B'C'. The welfare gain from a price reduction of long distance-calls is equal to CDE, which exceeds the triangle A'B'C'.

In addition, a shift of the aggregate demand curve for long-distance calls (DL) is likely. On the one hand, an inward shift of the demand curve is possible due to a reduced number of subscribers. On the other hand, because of lower transmission costs, one may expect new innovative value-added services to emerge, which could lead to a higher aggregate demand of long-distance calls (DL_1). In Fig. A8b the latter case is expected to lead to a welfare gain equivalent to the area BCDF.

Fig. A8: Cost-oriented Tariffing and Welfare Gains

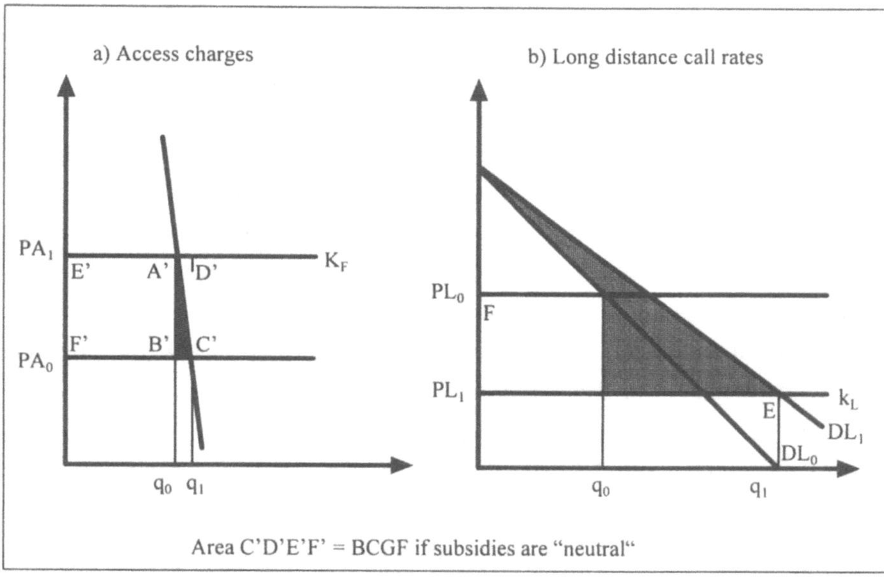

DA = aggregate demand for basic access
DL = aggregate demand for long-distance calls

This phenomenon can be observed in the US, since American consumers spend twice as much money on telecommunications services as most Europeans and Japanese. This difference can only be explained by the higher volume of calls and the greater use of advanced telecommunications services in the US since US long-distance call charges are rather low in comparison to Europe (OECD, 1993, p.21).

The potential economic welfare benefits from rebalancing are substantial. According to a study comparing the actual price structure in Britain with a possible rebalanced price structure, changes in relative prices that occurred between 1982/3 and 1990/1 led to a gain in economic welfare equivalent to about £ 580 mill. per year, or 0.1% of UK GDP (ATTENBOROUGH, 1993, p.6).

Price Cap Regulation or Direct Competition

If competition is not effective or is absent, price cap regulation is a strategy to rebalance prices and increase the efficiency of a national PTO. Whether competition or price cap regulation is more efficient is a question of market failure versus policy failure. Nevertheless, in the period of transformation price cap regulation can be a useful instrument to adapt prices to the changed circumstances. In Germany, a new price cap regulation is expected to take effect in 1996. The Bundesministerium für Post und Telekommunikation (BMPT) intends to reduce prices to 87% (including tax) and 75% (excluding tax) of the current level until January 1998. The aim is to bring down German prices to a European level before the 1998 liberalization of voice telephony to European service providers (BMPT, 1993).

In the UK, price cap regulation was introduced in addition to competition. OFTEL, the British regulatory body, pursued two aims: the reduction of phone call prices and the promotion of competition. To reduce prices for local and national phone calls, an initial RPI-3 price cap was introduced. In 1989, the price cap formula was increased to RPI-4.5; in 1991, a new RPI-6.25 formula took effect. Moreover, the price cap regulation has been extended to international phone calls. Since August 1993, a RPI-7.5 price cap regulation has been valid (OFTEL, 1993, p.5). This development raises the question of whether price cap regulation is a trial and error process due to asymmetrical information, or a planned process anticipating the short and long term potentials of cost reduction.

Price cap regulation in Britain, however, seems to be less efficient than one would expect. Even though tariffs in Britain have fallen in real terms over the last years, tariffs are still high in comparison to other European countries with more liberal price setting rules like the Netherlands (OECD, 1993, pp.57-59 and PTT NETHERLANDS, 1993, p.9). To facilitate the market entry of Mercury, the second British network provider since the telecommunications act of 1984, exchange line rental prices are subject to an RPI+2 price cap regulation. Therefore, the price distortions shown in Tab. A10 are partly a result of the regulatory policy by OFTEL.

In terms of economic efficiency, price distortions lead not only to welfare losses, but also to incorrect signals to potential entrants; this may result in inefficient entry (in principal, the problem could also be avoided by separating network ownership and operations). This raises the question of whether price regulation is necessary at all.

Given a starting point of low or no competition at all, price liberalization could lead to soaring prices as the incumbent operator attempts to maximize its profit. On the one hand, the regulatory body has to prevent excessive pricing. On the other hand, price increases are incentives for market entry. The period needed to establish workable competition depends on the existence of alternative networks or the time necessary to build up competitive networks. To avoid excessive prices in the first stage of transformation, regulatory measures like price cap regulation might be considered.

Competition and Productivity

Tab. A11: Mainlines per Employee in PTO 1980 - 1994

	1980	1990	1992	1994	CAGR 1980-1990	CAGR 1990-1994
Germany	105.3	141.4	139.8	174.2	2.99	5.35
France	98.7	157.0	185.8	207.1	4.75	7.17
Spain	102.4	167.1	185.3	203.4	5.02	5.04
Sweden	120.5	122.5	142.0	183.1	0.12	10.6
U.K.	73.5	111.8	152.7	196.9	4.28	15.2

Compounded annual growth rate
* Negative growth rate is a result of German unification

Source: OECD (1993), various annual reports, own calculations.

Tab. A11 shows that productivity has strongly increased during the last few years in comparison to the decade between 1980 and 1990. In general, this tendency is likely to be a result of national and international pressure to offer more reliable and sophisticated services at lower prices. The announced EC telecommunications policy for the opening-up of national markets for telephony services in 1998 might also play an important role since national PTOs will then no longer be protected by national market entry barriers.

Fig. A9a: Cost of International Calls, Average Single Line Business Customer, Average Cost per Call Minute, 1983-1994 (in 1994 UK £)

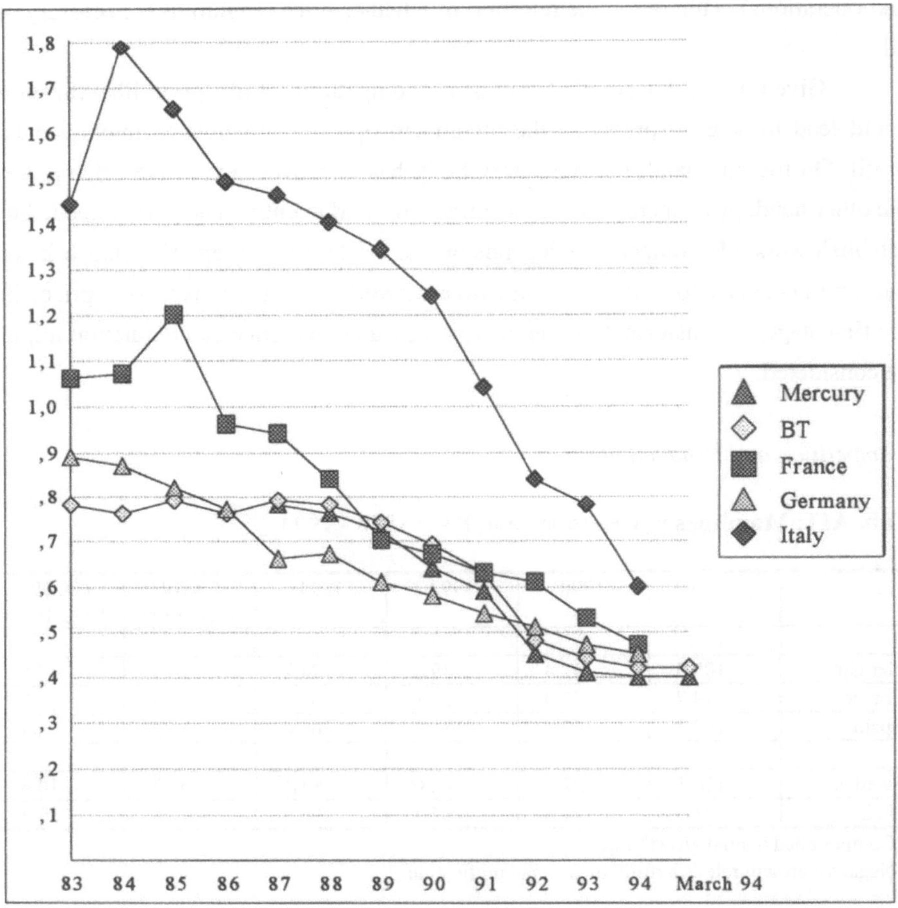

Source: Analysys (1994)

Fig. A9b: Total Cost of Calls, Average Residential Customer, Average Cost per Line, 1983-1994 (in 1994 UK £)

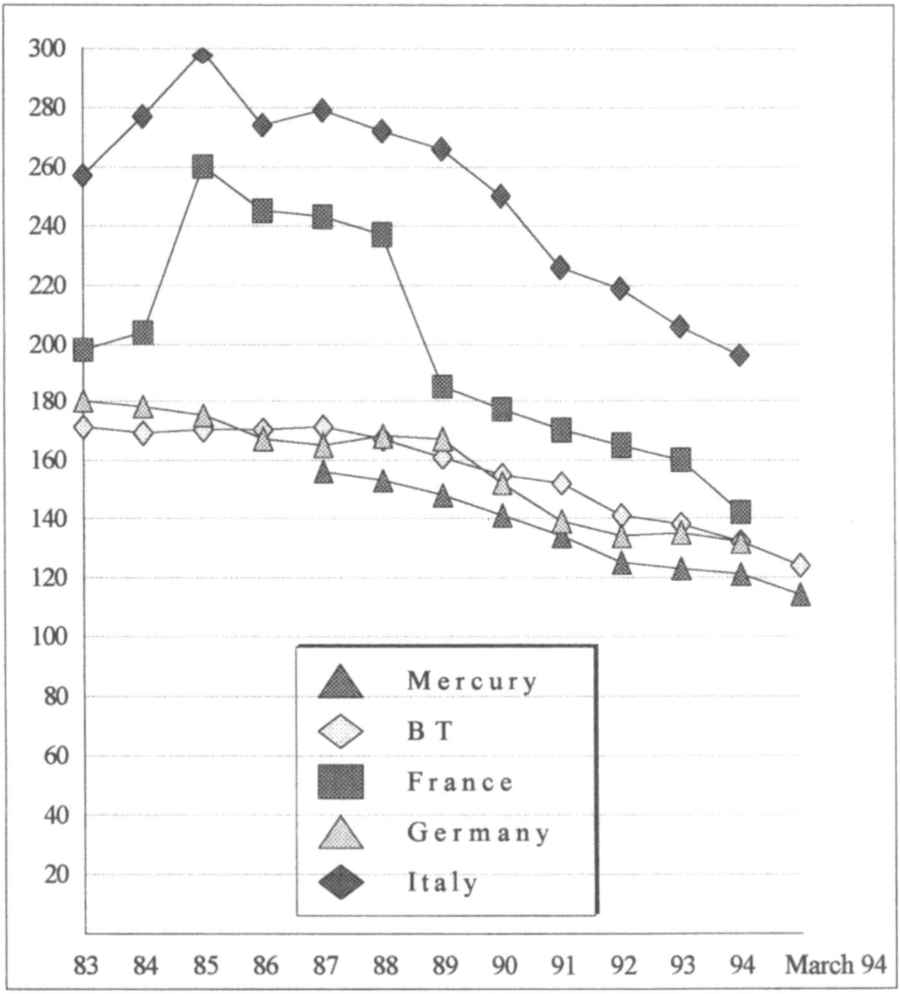

Source: Analysys (1994)

Sharp increases in productivity can especially be observed in Sweden and the United Kingdom, because incumbent operators are forced by competitors or potential competitors to reduce costs and X-inefficiencies. In 1993, Telia, one of the two Swedish network operators, reduced its number of employees from 41,771 in

1992 to 34,090 in 1993 causing an increase of labor productivity of 14.0% (TELIA, 1994, p.70). Nevertheless, further job cuts in the telecommunications sectors are likely to occur. Even if BT has already reduced its staff from 226,900 in 1990 to 137,500 in 1994/95, the number of mainlines per employee is still low in comparison to US standards. US Bell Operating Companies (BOCs) currently have an average of 265 mainlines per employee; 391 mainlines is the expected number for a BOC employee at the end of the century (FINANCIAL TIMES, 1994a, p.5).

Competition and Prices

Fig. A9a shows the average cost per call minute of international calls; Fig. A9b shows the average cost per line for an average residential customer for Italy, Germany, France and the UK (BT and Mercury). The figures indicate a decrease in cost of international telephone calls and total cost of telephone services for all countries. Reasons for this general trend can be found in national pressure to reduce cost and in network modernization. The lowest charges can be found in the UK - a result of competition. In international calls, the gap between the newcomer Mercury and BT has narrowed progressively in the past few years, as competitive and regulatory pressures have forced BT's tariffs sharply downwards (FINANCIAL TIMES, 1994c, p.II).

Fixed networks are typical examples of the problematic nature of sunk costs. The average cost curve of a monopolist is described by k in Fig. A10. The price set by the monopoly will be somewhere above P_0 to cover average costs. In the case of a market exit, the net value to be recovered is much less than the purchase price of the network equipment plus installation costs. Therefore, in the case of a newcomer's market entry, only the net realizable value (reversible costs) has to be regarded by the incumbent operator. The reversible costs are depicted by k_R in Fig. A10; the k_N curve describes the average total costs of a newcomer. In this case a market entry of a newcomer seems unlikely, since the monopolist can credibly threaten to decrease its price below the lowest possible price of the newcomer (P_1).

Fig. A10: Sunk Costs and Market Entry

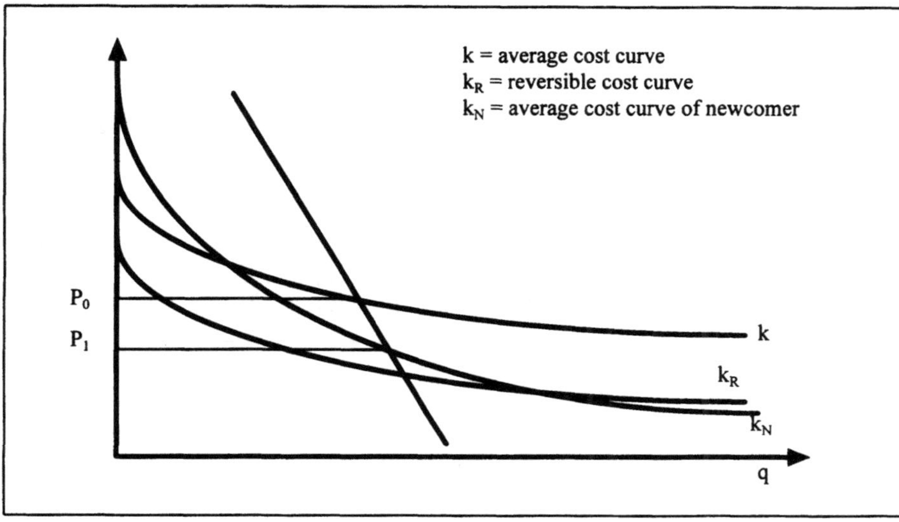

Market Entry Barriers

If alternative networks already exist, the sunk-cost problem is less severe. A cable television provider with its own network can enter the telephony market segment with lower costs than a total newcomer. Therefore, an inward shift of the k_N curve can be expected making a market entry more possible even without regulatory measures. In this case one could ask the question of whether a newcomer's market entry should be made easier.

In the Netherlands, the admission of a second network provider for the provision of data communications is planned before 1995. According to the Dutch minister of telecommunications, the new company should consist of different existing network providers like Dutch railways, electricity companies and cable-television network providers. The various infrastructure providers will in principle have the same rights and obligations. For the second provider, equal obligations include the supply of certain types of leased line on a nationwide and non-discriminatory basis. The draft of the new telecommunication act does not include asymmetrical legislation in favour of the new cooperative venture. Instead, special rights are given to the Dutch PTT (KPN), while the provision of voice communications is reserved for KPN until 1997. This transition period is seen as necessary for KPN in order to adapt to the changed circumstances and rebalance tariffs.

Tab. A12: BT Share of UK Public Switched Telephony Market in %

	1984/85	1990/1	1991/2	1992/3
All business lines and calls	100	94	90	87
All residential lines and calls	100	99+	99	98
Inland calls	100	96	94	93
• local	100	99	98	97
• long distance	100	94	92	90
International calls	100	89	83	80
District of London	100	70	n.a.	<50

Source: Attenborough (1993), p.3.

Even if the idea of a duopoly structure is tempting, the British experience shows relatively little success. In 1984 only two concessions for telephony services were given, one to British Telecom and the other to Mercury. Tab. A12 provides estimates of the change in BT's share of the public switched telephony market over the past 8 years.

The data shows that competition developed rather slowly, especially when compared to developments in the United States. Due to the divestiture of AT&T in 1984 and regulation policies that encouraged market entries, AT&T's share of interstate switched-services long-distance usage has fallen from 84% in 1984 to 63% at the end of 1991 (TAYLOR and TAYLOR, 1993, p.185).

The experiences of the duopoly strategy have led to a new regulatory framework in the UK. Since 1991, any provider can award a licence to a domestic service where adequate resources seem likely to be available and where environmental costs do not contest (CARSBERG, 1991a, p.3).

As estimates in Tab. A13 indicate, market shares of BT are predicted to decrease sharply in the UK in the 1990s. Apart from Mercury, cable TV operators will especially strengthen their future market position. By 2003, cable operators' television penetration rate will be at 47%, with telephone penetration rates of between 22% and 33% (FINANCIAL TIMES, 1994a, p.7).

Tab. A13: Market Shares of UK Telecommunications Carriers (per cent)

	1994	1995	1996	1997	1998	1999	2000	2001
BT	89	86	83	80	77	74	72	70
Mercury	11	12	13	14	15	16	17	17
Cable	0	1	2	4	6	8	9	9
Others	0	1	1	1	2	2	2	3

Figures based on estimates of total telecommunications service revenues, excluding entertainment revenues from cable television and customers' premises equipment from BT.

Source: Financial Times (1994a), p.7.

In the Netherlands, the British experiences were also taken into consideration. The decision for a duopoly was based on a McKinsey study, indicating that as of yet the market could not support more than two national infrastructure providers. The number of national infrastructure providers, however, will not be stipulated in the new Dutch telecommunications act (MAIJ-WEGGEN, 1993, p.6). A change of the future duopoly strategy is therefore possible and probably necessary, if the Dutch developments are seen in the context of European telecommunications policies in 1998.

To enable competition between various telecommunication operators, special attention must be paid to the regulatory framework's interconnection rules and universal service obligations.

Interconnection Rules

While the incumbent network operator offers all kinds of services, new market entrants or alternative network operators will limit their services to special market segments like trunk services (railways, electricity providers), local services (cable TV operators) and airwave services (satellite operators, GSM or PCN network operators). Interconnection rules are necessary to link the different networks to enable a full telecommunications service. Interconnections rules must contain technical compatibility and number portability, equal access and interconnection charges.

Technical compatibility is a necessary prerequisite of network connection. With respect to the European telecommunications market, technical minimum standards set by international organizations like the European Telecommunications Standardization Institute (ETSI) should be introduced to ensure a minimum of communication quality. Number portability can also be attached to the point "technical prerequisites"; the lack of number portability can be a significant barrier to the development of competition in the supply of local telecommunications services. In the UK, market research indicates that about 20% of people who could profit from cable TV telephony services find that costs for switching their custom to cable TV operators are too high; these costs include the changing of number and resulting loss of business (ATTENBOROUGH, 1993, p.12). Number portability should be introduced at an early stage to enable the change of network operators according to price differences.

Equal access to all network providers is another prerequisite of competition promotion. It enables customers to choose between two or more long-distance competitors according to prices and services. The selection is made at the local exchange and effective either by dialling additional digits or by a preselection for all calls (CARSBERG, 1991b, p.11).

Newcomers or alternative network operators are restricted to certain areas or services such as long-distance calls, local calls or mobile communications; access to other networks, especially to the incumbent PTO's network, is therefore crucial. Operators have to pay interconnection charges for the use of other networks, which should be cost-oriented from an economic point of view. Non-cost-oriented charges could restrict market entry if charges are disproportionately high. High interconnection charges can only be set in market segments with no or limited competition.

In most European countries, incumbent PTOs have a monopoly on local and long-distance phone calls. If a newcomer enters the long distance market segment, interconnections to local networks of the incumbent provider are indispensable. To prevent market entry, the incumbent PTO could increase interconnection charges, which are valid for both the incumbent and newcomer according to equal access principles. Even if interconnection charges have to be paid by both operators, only the newcomer is affected. Therefore, interconnection charges should be based on the principles of cost-orientation, transparency and non-discrimination as required by the directive of EC open network provision, and be monitored by the regulatory body.

Even if competition exists in all different market segments, interconnection charges still play an important role. In the local loop, competition between the incumbent PTO, cable network providers and airwave network operators is imaginable. Nevertheless, it is hardly economically advantageous for every household or enterprise to have access to every network. Given three different network providers A, B and C in an environment with no links or high interconnection charges, only the provider with the most new subscribers will survive. Assume that provider A is the incumbent PTO and newcomer B offers lower prices than A: Market entry seems unlikely since subscribers of B could not get a link to subscribers of A, or only at unreasonable prices.

Universal Service Obligations

Universal service obligations include obligations to provide general access to everyone at average prices. At present, those obligations are imposed on the incumbent PTO, which compensates for the losses of non-profitable areas with profits from other market segments. Since newcomers will only enter into profitable market segments (reducing prices and hence profits for the incumbent PTO), non-profitable segments can only be served at higher prices. Since higher prices for certain areas lead to several undesired side effects (especially for low income groups), new compensation rules should be introduced. Several different systems are imaginable: compensation via interconnection charges, via state subsidies to the operator or via direct transfers to low income subscribers.

Compensation via interconnection charges could lead to price distortions and market entry barriers. The efficiency of state subsidies depends on the award-system of subsidies. A once-and-for-all subsidy can be regarded as a disincentive for modernization and rationalization; a more market-oriented and incentive-emphasizing method should be introduced. This could be achieved by a licensing system granting fixed subsidy to that operator with minimum subsidy requirements. To enforce cost reductions, the licence should be limited to a fixed period, for example 15 years; after that period the licence will be newly auctioned. The money necessary for subsidies can be obtained by the revenues from licence auctions for profitable market segments, or via a levy on the turnover of all operators.

The abolition of average pricing will result in different access charges for customers in different regions. If those prices are cost-oriented, a more efficient allocation of scarce resources can be expected since real costs of a fixed network access will be compared to alternative networks (like airwave transmission). Rural areas can be expected to choose mobile communications. According to Tab. A10, an increase of access charges can generally be expected with regional differences due to the fixed cost bloc of a regional network operator. To prevent social disparities, direct payments could be given to low income groups, which is certainly more efficient than average pricing.

Mobile Network Competition

Interconnection rules must also be valid for mobile networks. In addition to these rules, another problem still has to be solved: efficient allocation of scarce frequencies.

Mobile network competition seems possible, as is shown by the example of France, Germany and the UK. Nevertheless, the correct form of competition, managed or totally liberalized, is still being contested. At present, licences given by national regulatory bodies are necessary to operate a mobile communications network. Even if economic aspects are taken into consideration, the number of licences seems to be arbitrary. Moreover, if competition is allowed, one or more licences is given to the incumbent PTO. In Germany, for example, a licence for an analogue and another for a digital network are given to Deutsche Telekom, while a third is granted to a private consortium; previously a fourth was also given to a second private consortium. This raises several questions.

The first question is whether a monopolist of one segment, like the Deutsche Telekom concerning fixed networks, should receive licences in market segments exposed to competition. From a theoretical point of view, this should only be allowed if welfare gains from economies of scope exceed welfare losses due to cross-subsidization and price distortions. According to the German competitors of Deutsche Telekom, several forms of cross-subsidization between Deutsche Telekom and DeTeMobil, a 100% subsidiary of Deutsche Telekom engaged in mobile communications, exist as credits at low rates of interest, lower costs of advertising and the use of telecom outlets at prices under costs (DEUTSCHE TELECOM E.V., 1993, p.3). Even if these criteria can hardly be verified due to different cost allocation methods, cross-subsidization is likely to occur.

Tab. A14: European Mobile Network Operators: GSM Digital Cellular Tariffs (in ECU)

Country	Operator	Connection charge	Monthly access	Call charges Peak	Off-peak
Austria	PTT Austria	33,25	28,82	0,39	0,25
Belgium	Belgacom	90,69	45,34	0,38	0,19
Denmark	Sonofon	74,31	8,07	0,28	0,14
	Tele Denmark	74,31	8,40	0,30	0,15
Finland	Telekom Finland	6,39	7,99	0,23	0,13
	Radiolinja	5,24	7,86	0,26	0,13
France	France Télécom	53,05	37,89	0,61	0,38
	SFR France	53,05	37,89	0,56	0,38
Germany	Deutsche Telekom D1	33,81	36,41	0,62	0,25
	Mannesmann D2	35,37	35,37	0,63	0,25
Greece	Panafon	86,32	34,53	0,32	0,16
	STET Hellas	86,32	34,53	0,32	0,16
Ireland	Eircell	63,74	25,50	0,32	0,22
Italien	Telecom Italia	106,57	27,48	0,36	0,11
Luxembourg	PTT	75,57	50,38	0,45	0,45
Norway	Tele-Mobil	9,82	4,91	0,33	0,29
	Netcom	0,00	3,93	0,35	0,29
Portugal	TMN	41,80	32,23	0,36	0,16
	Telecel	45,32	32,73	0,37	0,19
Sweden	Telia	33,54	6,71	0,30	0,13
	Comvik	32,96	11,18	0,22	0,12
UK	Cellnet	65,02	32,51	0,33	0,13
	Vodafone	65,02	32,51	0,33	0,13
	Mercury One-2One*	25,80	25,80	0,21	0,10
	Orange**	39,01	32,51	0,26	0,13

* PCN - network, expected prices;
** Individual charges exist for Talk 15, Talk 60, Talk 200, Talk 360, Talk 540; Talk „X" includes „X" free minutes of airtime. In this table, charges for Talk 60 (including 60 free minutes of airtime) has been choosen.

Source: Financial Times (1994c), Deutsche Telecom e. V. (1993), own calculations

The second question concerns the number of licensees. Since mobile communications markets are characterized by two-digit rates of increase, the optimum number of operators can hardly be estimated. From an economic point of view, licensing could hamper the development of new services as well as new and more efficient transmission networks. From a technical point of view, the necessity of licensing is often emphasized by scarce frequencies; frequencies given to a mobile network provider cannot be used by other competitors, causing a minimum number of frequencies per provider to be necessary. Nevertheless, sophisticated technology makes the use of allocated frequency bands more efficient and leads to the exploration of higher frequency bands. To enable an efficient allocation of frequencies, frequency rights should be tradeable and transferable. The first rights should be auctioned to promote optimum use of frequencies and should include obligations and commitments to prevent price speculation.

Tradeable frequency rights facilitate the market entry of newcomers since they can purchase frequency rights from incumbent firms. An incumbent mobile communications operator has to compare business profits with the opportunity costs resulting from sales values of frequency rights. If opportunity costs exceed business profit, the sale of frequency rights is economically wise. Therefore, tradeable frequency rights reduce market exit barriers since sunk costs are fully or partly compensated for by the sales profits of frequency rights.

Tab. A14 shows GSM (and two PCN) digital cellular tariffs in ECU in 1994. The individual chargeable elements of operator's tariffs (connection charges, monthly charges and call charges) show striking differences. Over the last twelve months, however, more than 60% of GSM operators have reduced prices for peak rate call charges (FINANCIAL TIMES, 1994c, p.V). Reasons for price reduction are different. Vodafone of the UK has reduced tariffs some 17.5% in the face of growing domestic competition. One of the new competitors in the UK is Mercury One-2-One, a system being based on PCN-technology. Tab. A14 shows the One-2-One charges for business subscribers; monthly charges are lower and call charges higher for private subscribers (DEUTSCHE TELECOM E.V., 1993, p.14). Even if competition between British analogue and digital mobile network operators is still restricted to London and other major cities, competition forces operators to decrease prices. In Germany, both GSM operators (Deutsche Telekom Mobil D1 and Mannesmann D2) have maintained high tariffs levels because of overwhelming demand for mobile communication, reinforced by German unification. The German example shows that two operators will not compete in a fast growing market. Yet, a sharp drop in prices could be observed in 1995 after the market entry of a third mobile network operator (E-Plus).

Conclusions

Fig. A11 and Fig. A12 summarize the main ideas of Section 3 and Section 4. Fig. A11 shows the influence of different kinds of market entry barriers and entry facilities on different market segments. An X indicates that market entry is hampered due to economic, technical or politico-social reasons; a double X indicates that these factors strongly prevent market entry. Fig. A11 shows strong market segment differences since no X can be found in the terminal equipment sector. In contrast, market entry barriers strongly influence the network segment, especially the fixed network segment. Concerning the latter, the most important barriers to entry are certainly economies of scale, especially in the local loop, and sunk costs. In Western European countries, rights of way are likely to play a minor role in comparison to Central and Eastern European countries, where uncertainties concerning property rights and competence exist. Since alternative networks heavily depend on their interconnection points to the network of the incumbent telecommunications operator, high interconnection charges are another strong barrier to entry. Interconnection rules are also very important to mobile networks. In mobile communications, scarce frequencies due to technical reasons (or political decisions) play a even more important role since a mobile network operator without frequency rights has no chance of entering a market.

Entry-easing factors (Z) open up markets. A double Z indicates that those market entry-easing factors have strong effects on the market segment. The total number of Z exceeds the total number of X; this could be seen as an indicator of deregulation and competition in telecommunications markets. In the fixed networks, the decrease of transmission costs by technological changes (micro-chip developments for switches, digitization and ISDN) and the demand increase of new services and capacity facilitate the market entry of potential network operators. Entry-easing factors strongly affect the mobile network segment since technological changes increase the use of scarce frequencies (in highly populated areas) allowing more competing operators and reduced transmission costs. Moreover, technical changes lead to a higher mobility that offers new opportunities to subscribers and service providers. The demand for capacity is therefore expected to increase.

Keeping these market entry barriers and entry facilities in mind, one can see Fig. A12 as a summary of potential competitors and necessary regulatory measures to create a competitive environment.

Fig. A11: Market Entry Barriers and Entry Easing Factors

	Equipment		Services		Networks	
	Terminal equip-ment.	Switches, Cables etc.	Value added service	Basic services	Fix	Mobile
ENTRY BARRIERS						
Economic reasons						
- Economies of scale		X			XX	
- Economies of scope					X	
- Learning effects		X	XX	X	X	X
- sunk costs		X			XX	X
- Interconnection charges					XX	XX
Technical reasons						
- Number portability					XX	X
- Scarce frequencies						XX
- Rights of ways					(X) X	
Politico-social reasons						
- Universal services			X	XX	X	
- Subsidies					XX	X
Entry Easing Factors						
Technological changes						
- Micro-chip developments	ZZ	ZZ	Z	Z	ZZ	ZZ
- Digitization + ISDN	Z	Z	ZZ	Z	ZZ	ZZ
- Fibre optic			ZZ		ZZ	
- Airwave transmission			Z	Z		ZZ
Falling Costs						
- Equipment	ZZ	ZZ			Z	ZZ
- Transmission			ZZ	ZZ	Z	ZZ
Increasing demand of						
- New services			ZZ		ZZ	ZZ
- Equipment	ZZ	Z				
- Capacity					ZZ	ZZ

X = prevents market entry
XX = strongly prevents market entry
Z = opens up markets
ZZ=strongly opens up markets

Fig. A12: Market Segments, Likely Competitors and Regulatory Measures

Market segment	Potential competitors	Regulatory measures
EQUIPMENT		
Terminal	national and international terminal equipment producers (market entry of Eastern European producers possible)	technical approvement by independent laboratories according to technical requirements
Switches, cables, etc.	specialized MNCs like Siemens, Alcatel, Ericson, etc. intermediate goods from Eastern European Countries / local production in Central and Eastern Europe	no regulation! if markets are not liberalized, PTO: incentive regulation or monitored procurement
SERVICES		
Value-added services	national and international service providers and network operators small firms as niche providers	national licence if reprocity rule in EU, equal access rules, ONP, national / international settlement rules
Basic services	national PTO other fixed network operators mobile communication opperators	minimum level of services (emergency call, area coverage requirements), compensation for uniform service obligations
Mobile communications services	national PTO mobile communication providers cable TV operators (PCN links?)	prevention of cross subsidization, auctions of scarce frequencies, tradable frequency rights, non-discriminatory access rights
NETWORKS		
Local loop	national PTO cable TV operators energy utilities mobile communications operators corporate networks (LAN, WAN)	interconnections rules, number portability, minimum technical standards, open interfaces, transfer payments to low income subscribers, subsidy for service obligations
National level	national PTO corporate networks railways, electricity utilities mobile communications operators satellite systems (?)	interconnection rules, equal access, compensation rules for non profitable areas
International level	national PTO alliances (Unisource, trans European networks) railway, electricity utilities mobile communications operators multinational corporations satellite communication systems	international interconnection rules, international technical standards, equal access, international frequency allocation, international number systems

5. Prospects and Consequences for Central and Eastern Europe

European telecommunications markets are currently in a period of transition from national monopolies to local, national and European competition. Here, EC regulation plays the most important role. The EC Commission intends to create a single European telecommunications market to promote competition within the European Community and strengthen the position of European providers in international markets. Since the 1984 Community action programme on telecommunications, several directives (binding EC legislation) have been passed. The progress towards liberalization in telecommunications equipment, value-added services and data transmission is striking. Even if effective competition in those fields seems relatively small from a consumer's point of view, the path to competition is open. The market for telephone sets is a positive example. After liberalization, the supply of different telephone sets has sharply increased at falling prices; moreover, telephone sets are becoming increasingly service- and user-oriented. The high rate of technological development in OECD countries offers new opportunities for Central and Eastern European countries wanting to modernize their telecommunications system and increase East-West trade in Europe.

In the EU, the decision to open up the market for voice telephony is probably the most contested but crucial step towards competition. This step requires a totally new regulatory framework concerning interconnection rules, universal service obligations and compensation measures. A new regulatory framework to solve these problems in a satisfactory manner can be found neither on the European nor national level. The experiences in Britain emphasize the problems resulting from interconnection charges, equal access and other market entry barriers (e.g. the non-existence of number portability). The British example, however, also demonstrates the general possibility of network competition, even if its success is limited under the duopoly strategy and the above-mentioned non-regulatory entry barriers. The new regulatory steps towards general and unrestricted competition are therefore relevant to other European countries embarking on telecommunications liberalization (GRAACK, 1995).

In the Netherlands and France, new regulatory conditions are currently under discussion. Since both countries intend to liberalize their voice telephony market, the discussions include the future role of the regulatory body and necessary flanking measures. The proposals emphasize the use of alternative networks like railways networks, electricity networks and cable TV operators. In the Netherlands, however, full liberalization of the voice telephony market cannot be expected before 1997. According to the Dutch Ministry of Telecommunications, a transition phase is needed for the Dutch PTT to rebalance its distorted price system. In France, a step towards a partial liberalization of alternative networks for voice telephony has been announced: The use of alternative networks will be limited to the second French mobile network operator to enable different base-station links via leased lines.

Future competition of network services will result from corporate networks. Corporate networks consists of local, national and international networks still reserved to a single firm or special user groups. Nevertheless, these networks are gaining importance since they offer voice, data and video service to the most attractive customer: the business sector. It is estimated that 14,000 private data networks existed in Europe in 1989. At the start of 1995, the number of data networks is expected to be 24,700 (BOHLIN and GRANSTRAND, 1991, p.465).

Some German developments (especially with respect to German unification) are of special interest; they concern both the role of corporate networks and the modernization of the East German telecom sector.

In Germany a corporate network operator was founded by Mannesmann, Deutsche Bank and RWE, an electricity utility. The consortium intends to invest several hundred million Deutsche Marks within the first phase to extend their already existing networks (BMPT, 1994, p.7). The creation of an efficient alternative network is of special interest to Mannesmann since it is a joint proprietor of the D2 consortium, a mobile communications network operator. It is therefore imaginable that after the liberalization of voice telephony services, the newly founded corporate network operator will also be used to create a link for transmitters of the E-Plus mobile communication network. The size of the firms and their financial potential clearly suggest that after 1998 those network operators will be able to penetrate into highly profitable markets underlining the necessity of regulatory measures with respect to universal services obligations.

The telecommunications situation in the GDR was similar to that of most ex-CMEA countries. After the unification of East and West Germany, the monopoly rights of Deutsche Telekom were spread to East Germany. By the end of 1992, DBP Telekom had invested approximately DM 17 bill. ($ 10 bill.) and created around

1.35 mill. new main lines (DBP TELEKOM, 1993, p.4). According to the average cost per main line of $ 1500, between $ 7 and 8 bill. have been invested in network modernization. Telekom is striving to install West German telecommunication standards in East Germany by 1997; a faster modernization and extension of the network is impossible due to financial constraints. The share of equity has fallen from 31% in 1990 to 24% in 1992, which indicates that financial requirements are being satisfied by credit markets. For ex-CMEA countries, financial constraint would be even more important rendering a monopoly strategy in network operation adequate.

Telecommunications markets in Central and Eastern European countries have to be viewed in the context of the above-described developments. The German experience makes it clear that a rapid modernization and extension of telecommunications network in Central and Eastern European countries is incompatible with a monopoly strategy for network operation. Even in Germany, critics have demanded a new and more competition-oriented regulatory framework to accelerate network extension (MONOPOLKOMMISSION, 1991, p.8).

The telecommunications infrastructure in Poland, Hungary and other ex-CMEA countries is obsolete; the penetration rate varies in a range of 10 to 15%; which is low even in comparison to relatively poor European countries like Portugal (24%), Ireland (28%), Spain (32%) and Greece (38%). The average penetration rate amounted to 41% in the EC.[7] From 1980 to 1990, the EC average growth rate of main lines was 4.75% per year, with considerable above average growth rates in Greece (5.69%), Spain (5.72%), Ireland (7.36%) and Portugal (9.17%).

Polish Perspectives

To achieve an EC average penetration rate of 41%, annual growth rates in Poland must be around 15% per year, even higher if an increasing average penetration rate and population trend is taken into consideration. However, the data indicates an enormous potential for demand since around 12 mill. more main lines are necessary to reach the EC average. The international average cost per main line is calculated around $ 1.500, which leads to financial requirements of $ 18 bill. or so, plus investment for the modernization of existing networks. Between 1980 and 1990, 23.35% of total telecommunications investment has been spent on network modernization in Spain, 48.22% in Portugal and 65.36% in Greece (see Tab. A3).

The investment requirements clearly show that a single network operator is overburdened.

As an associate member of the European Union that has applied for full membership, Poland should adopt EC legislation at an early stage. The adoption of telecommunications legislation could give positive signals to European service and network providers to invest in Poland. In addition to EC legislation, network competition and the commissioning of alternative network providers (electricity utilities, railways networks) are likely to reduce bottlenecks in long distance-calls. To link these networks, new interconnection rules under the principles of transparency, cost-orientation and equal access are important prerequisites. Access charges and call tariffs should be cost-oriented to avoid price distortions and to promote business activities via telecommunication.

To accelerate network expansion, the scarce basic access has to be allocated according to potential subscribers' marginal willingness to pay. This will certainly create social conflicts since the business sector and wealthy people are privileged. Nevertheless, the telecommunications sector will only prosper if profits that can be reinvested in network expansion and modernization are made. To promote network expansion, investment obligations or levies on the network providers' turnover, to compensate poor or non-profitable network operators, are imaginable. These measures should be temporary limited and regularly examined to avoid wrong investment incentives and a waste of scarce financial resources. Fig. A13 concludes with the consequences for Central and Eastern Europe.

Fig. A13: Consequences for Central and Eastern Europe

Many countries in Eastern Europe consider EU membership a major political goal. Poland, Hungary and the ex-CSFR are looking Westward for economic integration, technological modernization and useful experiences in regulation. The developments in the EU offer several broad messages or lessons:

1) The telecoms market ranks among the most dynamic markets in the world economy; a steady pace in the modernization race is particularly important for Eastern Europe. Long term modernization requires that strategic decisions are made for the telecoms business.
2) Internationalization of the telecoms industry is a typical trait in Western Europe and could be even more useful in Eastern Europe. International consortia are working toward bundling know-how and rapid access to advanced technologies as well as forward easier connection with foreign networks. US and European network operators which become part of a consortium in Eastern Europe have an incentive to promote international links and reinforce the attractiveness of the networks by opening up to cheaper foreign service providers.
3) An expanding outward-oriented private business sector and an expanding service sector are two major proponents of telecoms liberalization. Progress in privatization in Eastern Europe will thus raise the demand for liberalization.
4) A decentralized regulatory structure is useful in the stage of state telecoms networks: thus deregulation can more easily be achieved.
5) The EC shows: A lack of competition in network operation implies having a hardware industry which is not internationally competitive and therefore yields low profits (in an East European context this could imply high subsidies); small tax receipts and the foregone tax revenue from other sectors which would expand in the presence of a competitive telecoms industry are certainly more important than the monopoly rent a government could extract from a continued dominance of a national network operator, a wide range of reserve services and a mainly national equipment industry.
6) The technological developments in the computer and telecoms industry imply that it will not be feasible to maintain a national network operator with a dominant realm of reserved services. In an environment of rapid technological change a state monopoly would require ever-expanding complex regulations due to progressively increasing regulation costs and increasing opportunity costs of the monopoly.
7) Increasing market integration in Western Europe and new telecoms alliances raise the question of whether countries in Eastern Europe should not consider joint regulatory frameworks and cross-border alliances in the telecoms business.
8) Small open economies like the Netherlands and Sweden, and not big countries like France, Germany, or Italy, are leading prospective liberalization in telecoms. The UK is an exception in this respect: BT's sustained dominance shows that a simple privatization of a state monopoly is insufficient to achieve true competition. In France some network competition is being considered, but political determination seems unclear.
9) Big countries (the UK) as well as small countries (the Netherlands) can have competing long-distance networks. Most countries in Central and Eastern Europe could envisage competition in trunk networks.
10) Mobile telecommunication offers the fastest way to raise the telephone penetration rate. At least two mobile network operators are necessary to obtain a minimum of competition.
11) Building-up a network requires a different regulation than a mature state of network saturation.
12) Price structures in Western Europe are grossly different from the law of one price, suggesting inefficiencies and lack of competition: Furthermore, compared to the US and Canada telecoms services in Western Europe are quite expensive, underlining the inefficiency of having a national network operator.
13) Cross-subsidization should be avoided in the telecoms industry. Prices based on (least) costs are the appropriate market signals.
14) Instead of imitating some poor West European models of state monopoly cum regulation, countries in Central and Eastern Europe could fully anticipate a liberalized West European telecoms sector and pursue deregulation early on.

Endnotes:

---

[1] A thorough discription of the telecommunications sector can be found in NOAM (1992). The book not only describes the dynamic changes and the forces that transform the traditional PTO system but also discusses them.

[2] It is no coincidence that the software company Microsoft has created a joint venture with McCaw, a cellular phone company.

[3] For radiopaging POCSAG (Post Office Code Standards Advisory Group) and ERMES (European Radio Messaging) are widely used, but a series of new technologies will be launched in the 1990s.

[4] For the same business tariff basket in January 1992, Mercury charged 9.9% less than BT, with a striking difference of 22.1% for the long distance call basket. OECD (1993), p.61.

[5] CCITT: International Telegraph and Telephone Consultative Committee.

[6] PABX stands for "private automatic branch exchange". Further explanations are given in: BOHLIN and GRANSTRAND (1991), p.455.

[7] Data from 1990 and 1991. OECD (1993), p.41 and BERLAGE and SCHNÖRING (1992), p.5.

Appendix A: Position of the Community and Community Action

POSITION OF THE COMMUNITY:
The telecommunications sector is the largest potential investor in new information technologies, using large numbers of mirco-electronic devices and software development. Telecommunication has a large potential effect on employment, productivity and competitiveness. The Community's traditional trade surplus in telecommunications equipment is decreasing rapidly. It now has substantial deficits with the USA and Japan. The Community faces a number of handicaps such as: fragmented national markets, uncertainties of companies over what development strategies to choose; backwardness of less-favoured regions in Europe with respect to networks; regulatory inconsistencies among the different Member States concerning tariffs, access conditions to the networks, public procurement, etc.

The economic benefits for Europe of 1993 will depend in large measure on completing the internal market for the communcations sector, as shown in the Commission's study on the economics of 1992 (Cecchini Report of 1988).
As a result of a combined equipment and service market around ECU 80 billion in the EC, gains of up to ECU 10 billion are to be realised from a single market.

COMMUNITY ACTION:
The objective is to create conditions which reverse the trends towards a relative decline of the Community's telecommunications industry and to permit economies of scale. The Community has taken action in several areas:

GREEN PAPER on the development of the common market for telecommunications services and equipment (COM(88) 0290).

The Green Paper was widely welcomed and has been followed up by a time-schedule of actions (COM(88) 0048) and a progress report on the implementation of a Community telecommunications policy (COM(88) 0240). One of the first actions resulting from the Green Paper was the Commission Directive on competition in the markets of telecommunications terminal equipment (OJL 131, 27.5.1988). The Council resolution of 30.6.1988 supported the Commission proposal to create a telecommunications market by 1992. It invited the Commission to pay special attention to the external effects (negotiation with third countries) in its liberalisation efforts.

SERVICE AND OPEN NETWORK PROVISION
In June 1989, the Commission adopted a Directive under Article 90(3), designed to abolish the exclusive rights granted by the Member States in the telecommunications service sector, with the exception of voice telephony. The Commission decided that this Commission Directive will take effect on the day on which the Council adopts the Directive on Open Network Provision (ONP). ONP lays down the ground rules for harmonisation of the conditions of access to and use of public telecommunications network, their interfaces and tarrifs.

- *Framework Directive 90/387/EEC/ (OJL 192, 1990) on Open Network Provision;*
- *Directive 92/44/EEC (OJL 165, 1992) on the application of ONP to leased lines;*

- *Proposal for a Directive on the application of ONP to voice telephony (OJC 263, 1992).*

ISDN (Integrated Services Digital Network)
Recommendation 86/659/EEC (OJL 382, 1986), on the introduction of ISDN with timetable for the introduction of ISDN services.

TERMINAL EQUIPMENT
Directive 86/361/EEC (OJL 217, 1986), amended by Directive 91/263/EEC (OJL 128, 1991) introducing a single approval procedure for the interconnection of terminals to the networks.

STANDARDIZATION
Directive 83/189/EEC (OJL 109, 1983) laying down a procedure for the provision of information in the field of technical standards and regulations;
Recommandation 84/549/EEC (OJL 298, 1984) on the implementation of a common approach in the field of telecommunications;

Decision 87/95/EEC (OJL 37, 1987) on standardization in the field of information technologies (IT) and telecommunications.

One of the goals of the action programme to implement the Green Paper was to establish a European Telecommunications Standards Institute (ETSI). ETSI, founded in 1988, is now responsible for preparing common telecommunications standards.

MOBILE COMMUNICATIONS
Directive 87/372/EEC (OJL 196, 1987), reversing certain frequency bands for mobile communications;

In 1991 the Member States began installing a single standard for mobile radiotelephony (GSM). This will replace the different incompatible systems in force in the EC (Recommendation 87/371/EEC, OJL 196, 1987). The Council adopted Directive 91/288/EEC to promote the introduction of DECT (Digital European Cordless Telecommunications) from January 1992. Finally, from 1992 a single European radio messaging standard (ERMES) was due to replace the five systems in force in the Member States (OJL 310, 1990).

SATELLITE COMMUNICATIONS
To define a European position on satellite communications, the Commission has issued a Green Paper on the Common Approach to Satellite Communications in the EC (COM(90) 0490). The Commission has proposed:

- *the full liberalisation of the earth segment;*
- *unrestricted access to space segment capacity;*
- *harmonisation and full commercial freedom for space segment providers.*

Quoted from: European Parliament (1994), Fact Sheet on the European Parliament and the Activities of the European Union, pp.239-240.

Appendix B: Liberalization

	Customer premises equipment				Telephone services		Other	
	Telephone	PBX	Telex	Modem	National	Intern.	Mobile	Data
EC & EFTA								
Austria	Y	Y	Y	Y	N	N	N	N
Belgium	P	P	P	P	N	N	N	N
Denmark	Y	Y	Y	Y	N	N	P	Y
Finland	Y	Y	Y	Y	N	P	P	Y
France	Y	Y	Y	Y	N	N	P	N
Germany	Y	Y	Y	Y	N	N	P	P
Greece	P	P	P	P	N	N	N	N
Iceland	Y	Y	Y	Y	N	N	N	N
Ireland	Y	Y	Y	Y	N	N	N	N
Italy	Y	Y	Y	Y	N	N	N	N
Luxembourg	Y	Y	Y	Y	N	N	Y	Y
Netherlands	Y	Y	Y	Y	N	N	N	N
Norway	Y	Y	Y	Y	N	N	Y	Y
Portugal	Y	Y	Y	Y	N	N	Y	Y
Spain	Y	Y	Y	Y	N	N	P	P
Sweden	Y	Y	Y	Y	Y	Y	Y	Y
Switzerland	Y	Y	Y	Y	N	N	N	Y
UK	Y	Y	Y	Y	Y	Y	Y	Y
Southern								
Cyprus	Y	Y	Y	Y	N	N	N	N
Malta	Y	Y	N	Y	N	N	N	N
Turkey	N	N	N	N	N	N	N	N
C & E								
Albania	N	N	N	N	N	N	N	N
Bulgaria	Y	Y	Y	Y	N	N	N	N
Croatia	Y	Y	Y	Y	N	N	N	N
CSFR	Y	Y	Y	Y	N	N	P	P
Hungary	Y	P	Y	N	Y	Y	P	P
Poland	Y	Y	Y	Y	Y	N	P	Y
Slovenia	Y	Y	Y	Y	N	N	P	P
Romania	Y	Y	Y	Y	N	N	Y	Y
Yugoslavia	P	P	P	P	N	N	N	N

Note: Y=Competition allowed; P=Partial competition allowed; N=Competition not allowed

Source: ITU (1992), European Telecommunications Indicators, p. 2.

References:

ANALYSYS (1994), Cutting the Cost, Cambridge: Analysys Publications.
ATTENBOROUGH, N. (1993), Regulation of Competitive Telecommunications Markets, n/e/r/a topics no.12 ,London: n/e/r/a.
BARTH, H. (1992), Moderne Telekommunikation: Netze, Dienste, Instrumente, Normen und praktischer Einsatz, München: Franzis.
BAUMOL, W. and SIDAK, G. (1994), Toward Competition in Local Telephony, Washington DC: American Enterprise Institute and MIT Press.
BERLAGE, M. and SCHNÖRING, T. (1992), Telekommunikation in Osteuropa Bestandsaufnahme und Entwicklungstrends, Diskussionsbeitrag Nr. 93, Bad Honnef: WIK.
BLANKART, C.B. and KNIEPS, G. (1989), What Can We Learn from Comparative Institutional Analysis? The Case of Telecommunications, Kyklos, 42, p.579-598.
BMPT (1993), Genehmigungskonzept des BMPT für die Entwicklung der Telefontarife der DBP TELEKOM, Bonn: Bundesministerium für Post und Telekommunikation.
BMPT (1994), Wirtschaft steigt ein, in: Post Politische Information, Bonn: Bundesministerium für Post und Telekommunikation.
BOHLIN, E. and GRANSTRAND, O. (1991), Strategic Options for National Monopolies in Transition: the Case of Swedish Telecom, in: Telecommunications Policy, Vol. 15, no. 5, p.453-476.
CARSBERG, B. (1991a), OFTEL: Competition and the Duopoly Review, in: Cento Veljanovski, ed. (1991), Regulators and the Market, London: Institute of Economic Affairs, p.98-106
CARSBERG, B. (1991b), The UK Duopoly Review - What it means for business users, London: OFTEL.
DBP TELEKOM (1992), Geschäftsbericht 1991, Bonn.
DBP TELEKOM (1993), Geschäftsbericht 1992, Bonn.
DECKER, P. and WALKE, B. (1993), Mobile Datenkommunikation - eine Übersicht, in: Informationstechnik und technische Information (IT+TI), No.5, p.12-25.
DEUTSCHE TELECOM E.V. (1993), Telecommunications Info, Nr. 95, Overath: Deutsche Telecom e.V.
EC COMMISSION (1987), Grünbuch über die Entwicklung des Gemeinsamen Marktes für Telekommunikationsdienstleistungen und Telekommunikationsgeräte - KOM(87) 290 endg., Brussels.
EC COMMISSION (1990), Richtlinie der Kommission vom 28. Juni 1990 zur Verwirklichung des Binnenmarktes für Telekommunikationsdienste durch Einführung eines offenen Netzzugangs (Open Network Provision ONP) (90/387/EWG; ABl.L 192/1, 24.07.90).
EC COMMISSION (1992), Bericht der Kommission an den Rat und an das Parlament zur Lage im Bereich der Telekommunikation, Ausgabe 10-1992, Ziff.1.3.67.
EC COMMISSION (1993a), Wachstum, Wettbewerbsfähigkeit, Beschäftigung, Weißbuch, KOM(93) 700 endg., Brussels.
EC COMMISSION (1993b), Panorama of EC Industry, Brussels.
EC COMMISSION (1993c), Entschließung des Rates zum Bericht über die Lage im Bereich der Telekommunikationsdienste, Ausgabe 6-1993, Ziff.1.2.122.

EC COMMISSION (1993d), Developing Universal Service for Telecommunications in a Competitive Environment, COM(93) 543 final, Brussels.
EUROPEAN PARLIAMENT (1994), Fact Sheet on the European Parliament and the Activities of the European Union, Luxembourg.
ECONOMIST (1993), Telecommunications, October, 23rd, 1993, London.
FINANCIAL TIMES (1994a), Telecom Markets, issue no. 243, March 30, 1994, London.
FINANCIAL TIMES (1994b), EU May Liberalize Telecoms Infrastructure Soon, June 22, 1994, London.
FINANCIAL TIMES (1994c), Telecommunications in Business, Financial Times Survey, June 15, 1994, London.
FOREMAN-PECK, J. and MÜLLER, J., eds. (1988), European Telecommunications Organisations, Baden-Baden: Nomos.
FRANCE TELECOM (1992), Rapport d'activité 1991, Paris.
FRANCE TELECOM (1993), Rapport annuel 1992, Paris.
FROHBERG, W. (1993), Künstliche Intelligenz und Telekommunikation, herausgegeben von Fridhelm Bergmann, München: Verlag Technik.
GRAACK, C. (1995), EU-Telecom Markets and International Network Alliances: Developments, Strategies and Policy Implications, Discussion Paper No.8, European Institute for International Economic Relations, Potsdam.
GRUPP, H. and SOETE, L. (1993), Analysis of the Dynamic Relationship between Technical and Economic Performances in Information and Telecommunication Sectors, Karlsruhe.
HARPER, J.M. (1991), A 21st Century Structure for UK Telecoms, London: CommEd.
HAUSMAN, J., TARDIFF, T. and BELINFANTE, A. (1993), The Effects of the Breakup of AT&T on Telephone Penetration in the United States, in: The American Economic Review, Vol.83, p.178-184.
HDTP (1993), Review of the Telecommunications Act, Outline Memorandum, Provisional point of view of the Cabinet, 15. July 1993, The Hague: Telecommunications and Post Department.
HENRY, C. (1993), Public Service and Competition in the European Community Approach to Communications Networks, Oxford Review of Economic Policy, Vol. 9, No. 1, p.45-66.
ITU (1992), European Telecommunications Indicators, Geneva.
ITU (1994), World Telecommunication Development Report 1994, Geneva.
KNIEPS, G. (1993), Europäische Telekommunikationspolitik, Freiburg i.Br., mimeo.
LAFFONT, J.-J. and TIROLE, J. (1993), A Theory of Incentives in Procurement and Regulation, Cambridge, Mass: MIT Press.
LUTZ, E. (1993), Mobilkommunikation über geostationäre (GEO) und umlaufende (LEO) Satelliten, in: Informationstechnik und technische Information (IT+TI), No.5, p.26-34.
MAIJ-WEGGEN, H. (1993), Definitve Government Standpoint on the Broad Outlines of the Review of the Telecommunications Act, The Hague: Ministry of Transport, Public Works and Water Management, November 17, 1993, (not fully authorized version).
MERCER (1994), Future Policy for Telecommunications Infrastructure and CATV Networks, Brussels: European Commission.

MEYER, J.R., WILSON, R.W. and ALAN, W. (1980), The Economics of Competition in the Telecommunications Industry, Cambridge, MA: Oelschlaeger, Gunn & Hain.
MONOPOLKOMMISSION (1991), Zur Neuordnung der Telekommunikation, Sondergutachten 20, Baden-Baden: Nomos.
NATA (1990), EuroTelecom 1992, Washington: North American Telecommunications Association.
NOAM, E. (1992), Telecommunications in Europe, New York and Oxford: Oxford University Press.
OECD (1993), Communications Outlook 1993, Paris: OECD.
OECD (1995), Communications Outlook 1995, Paris: OECD.
OFTEL (1993), BT's Price Changes, in: OFTEL NEWS, issue no.25, 25. Dezember 1993, London: OFTEL.
OMSYC (1991), Les chiffres-cles des télécommunications mondiale, Paris: OMSYC.
POSPISCHIL, R. (undated), Frankreich, in: F. Arnold, ed. (undated): Handbuch der Telekommunikation, (Loseblatt-Werk), Grundwerk, Kapitel 11.1.2.4., Köln.
PTT Netherlands (1993), Annual Review 1992, The Hague.
SCHNÖRING, T. (1992), Entwicklungstrends auf den europäischen Telekommunkationsmärkten, WIK Discussion Paper, No. 102, Bad Honnef: WIK.
TAYLOR, W.E. and TAYLOR, L.D. (1993), Postdivestiture Long-Distance Competition in the United States, in: The American Economic Review, Vol.83, No.2, p.185-190.
TELIA (1994), Annual Report 1993, Farsta.
UNGERER, H. and COSTELLO, N. (1990), Telekommunications in Europe, Luxembourg: Office for Official Publications of the European Communities.
VAN AS, H.R. (1993), The Evolution Towards High-Performance LANs and MANs, in: Informationstechnik und technische Information (IT+TI), Nr.4, p.9-19.
WEIZSÄCKER, C.C. von (1984), Free Entry into Telecommunications?, in: GIERSCH, H., ed., New Opportunities for Entrepreneurship, Tübingen: Mohr, p.201-239.
WEIZSÄCKER, C.C. von (undated), Die Bedeutung der Mehrwertdienste aus volkswirtschaftlicher Sicht, in: F. Arnold, ed. (undated): Handbuch der Telekommunikation, (Loseblatt-Werk), Grundwerk, Kapitel 11.1.2.4., Köln.
WELFENS, P.J.J. and C. GRAACK (1996), Telekommunikationswirtschaft. Deregulierung, Privatisierung und Internationalisierung, Heidelberg: Springer.

B. Telecommunications in Systemic Transformation: Theoretical Issues and Policy Options

Paul J.J. Welfens

1. Introduction

In OECD countries there are four major forces driving the liberalization of telecommunications: (i) national liberalization initiatives (e.g. US, UK, Sweden) which set a role model for other countries; (ii) the European Commission's liberalization scheme with the magic date of 1998; (iii) US pressure on EU countries to open up their national market as a prerequisite for EU consortia to invest in the USA; (iv) strong technological dynamics in telecoms technologies and computer network development. Already in the 1990s the local loop is no longer a natural monopoly as wireless technology allows to bypass the national operator's local terrestrial network. Moreover, with the expansion of the cable TV business and the option that electricity firms (grid companies) enter the telecoms market, there exist potential newcomers for the local network. In some countries the electricity grid companies will largely remain in the trunk (long distance) business, where they can easily add to competition. At the same time this will create new problems with regard to cost allocation because it is not clear to which extent the cost of adding optical fibres to the existing electricity network will be taken as the only marginal costs of telecoms network expansion, while an opportunity cost approach would include at least part of the depreciation of the existing pylons (WELFENS, 1995).

It is clear that the initial excess demand for telecoms services in transforming countries could lead to a rapid expansion of the existing and new networks. Such a rapid expansion will, however, not be cheap, unless competition helps to achieve efficient network development. The excess demand situation could temporarily be aggravated as the creation of new firms will add to the existing demand overhang.

Transforming economies that aim to become OECD member countries or want to join the EU or the European Economic Area naturally will monitor the liberalization drive in OECD countries. In transforming economies there are also some fundamental reasons why modernizing the telecoms sector is of strategic importance:

- An efficient telecoms system creates greater transparency in the goods sector. Telecoms modernization will stimulate arbitrage processes and thereby generate economic benefits.
- Even big firms can be organized in a rather decentralized way if there is a modern, efficient and innovative telecoms system. The legacy of postsocialist countries with their initially oversized firms makes a modern telecoms system particulary valuable. While dismemberment of firms sometimes might not be feasible because of a shortage of managers or because of political reasons, modern telecommunications technology allows firms to be organized in a decentralized manner. Indeed, the overall degree of decentralization in a market economy is determined by both the scope of the market and the degree of decentralization in firms.
- Telecommunications is important for the expansion of trade and the development of financial services (in turn a basis for domestic and international trade). By stimulating trade a modern telecommunication system - allowing exporters and importers to learn about new markets and to adjust to international price trends - becomes a valuable ingredient in the process of economic opening up in ex-CMEA countries.
- Telecommunications facilitates the process of building alliances among firms in the goods producing sector and in the service industry. This will generate economic benefits to the extent that cooperation raises factor productivity and that structural change towards a service society is facilitated.
- Foreign investors always place a particular emphasis on a modern telecommunication system which allows the subsidiary to be quickly linked up with the parent company and other subsidiaries. Given that ex-CMEA countries have to catch up with OECD countries in the internationalization of production, it is clear that a modern telecoms system is important: for attracting more foreign investment in the medium term and for generating firm-specific advantages for domestic firms that will allow successful foreign investment abroad. While total factor productivity as the residual in the estimation of a production function, is difficult to associate with visible technology, one may note that the presence of foreign MNCs as well as the

share of domestic MNCs in global outward foreign investment in manufacturing could be a proxy for total factor productivity in the home economy. This is a plausible hypothesis because high technology and most international technology transfer is related to MNCs (KLEIN and WELFENS, 1992).

- The OECD countries are moving towards the information society and higher R&D intensities, respectively. In such a knowledge based economy information (T), especially know-how and technology-related information, is a genuine input - with some public good properties - into the production function of each firm and each sector. Y=f(K,L,N,T,Z). Output level Y will be higher the higher the level of services extracted from the capital stock K and labor L, the greater the natural capital stock N [environment] and the higher the level of technology Z. Finally, output will increase if the exchange of information T - via telecoms networks - is increasing.

In Eastern Europe socialist systems were associated with national telecoms monopolies, and in no ex-CMEA country was the national monopoly really interested in expanding the telecoms system or in fully exploiting the network's capacity to transmit data and information. There was an enormous excess demand for telephone lines on the one hand, while on the other hand poor technological standards made especially international telecommunication cumbersome and expensive - this held even for communication within the CMEA. Interestingly, the telecoms industry was open for trade in equipment to a certain extent: Polish companies, for example, were active in expanding the Soviet telecoms network in the 1980s.

Governments in socialist countries did not invest much in the telecoms network, which was tightly controlled not only for political reasons. Subscriber charges and basic telephone service prices were low in CMEA countries, however, and the huge excess demand among potential telephone subscribers effectively invalidated the existence of cheap (uniform) domestic call rates. International telephone traffic was also impaired by a poor network, low quality and insufficient digitization, which reduced not only personal international communication but prevented the expansion of trade, international investment and global business communication.[1] While political resistance to the increase of East-West economic and communication links was a natural by-product of the socialist CMEA, the post-1989 switch towards a market-oriented decentralized economic system in Central and Eastern Europe renders the lack of national and international communication a critical problem of transformation. In a decentralized economy with autonomous decisions of millions of firms and households, there is a much bigger need for communication and information than in a centralized command economy. Moreover, the desired economic opening-up to OECD countries as well as to Newly Industrializing Countries - using telecoms networks and telecoms services on a broad basis - is difficult; an underdeveloped telecoms sector is a major bottleneck for learning about new technologies, financing conditions and possible international alliances, as well as for changing preferences in world markets.

Globally, the rising research & development intensity plus the rapidly growing stock of knowledge, in combination with increasing scientific specialization, also raise the shadow price of communication capacity. Through the rapid progress in technology, prospective liberalization of the telecoms sector in (Western and Eastern) Europe and the growing importance of information in modern societies, the telecommunication industry becomes a strategic industry (BAR and BORRUS, 1989). A modern telecoms sector is vital for international competitiveness in the tradables sector and for economic growth. Those reforming countries in Central and Eastern Europe which can quickly reorganize and expand their telecoms networks and improve telecoms services will face improved opportunities for economic growth and prosperity. The telecoms sector's expansion is being pulled by economic growth, but at the same time the sector itself is a driving force of economic expansion. Therefore, it is crucial that governments in Central and Eastern Europe adopt policy strategies which will lead to a viable and efficient telecoms sector. This concerns terrestrial, mobile and satellite networks in general; in particular it concerns local subscriber access, interconnection charges and conditions (third party access), trunk network operation for long distance as well as the equipment industry, and telecoms service provision. Policymakers are facing important decisions with respect to organizing, privatizing and restructuring networks operation, telecoms services and the equipment industry.

Through systemic transformation almost everything in the ex-CMEA countries has changed; in Poland the changes adopted were particularly radical and sweeping, obviously with some success as GNP growth rates of 4-6% for 1993/95 indicate. Sustained economic growth is a key ingredient for successful systemic transformation since aspiration levels in Central and Eastern Europe are high on the side of the population and because political stability can probably be obtained only if influential losers of the transformation process can be compensated via higher expected income growth. High growth is crucial for systemic transformation because it is associated with lower government deficits, limited unemployment and better opportunities to exploit static and dynamic economies of scale. Prospects for Central and Eastern Europe are modest for the mid-1990s (HUNYA et al., 1994; UNCE, 1994), although medium-term growth rates could increase strongly if adequate policies were adopted and favorable international conditions can be maintained.

Telecommunication is facing a thorough reorganization in all ex-CMEA countries and a host of questions arise for the transforming economies: Which problems emerge for establishing a competitive telecoms industry? Which strategies are most success-promising to fully use the general growth momentum associated with the expansion of a modern telecoms industry? The equipment sector, the infrastructure networks, as well as the developed value-added services could become expanding industries to generate income and hard currency revenues, raise the aggregate technology level, directly create employment and indirectly stimulate the expansion of all other industries.

The following analysis highlights the importance of a modern telecoms sector for former socialist economies and discusses which policy options are favorable for transforming economies. The sequel primarily focuses on the points of departure in Central and Eastern Europe, presenting some key figures of telecoms in Central and Eastern European countries. Moreover, the fundamental economic significance of the telecoms industries for transformation and growth is addressed in Section 2. Section 3 presents a basic theoretical analysis in order to highlight basic problems of the telecoms industry and of efficient regulation of the industry. Section 4 is a discussion of alternative privatization and liberalization schemes for post-socialist economies in general and for Poland in particular. The basic idea of the paper is to review theoretical as well as regulatory policy issues related to telecoms, and to put standard analysis as well as some new ideas into the context of systemic transformation.

2. Points of Departure in Eastern Europe

In Europe, the telecommunications industry traditionally consisted of a national network for voice and data communication, which was a government monopoly. Socialist countries in Eastern Europe were not much different from Western Europe in this respect, except that the number of main lines per inhabitant was markedly behind even poor EC countries. Since socialist systems were hierarchically organized, vertical communication lines in industry were provided by state PTOs, but the individual horizontal communication links that are so vital for competing firms in market economies were quite neglected.

Telecoms revenues $P^T T$ as a percentage of national output Y - the indicator ß - were markedly lower in socialist countries than in OECD market economies (see Tab. B1); in market economies with competitive networks (US, UK and Sweden) it exceeded the value ratio of most countries with a monopoly operator. This is remarkable because it suggests that the price elasticity of network demand exceeds unity, or that the output promoting effect of cheaper telecoms services is considerable. Within the ex-CMEA countries Bulgaria and the CSFR had relatively high revenues as a percentage of net material product; this is partly the result of much higher per capita investment in the telecoms network in these two countries than in other ex-CMEA countries. Insufficient revenue can itself become the cause of insufficient investment; however, in a functional market economy with an initial excess demand for telecoms services, financing of telecoms investment via capital markets cannot be a major problem.

Part of the East-West difference in Europe might be attributable to the fact that the income elasticity of the demand for telecoms services exceeds unity. However, since revenue-per-main-line in poor EU countries was much higher than in rich CMEA countries, the low ß-ratio indeed points to unexploited business potential. This perception is reinforced by the high excess demand with respect to network access in socialist economies: Waiting lines used to be very high in Central and Eastern Europe. Even if the Visegrad countries have made progress in reducing the inherited socialist waiting list, one may not overlook that the demand for telecoms services is growing faster than GNP. Poland, as a country with considerable economic growth in the early 1990s, is thus facing a major challenge with respect to quick network expansion.

The number of main lines was low in Central and Eastern Europe compared to Western Europe. No CMEA country had a line density high enough to bring it in the international top 20 league. Moreover, the revenue per main line was poor compared to EC countries: e.g. in Portugal this figure was five times as high as in Bulgaria (see Tab. B2).

Tab.089: Telecoms Revenue ($R^T/Y = ß$) as Percentage of National Output and Investment per Capita I/P in US$, 1989

	ß	I/P
New Zealand	3.38	24.32
UK	2.50	75.23
US	2.32	77.16
Australia	2.23	99.85
Sweden	2.13	24.77
Portugal	2.08	27.87
France	1.72	77.72
Italy	1.61	90.68
Germany	1.66	134.63
Spain	1.63	113.24
Japan	1.54	95.39
OECD	1.97	83.87
Bulgaria*	2.23	22.37
CSFR*#	2.07	16.99
Poland*#	0.82	14.83
Romania*	0.72	2.05
Yugoslavia*	0.96	4.36
Eastern Europe*	1.19	19.50

* telecoms revenue relative to Nat Material Product which is about 10% smaller than GDP figures so that the *-figures times 0.9 would be a suitable adjustment to get comparative figures with Western Europe; # 1990.

Source: YPSILANTI, D. and KELLY, T. (1993), Fostering Telecommunications Development. The Role of the OECD, in: STEINFIELD, C.; BAUER, J.M. and CABY, L., eds., Telecommunications in Transition, London: Sage, 118-139, here 138

Tab. B2: Selected International Telecoms Data

	I/GDP	Main Lines per 100 Inhab. 1992	$-Revenue per Main Line	Growth Rate of Main Lines			GNP per Capita
	1992	1992	1992	1991	1992	1993	1991*
Sweden	0,58	69,00	907,6	1,7	0,1	n.v.	79,0
Switzerland	0,92	59,48	1.241,1	3,5	2,6	1,9	98,4
Canada	0,62	58,52	853,1	3,4	(2,6)	3,0	87,3
Denmark	0,41	57,71	809,3	2,1	1,8	1,8	80,8
USA	0,44	55,27	1.525,5	2,4	2,6	3,7	100,0
Finland	0,58	54,24	832,3	1,8	0,9	0,7	72,9
Iceland	0,27	52,14	647,5	5,2	3,3	2,5	n.v.
Norway	0,48	51,60	1.200,7	3,1	3,2	2,9	77,6
Luxembourg	0,62	51,14	858,4	4,4	7,7	4,0	n.v
France	0,51	50,15	673,4	3,5	(3,2)	(4,7)	83,3
Australia	(0,78)	(47,87)	1.083,3	n.v.	3,4	3,7	75,4
Netherlands	0,58	47,63	789,7	3,4	3,1	3,2	76,0
Japan	0,54	(46,06)	812,3	(3,1)	(2,6)	(2,0)	87,6
UK	0,51	(45,74)	919,7	1,0	2,4	(3,1)	73,8
Hongkong	0,47	44,82	n.v.	6,8	6,7	6,1	83,7
New Zealand	1,00	43,91	927,2	1,6	2,7	3,8	63,1
Austria	0,76	42,74	875,4	3,8	3,7	3,3	79,9
Germany	0,82	41,81	837,8	7,1	5,5	4,2	89,3
Belgium	0,42	41,63	688,4	4,7	4,1	3,1	79,1
Greece	0,65	40,80	327,4	6,1	7,3	5,5	34,7
Italy	0,78	39,93	745,7	3,2	2,8	1,9	77,0
Singapore	0,65	36,53	n.v.	5,0	5,6	6,5	71,2
Taiwan	-	34,88	n.v.	8,4	7,4	7,1	n.v.
Korea, Rep.	1,24	34,45	n.v.	9,8	7,0	7,0	37,6
Israel	0,76	34,27	n.v.	4,7	5,9	8,5	60,8
Spain	1,26	33,39	663,6	5,3	4,0	3,3	57,3
Ireland	0,65	29,77	1.314,9	6,7	6,2	5,1	51,6
Portugal	1,18	25,46	622,3	13,2	11,9	8,2	42,7
Bulgaria	**-**	**25,01**	**118,8**	**6,5**	**6,1**	**3,0**	**22,5**
Belarus	**-**	**16,28**	**n.a.**	**7,7**	**4,2**	**4,0**	**n.a.**
CSFR	**0,37**	**15,65**	**220,5**	**5,6**	**-**	**-**	**28,4**
• **Czech Rep.**	**-**	**-**	**-**	**-**	**6,6**	**7,8**	
• **Slovak Rep.**	**-**	**-**	**-**	**-**	**8,4**	**8,8**	
Yugoslavia	**-**	**(15,14)**	**n.a.**	**n.a.**	**n.a.**	**n.a.**	**n.a.**
Uruguay	*-*	*14,52*	*n.a.*	*8,7*	*9,0*	*7,8*	*30,1*
Turkey	*0,69*	*14,13*	*n.a.*	*18,8*	*15,4*	*16,2*	*21,9*
Russia	**-**	**(13,75**	**n.a.**	**(7,3)**	**2,2**	**3,1**	**31,3**
Ukraine	**-**	**(12,98)**	**n.a.**	**(7,4)**	**(8,1)**	**3,1**	**23,4**
Kazakhstan	**-**	**11,12**	**n.a.**	**n.a.**	**4,3**	**(2,3)**	**20,3**
Argentina	*-*	*11,04*	*n.a.*	*2,6*	*7,8*	*13,8*	*23,1*
Hungary	**0,76**	**10,69**	**463,4**	**13,4**	**14,4**	**16,0**	**27,5**
South Africa	*0,61*	*9,63*	*na.a*	*5,0*	*2,6*	*3,9*	*n.a.*
Romania	**-**	**10,74**	**192,5**	**n.a.**	**3,0**	**11,0**	**31,2**
Poland	**-**	**9,29**	**157,9**	**8,3**	**10,5**	**12,1**	**20,3**

* PCC estimates (US=100)

Source: Siemens (1993, 1994 and 1995), International Telecom Statistics; World Bank (1993), World Development Report, New York, 296-297.

Tab. B3: East European Cellular Telephone Subscribers (January 1993)

	Subscribers	Penetration (per 1000 of population)
Czech Rep. / Slovakia	6,188	0.29
Croatia	6,000	1.28
Estonia	2,400	2.59
Hungary	22,010	2.08
Latvia	1,200	0.37
Lithuania	267	0.08
Poland	9,000**	0.24**
Russia:		
Moscow	2,415	0.24
St. Petersburg	1,057	0.21
Slovenia	4,000	2.00
Total	48,537	0.52
Total W. European*	5,985,146	16.25

* as of March 1, 1993, figures for analogue networks; ** estimate for end-1994

Source: FT, Sept. 7, 1993/FT Mobile Communications Newsletter

All ex-CMEA countries have started to modernize their telecoms sector (even Russia will liberalize telecommunications), but there is limited progress in privatizing and liberalizing the telecoms sector in Central and Eastern Europe (BERLAGE and SCHNÖRING, 1992; CSIS, 1992).

Mobile Telecoms Expansion

Mobile telecoms is rapidly expanding in Central and Eastern Europe. Mobile telecoms helps to cover the exploding communication needs in major cities in the short run and in the long term this may help to raise penetration rates in peripheral areas. Mobile telecoms generates additional business for the main network operator since leased lines have to be used. Hence capacity utilization of the main network is increased.

Hungary awarded two licenses for GSM tender in August 1993: two licenses for digital mobile telecom were won by US West, and a consortium of national telecom companies from the Netherlands, Denmark, Sweden and Finland. Winners paid about $ 100 mill. as a concession fee and committed themselves to network development, which could cost some $ 400 mill. over a decade. Hungary's penetration ratio is only 1/8 of the West European average, showing that there is huge long term market potential even in the most advanced telecom reforming country. In Poland, Ameritech of the US, France Telecom and local partners have invested $ 120 mill. in their Centertel joint venture. Poland enjoys the advantage of rising trade links with many OECD countries, most notably Scandinavia and Germany, so that Polish businessmen could even subscribe to West European networks and service providers. If Poland (GSM licence 1995) does not rapidly build up a host of reliable and cheap radio telephone networks and innovative service providers, this booming market segment could markedly contribute to a Polish balance of payments deficit. In the short run telecoms modernization will generally contribute to a deficit as hardware bottlenecks can quickly be reduced only via imports. However, the long term impact of modernization could be the strong increase of exports of telecoms services. A prerequisite for this is the expansion of services, nurtured by a competitive domestic environment.

2.1 Structure of the Telecoms Industry in an East-West Perspective

The telecoms industry consists of three sub-industries: (1) the telecoms network operators which create the basis for exchange of data and communication at the national and international level; (2) the telecoms equipment industry, and (3) the telecoms services industry which provides differentiated products to private households and industry (Fig. B1). In Western Europe monopsonist PTOs dominated the competitive equipment industry. In contrast, in socialist CMEA countries a very limited number of state equipment firms strongly influenced the PTO, which did not have the alternative to rely on imports form world markets. Since governments in transforming economies are naturally eager to privatize their respective telecoms industries - before considering the complex privatization of the network operator - the historically rooted influence of equipment makers over the network operators in Central and Eastern Europe could be prolonged. Moreover, the equipment firms which are interested in defending exclusive relationships with the national network operator are likely to support the survival of an exclusive government-owned network operator. The main reason for this is that by capturing the network operator, they can indirectly control market entry into the equipment industry. As long as the network operator is government-owned, domestic firms can hope to ensure their benefit from privileged government procurement contracts. Competing private networks would mainly try to obtain the cheapest and most modern equipment from domestic or international equipment suppliers. As soon as network competition is allowed or feasible, the equipment industry will face a new regime as well. There is a potential danger that large, western hardware manufacturers might seek to share monopoly positions with local network operators, thereby inhibiting a liberalization of the market.

Fig. B1: Structure of the Telecoms Industry

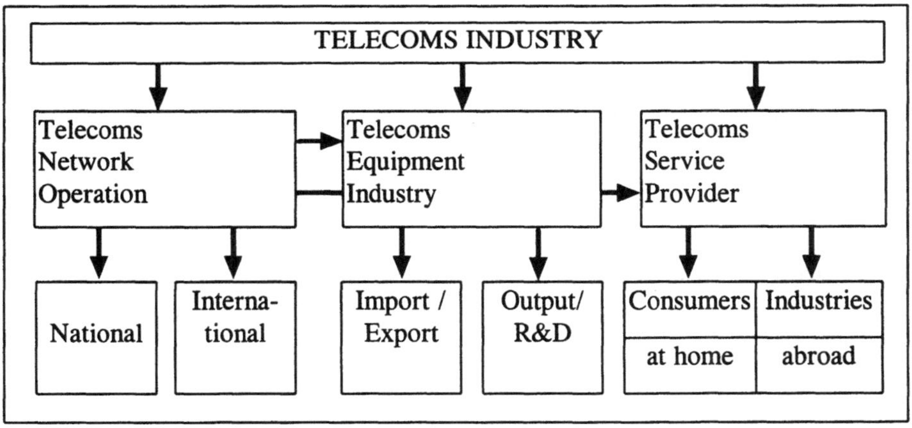

Facing network competition, the equipment industry will enjoy an offsetting effect coming from the service providers. In the early stage of market development, service providers will enjoy network externalities such that the marginal utility for existing users is increased by new users; this is the reason for the willingness of competing service providers effectively to subsidize access to networks, namely by offering a package of standardized services plus the hardware sold at a discount price. Network externalities in the early stage of network development imply that there will be a rapid expansion of both the hardware (telephone) and the service business, until some saturation level has been reached. Static and dynamic economies of scale in hardware production will help to keep production costs low even after networks have been completed.

In transforming economies the subsidization of subscribers/telecom hardware will become a problem if the equipment industry is not profitable. If national suppliers are not competitive at world market prices the expansion of the telecom network will in effect increase the pecuniary and welfare losses in the equipment industry. Therefore, an important prerequisite for fast efficient network expansion is that the equipment industry be reorganized and modernized in a way which will make it internationally competitive. The long-term benchmark for this is the tariff-free world market price at the equilibrium exchange rate. If tariffs exist in the initial transformation stage, they can be economically justified only if they are reduced over time.

Economic theory suggests that the equipment industry can be competitive only if several suppliers compete in a given market. If regional market integration is desired for political reasons, it is not necessary to establish several competing equipment suppliers at the national level; rather, one would have to make sure that integrated markets are served by competing suppliers from transforming economies. We assume here that market integration takes the form of a free trade area or a customs union. If the Visegrad countries were to form truly integrated markets, multinationalization of the equipment industry could be useful. This means that firms from Poland, Hungary or the Ex-CSFR would merge, and that the few firms thus created would each possibly accept a strategic overseas investor or an investor from the EU; looking for EU investors might make more sense because the Visegrad countries aim to belong to the EU in the future. If each country privatizes its equipment firms and immediately looks for a foreign investor, foreign investors' rivalry will make it quite difficult later to establish efficient company structures in the integrated region. A counter-argument would be relevant only in the case that foreign investors have a controlling interest and face no restrictions for divestment. In this case foreign multinational companies (MNCs) could swap investments, but it is clear that this reorganization of industry in Central and Eastern Europe is time-consuming. At the same time one might argue that expecting national privatization schemes to promote the emergence of regional multinationals is too bold to come true in a period of rising economic nationalism. However, at least the World Bank with its telecom aid rules could encourage competition within a broader regional framework. Existing World Bank telecom aid rules offer cheap loans only if competition is established at the national level. While this might not be an economic problem in some big countries of the ex-CMEA, it is doubtful that this approach is useful for small countries. The European Investment Bank and the EBRD are other important sources of external finance for modernizing the telecoms sector in transforming economies.

2.2 Telecoms Industry as a Strategic Industry for Systemic Transition

The dissemination of information and communication is strategically important for the functioning of a market economy. Developing an efficient telecoms industry is even more important in emerging market economies. Creating data banks for foreign investors and traders would be quite useful for both the transforming countries as well as for the rest of the world. Data coverage, storage and dissemination have to be reorganized, and much emphasis on commercial aspects added to the traditional technical and statistical data coverage in Central and Eastern Europe. The government might co-finance some specialized data banks, which are considered to be of special importance for the country. Finally, the revenue from foreign data bank users will be most valuable for financing an improved use of data banks abroad, which offer a wide range of valuable but expensive information. Experiences in Hungary show that both the use of (underdeveloped) domestic data banks as well as the use of foreign data sources is still limited in the first transformation stage and leaves many economic opportunities unexploited (HABUDA/STOCK, 1994). In the first stage, improved data coverage and data access is mainly important as a tool to inform and attract foreign investment and traders; in a second stage domestic firms with a firm-specific advantage could increasingly use data about investment opportunities abroad.

Telecommunication is not only a dynamic industry in itself but also a valuable input in every industry, namely an input which can add to international competitiveness. Creating efficient seamless telecoms networks means reducing information and transaction costs in all markets, thus raising the amount of profitable output and employment in the economy at large. The reason for this is straightforward: In markets with information and transaction costs, consumers' marginal willingness to pay will include these costs in addition to the price obtained by the firm. The consumers' net willingness to pay - excluding transaction and information costs - faced by the firms is given by the net demand schedule DD^n_0, while gross demand is DD_0 (Fig. B2). If S_0 is the supply curve equilibrium, output is given by q_1 and sold for producer price p_0. Consumers face nonpecuniary costs of AC, so that gross price is p_1. Transaction and information costs indeed drive a wedge between the effective price paid by the consumers and the average revenue obtained by firms - similar to the case of a value-added tax. Yet, if transaction and information costs could be reduced to zero, the gross demand curve DD_0 would coincide with the net demand curve such that point E gives market equilibrium; both output q_2 and price p_2 are higher than the initial values p_0 and q_1, which means that both value-added and employment are raised by the reduction of transaction and information costs. Welfare, as expressed by the consumer rent plus producer rent, will be increased by the area P_0CEAP_1. With the introduction of a modern telecoms sector, transaction and information costs will radically fall, and both output and average revenues in all markets will increase.

Fig. B2: Effects of Transaction Cost

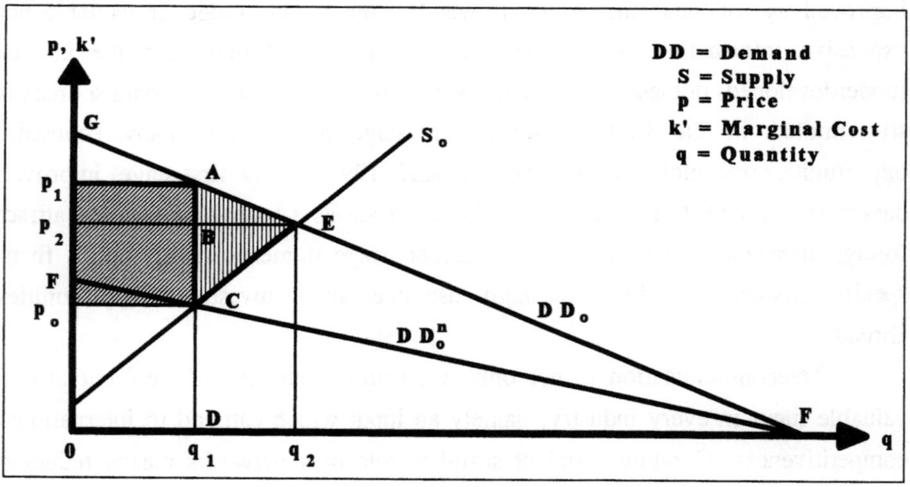

Finally, in sectors which intensively use information and communication, the modernization of the telecoms sector will mean a reduction of production costs such that the supply curve is shifted downwards (from S_0 to S_1). Prices will fall and output will be increased (Fig. B3). Hence the modernization of the telecoms sector will increase real output across all other industries. The effect on the officially measured price level is ambiguous, but the effective price level - including transaction costs - will fall.

Fig. B3: Effect of Reduction of Information Costs in Information-intensive Industry

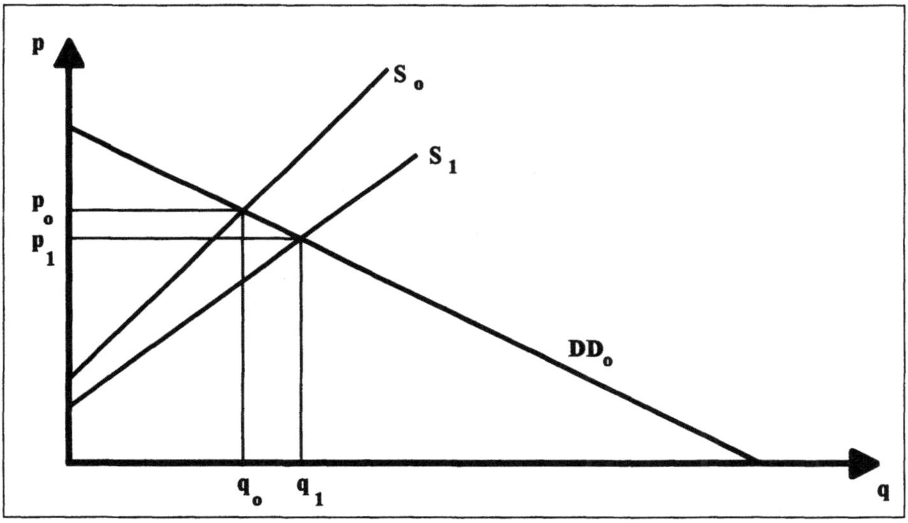

3. Theoretical Aspects of the Telecoms Industry

3.1 Some Problems of Uniform Subscriber Pricing

Traditionally, there are uniform prices across space in telecoms markets, although the marginal costs of network access are obviously low in centers of business activity when compared to peripheral regions. Let us assume for simplicity's sake two regional markets, both characterized by the fact that consumers at more distant locations can be connected to the network at rising marginal costs (Fig. B4). Demand is highest at the center O, but lower at more distant locations. If demand in market A) is DD_O, while a higher demand DD_O* is characteristic for market B), uniform pricing will lead to price p_O and an equilibrium distance of

M_0 in market A) and M^*_0 in market B). If each market were operated by separate networks, prices would be different: p^*_0 would hold in the market with higher demand but this market area would only be covered by the distance M^*_0. With a uniform network operating at a uniform regulated price P_0, consumers of market A) are effectively subsidizing users in market B), where a loss equivalent to the triangle FGH will occur (the welfare loss amounts to FE*G). This has to be covered out of region A's profits, which amount to the triangle DEP_0. The results obtained implicitly assume competitive behavior.

If one assumes monopolist behavior (Fig. B5) under a regulation requiring uniform prices, one would generate pressure on the company facing the relatively higher marginal revenue to reduce prices, down to the level of the competitor. This holds if subscribers are mobile to some extent so that neighboring regional markets cannot be separated by the two firms in regions A and B (otherwise each firm will set marginal costs equal to regional marginal revenue). Thus the firm facing the relatively higher market demand will be forced to expand its services beyond the radius established under monopoly pricing. This idea suggests splitting up the network into comparable regional network operators, where comparable refers to initial marginal costs. Since each operator will have at least one neighboring operator, competition in the overlapping boundary could be allowed. Moreover, the operator which is quicker in reducing marginal costs would force - even in a duopoly - the competitor to follow suit with price reductions and costs reductions.

In Europe, national PTOs engaged in all kinds of cross-subsidization activities. It is doubtful whether the size of the network, the scope of services provided and the equipment used were optimal from an economic point of view. In Europe state-owned PTOs and government regulation of the telecoms industry caused high equipment prices and weak technological progress, insufficient services and excessive prices when compared to the US with its liberalized and competitive telecoms market.[2] In California one even can find an efficient wholesale business in the telecoms network industry with differentiated pricing (MITCHELL and VOGELSANG, 1991), namely firms which offer time slots on the network at bulk rates to industry, universities and government agencies. This is a means to increase capacity utilization of an existing network, which normally

Fig. B4: Network - Competitive Telecoms Market with Different Demand

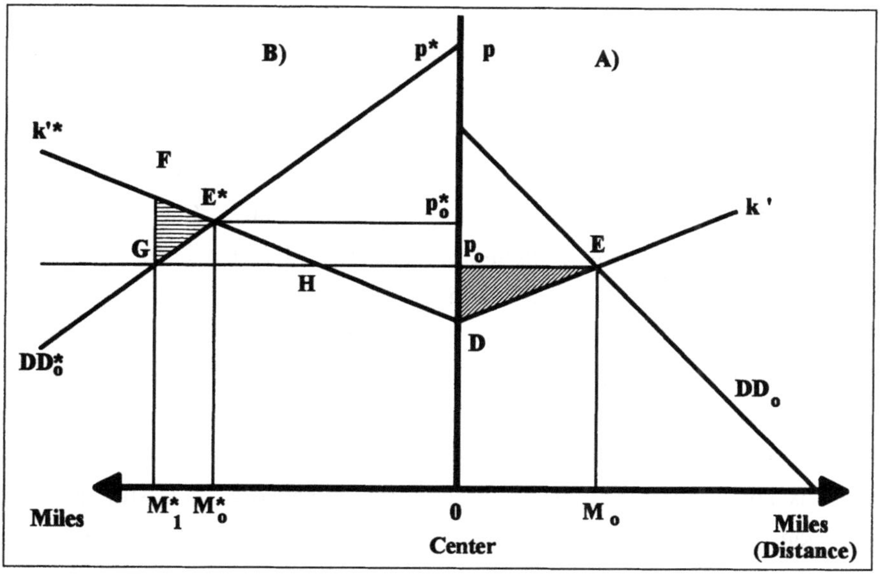

Fig. B5: Telecoms Market with Different Demand under Duopoly

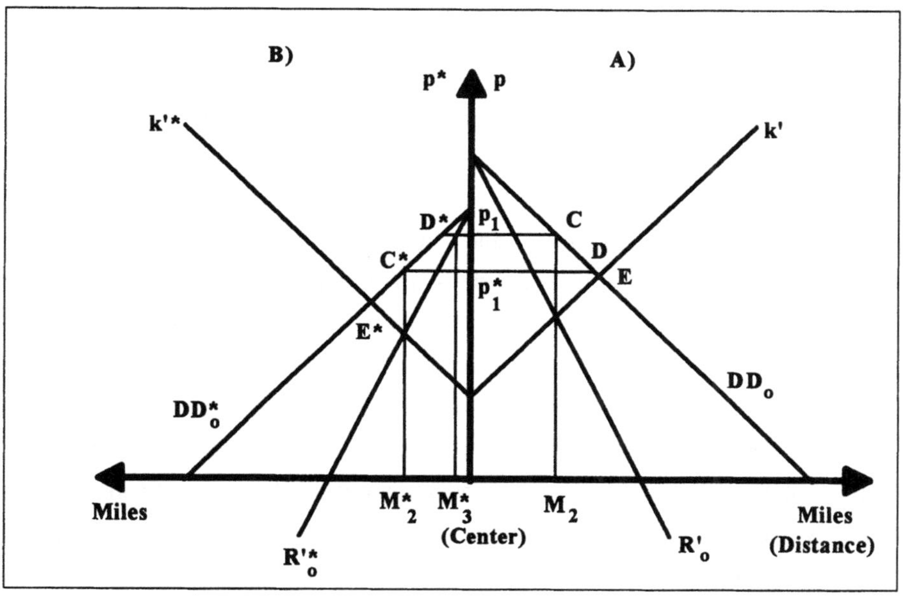

would increase should demand reach an excess at peak time and a low at off-peak periods.

In the 1980s, a phone call from Europe to the US typically cost two or three times as much as a phone call from the US to Europe. This has started to change only with the emergence of call-back services, which basically require European users to subscribe to a computerized call-back service provider allowing them to make international phone calls at foreign rates. For multinational companies the circumvention of high-cost European PTOs has been increasingly important in the 1980s. New technologies made service users effectively more mobile across space, and this meant increasing competitive pressure for European PTOs. Moreover, with envisaged satellite systems that could create a global telephone network operated from any country, the traditional demarcation of national telephone markets will be blurred after the year 2000. While a satellite network is not a full substitute for a fibre-optical network, it is still true that competition in the telephone network business will sharply increase in the long term. This is also true because national or regional grid companies could easily add fibre-optical cables to the electricity distribution network - the marginal costs will be rather low. Even the railway system, which in all countries operates its own internal communication network, could enter the telecoms network market by adding network capacity to its existing system.

3.2 Competition, Natural Monopoly and Economies of Scope

Traditionally, part of the telecoms network operation is a monopoly. However, new technological developments and innovative regulation policies could introduce competition. It is useful briefly to highlight the principal difference between monopoly and competition for the simple case of constant marginal costs k'. For ease of exposition we assume that marginal costs k' and average costs k are identical (Fig. B6). A monopoly typically charges higher prices and produces a lower output than firms under competition. If marginal costs are k'^{M}, competition would lead to price p_0 and output q_0, while a profit-maximizing monopoly would charge p_1 and supply a much lower quantity q_1. Consumer surplus - a measure of the extra benefit accruing to consumers - is equal to the triangle ZP_0C under competition, but boils down to ZP_1N in the monopoly case;

the switch from competition to monopoly means a deadweight loss of NMC (which is half the monopoly profit) and a redistribution of consumer rent P_1NMP_0 in favor of the monopolist firm. If demand is implicitly characterized by a constant elasticity function $p(q) = Aq^{\beta}$ $(-1<\beta<0;\ A>0)$, consumer rent is a monotonically increasing function of output q_0 and a monotonically decreasing function of price p.[3]

With a linear demand curve the consumer surplus under competition is a monotonically decreasing quadratic function of the price p_0; this is particularly important if one assumes that competition means faster process innovation and hence lower average and marginal costs k^C than in the monopoly situation. If k'^C holds under competition and higher marginal costs k'^M under a monopoly, the switch to competition would mean an increase in consumer surplus by the equivalent of $P_0CC'P_2$.

Fig. B6: Monopoly Profit and Cost of Reduced Innovativeness

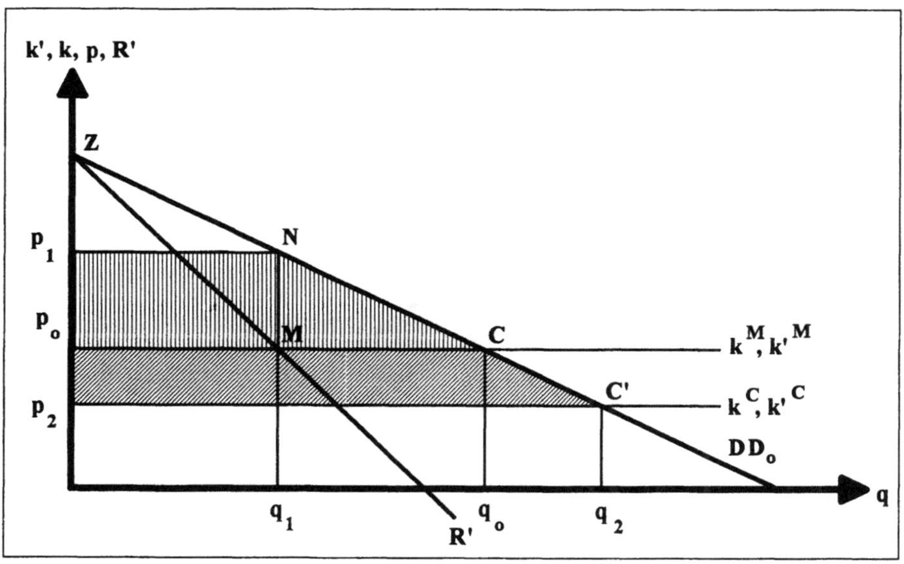

When comparing monopoly and competition under the assumption that process innovation is faster under competition, the costs of a monopoly in the sense of lost consumer surplus loss would be equal to the two shaded areas, namely P_1NMP_o plus $P_oCC'P_2$ plus the deadweight welfare loss NMC. A switch from a monopoly to competition would therefore bring enormous welfare benefits. These benefits will be further increased if one assumes that competition generates a higher rate of product innovations. Product innovations increase consumers' marginal willingness to pay, and thus the demand schedule will be steeper than initially. Product innovation means additional consumer surplus and a higher output q_3. The telecoms industry's switch to competition would thus bring another benefit of B'C"C'Z (Fig. B7). In general, welfare analysis would focus both on the consumer surplus and the producer surplus, but for political reasons one might strongly emphasize consumer surplus.

Fig. B7: Competition-induced Product Innovation

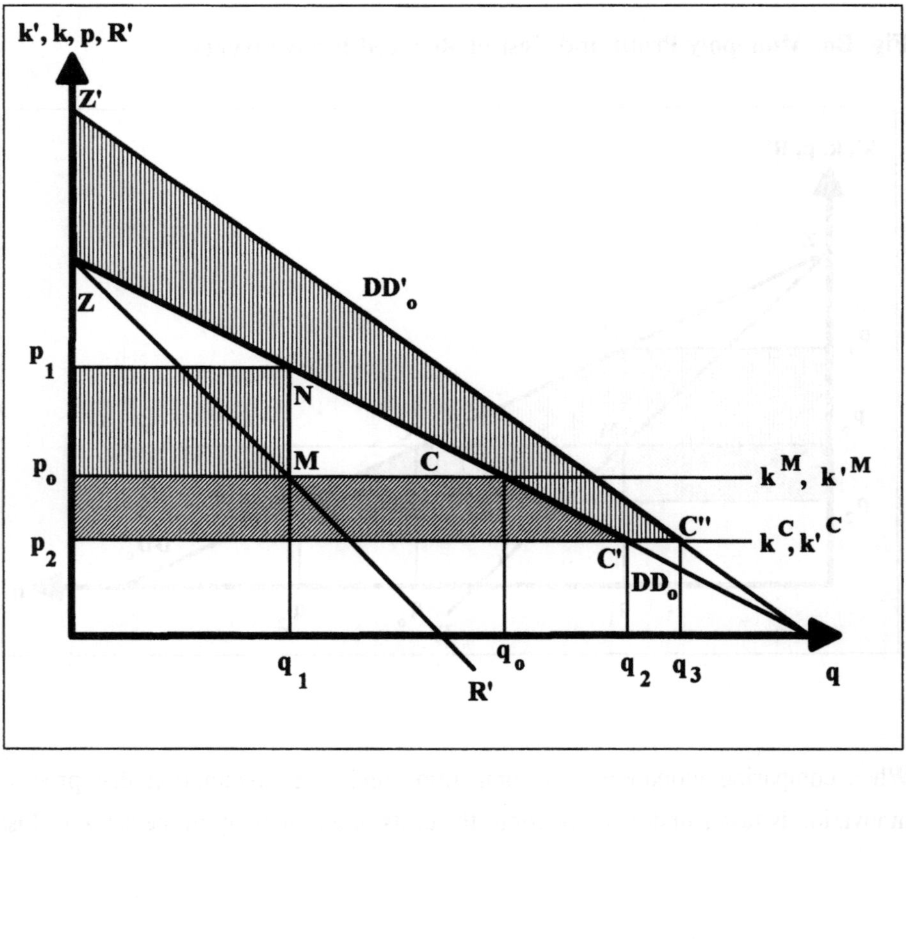

Natural Monopoly

In telecoms network operation there are economies of scale which could lead to a state or private monopoly; no stable set of competing firms can profitably exist due to technological traits associated with falling marginal costs and high barriers to entry, or if the telecoms industry is regulated inappropriately. The telecoms industry certainly enjoys network externalities (demand side) and economies of scale (supply side). The more subscribers to a network there are, the lower are the costs of providing an additional communication unit. This holds at least during early network expansion, which is the relevant case in transforming economies. Telecoms is the typical example of an industry with falling marginal costs k' and falling average costs k, namely a natural monopoly. In a natural monopoly situation, an initial set of competing suppliers will boil down to one big monopoly supplier, because raising the size of the individual firm always pays off in the form of cost advantages, causing the most aggressive firm to swallow all competitors. In a strict sense falling marginal costs are not required for a natural monopoly while sub-additivity of costs is.

In Fig. B8 the demand schedule is DD_0, and both the marginal and the average cost schedules are shown. The efficient economic solution requires equating the price p with marginal costs, here point E. However, a private firm supplying the amount q_0 would at price p_0 suffer a per unit loss equivalent to the distance DE. The shaded area is the total loss occurring with output q_0 at price p_0. A loss could be avoided if prices were set in accordance with average costs so that price would have to be raised to p_1. An efficient outcome is obtained if government pays a per unit subsidy of DE to cover the fixed costs, which are the reason for the firm's loss at price p_0. Alternatively, the firm could be allowed to price according to its average costs, so that a reduced output q_1 would be sold at price p_2. The consumer surplus lost under such an arrangement is the triangle BEF. However, the more important problem with a single firm is that lack of competitive pressure will allow the firm to postpone possible process innovations - here equivalent to a downward shift of the marginal cost curve. Moreover, if there is only one firm, this firm will act as a monopoly and maximize its profits by equating marginal revenue R' with private marginal costs k'^{P}, such that output is q_2 and the high monopoly price p^M will be charged (Fig. B9). Monopoly output is much lower than q_1 (or q_0 under hypothetical competition). The monopoly situation causes an enormous loss of consumer rent when compared to the efficient solution at point E. Average fixed costs DE could also be recovered by a monthly rental charge which is part of a two-part tariff (rental charge plus price for using the network).

Fig. B8: Average Costs Pricing as a Problem in Industry with Falling Marginal Costs

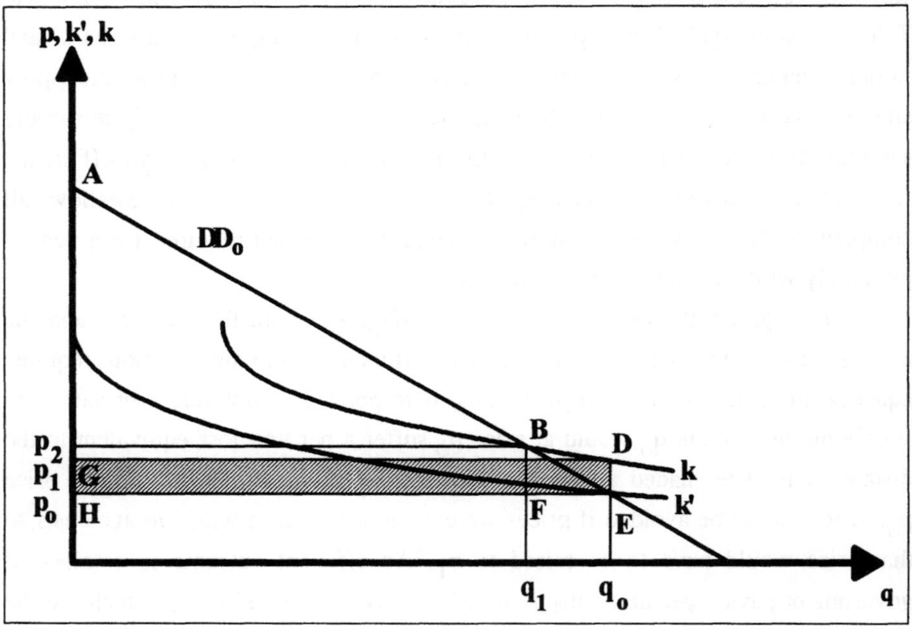

Output at q_1 implies a consumer rent equivalent to the triangle DGZ. Under monopoly pricing, however, the consumer rent is down to ZCM and the triangle EHE represents a deadweight loss to society. This situation should definitively be avoided in post-socialist countries. Only in the case of social marginal costs being much higher (k'^S) than private marginal costs could a monopoly be considered, at least in principle, as economically useful, since its tendency to price higher than competitive markets will automatically reduce demand and output. Negative external effects are rare: e.g. aggressive TV emissions.

Fig. B9: Monopoly Pricing with Falling Marginal Costs and Tax Credits for Users

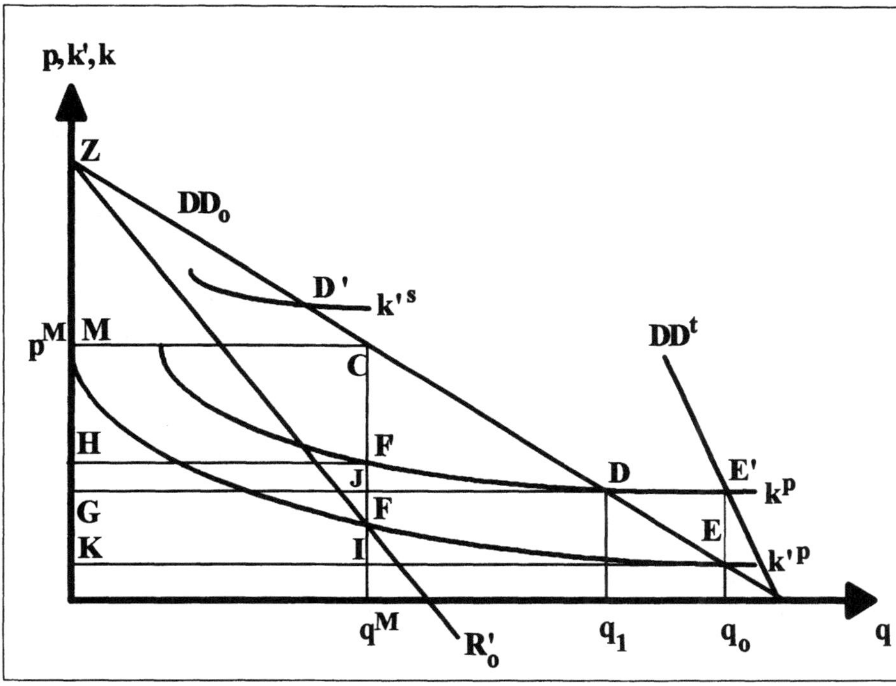

Tax Credits for the Demand Side Superior to Producer Subsidy

Since so many industries in transforming economies are calling for subsidies, any policy based on paying subsidies to producers is doubtful. The more common subsidies are the more firms and alliances will invest in lobbying for more subsidies; scarce energies and ideas are diverted from the main problem in Central and Eastern Europe, namely how to raise productivity, quality and market shares.

Paying subsidies to a telecoms operator - even if this were justified by theoretical economic reasons - is bound to find a multiple echo in many industries for which theoretical reasoning could not provide any convincing arguments. Hence, if social benefits exceed private benefits or if marginal cost pricing with normal demand would not allow the covering of total costs, one may consider subsidizing demand in such a way so that average-costs-pricing results in the optimal output. A tax credit for telephone users would solve the problem: it would replace the DD_0 demand curve by the DD^t demand curve. This unfortunately still leaves the unsolved problems of how optimally to promote competition and recover investment costs. A two-tariff policy also solves the problem.

Economies of Scope and Network Externalities

A phenomenon widely encountered in the telecoms business is the existence of economies of scope. Economies of scope in a two-industry case imply that expansion of industry A reduces marginal costs of an expansion of industry B. In the telecoms business an expansion of the network causes positive network externalities, in the sense that the usefulness of being hooked up to a network will increase for each user as new users - and potential sources of information/points of communication - join the network. This phenomenon is well-known in the telephone equipment industry, letting hardware suppliers form a strategic alliance with network operators and service providers. Equipment suppliers could offer telephones at subsidized prices in order to stimulate network expansion; but building a mobile telephone network would require a critical minimum of subscribers with the allied service provider and implicitly or explicitly with the respective network operator (e.g. in the US and Germany). The service provider or the network operator will then transfer part of the extra profits from network enlargement to the equipment supplier, who can thus recover his subsidies for equipment sales. For network operators such an alliance with an equipment supplier is particularly attractive because network enlargement means that network services can be increased at falling marginal costs; with given unit prices, this will lead to an increase in profits from operating the enlarged network. If network enlargement means a cheap opportunity to recruit a critical minimum mass for profitably providing specialized services (e.g. communication services among "minority clubs"), service providers could also transfer part of their extra-profits to equipment providers. The most important phenomenon in the telecoms industry is probably the potential combination of static economies of scale/network externalities for the network operator and dynamic economies of scale for high-tech equipment makers. Dynamic economies of scale imply the relevance of learning effects which positively depend on the cumulated output of certain equipment. For example, if network enlargement allows the operator to realize static economies of scale which translate into falling prices and hence higher demand for services by actual and new users, there will be a rising demand for equipment and improved potential for "learning by doing" effects. A strategic alliance between a network operator and equipment makers could thus be explained.

The real problem in the telecoms industry could be high sunk costs, mainly in form of idiosyncratic capital investment (terrestrial network), which drives a wedge between cost perceptions of the incumbent and potential entrants. Market entry is not free in terrestrial network operation. Monopolistic pricing and cross-subsidization could therefore play an important role in distorting competition.

3.3 External Effects of Telecoms Network Expansion

Markets for information are not functioning well in market economies. Markets for technological know-how are particularly imperfect for well-known reasons: those wanting to sell information have to disclose part of their know-how for free in order to put a clear offer on the bargaining table. The exchange of commercial and technological know-how as well as the organization of production and sales in space would be facilitated by an efficient telecoms network. Positive external effects of information would imply that the marginal social utility of information/the telecoms network is higher than the level expressed by market demand DD_0. Adding the positive external effects to the demand schedule DD_0, we obtain the demand schedule DD_1 relevant for optimum allocation (Fig. B10). Assume for simplicity's sake that DD_1 intersects the abscissa at twice the value of the intersection point of the DD_0 curve, such that DD_0 coincides with the marginal revenue curve relevant for DD_1. Without external effects on the demand side, optimum output is q_1; taking into consideration the full social utility of information leads to an optimum output of q_2, which is clearly much higher than q_1 and very much higher than the monopoly solution q_0. The extra benefit for society in terms of consumer rent is the equivalent of the area ZHJ minus the triangle HGE_2. Optimal allocation could require state subsidies for the operator of the network(s). The subsidy per unit of output should be equivalent to GE_2 in this case, where we assume that adequate price regulation or effective potential competition could induce the firm to supply q_2 if a corresponding government subsidy is guaranteed. Moreover, there will be positive network externalities in the early stage of network expansion. If the network operator is moving on the marginal cost curve from K to E_1, the existence of positive network externalities means that the initial demand curve DD_0 is shifting outwards (DD_1) since positive network externalities exist on the demand side; with a rising number of line subscribers the marginal utility of communication will increase. In a monopoly the interaction of falling marginal costs and network effects will bring about a lower price P_1 (instead of P_0) and a rising output q_1.

The telecoms equipment industry is highly innovative and represents a high R&D intensity. One may therefore assume that a local telecoms equipment industry has some positive technological spillover effects to other industries. In the presence of a viable equipment industry with local R&D activities, the expansion of the telecoms network indirectly generates positive external effects as long as a considerable part of equipment supply comes from domestic firms. Orders that are covered by domestic firms mean smaller output volumes for leading world suppliers, which then face smaller opportunities to exploit static economies of scale and dynamic scale economies ("learning by doing", which depends on cumulated output volumes). If there is import discrimination via tariffs, government procurement and other measures leading to international firms' exploitation of such scale economies will thus impair opportunities for import price reduction. If domestic supply prices in the equipment industry are not considerably above world market prices, and if crucial technology spillovers are expected, some import protection cum FDI promotion in the equipment sector might be useful in transforming economies. If prices are high above world market prices or if markets are too small and technology levels insufficient to attract foreign investors, the country should fully rely on imported equipment (within a regime of strong import competition).

Since demand for telecoms depends on relative prices and income, positive external effects on the demand side could be solved in a two tiered approach: For private households one would have to offer a tax credit taking into account the number of telephone calls made; for those paying no income taxes, tax credits could be applied to a fictitious minimum tax base, which would result in a small direct- income transfer.

Fig. B10: Natural Monopoly with Positive External Effects on the Demand Side (Network Effects)

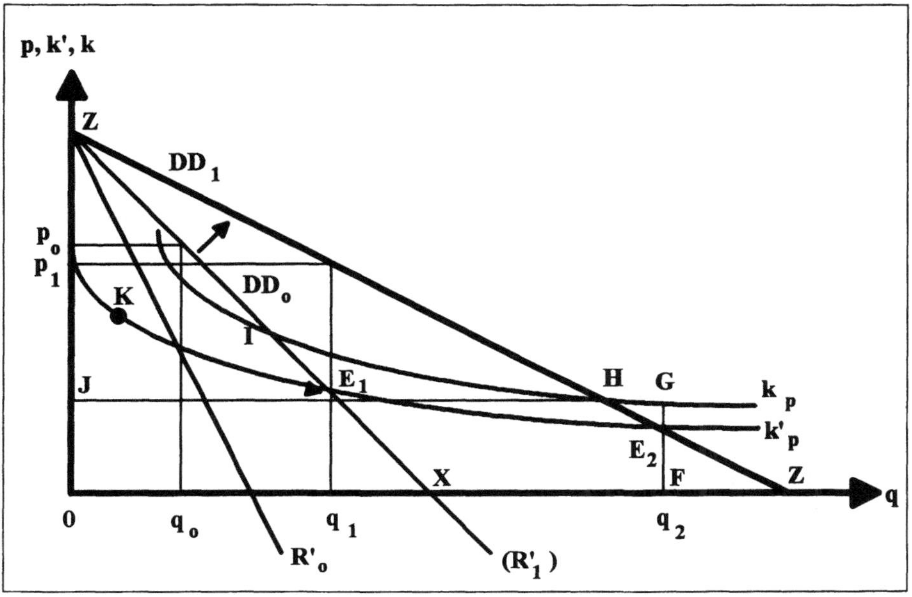

3.4 Oligopolistic Telecoms Industries

Kinked demand curve

There is only a small deviation between a full cost recovery policy (based on pricing according to average costs) and the quantity at optimal supply if a kinked HITCH-SWEEZY demand curve (line ABC in Fig. B11) holds. Such a demand curve assumes an oligopoly with the following behavior: At prices above the market-clearing price p_0, a rise of one firm's offer price will cause a large loss of demand for the firm so that the demand curve is very price-elastic. If, however, a firm were to lower its offer-price below the initial market-clearing price p_0, all other firms in the oligopoly will follow suit such that the demand curve appears to be relatively inelastic below p_0. It is easily recognized that in such a situation the deviation between output q_0 and the optimum economic level q_1 is quite small. It follows that a possible solution to the natural monopoly problem is to allow and foster the emergence of an oligopoly in the telecoms industry, while trying to ensure the survival of several competitors by allowing them extra profits in the provision of complementary services for which demand is inelastic - e.g. TV broadcasting. The condition that demand for the complementary good be inelastic makes sure that extra profits gained in the complementary industry do not much change the output in this industry and therefore do not cause serious output distortions. It is unclear to what extent vertical integration is a necessary ingredient for maintaining an oligopolistic network structure.

Fig. B11: Small Deviation Between Full Cost Recovery and Optimum Supply

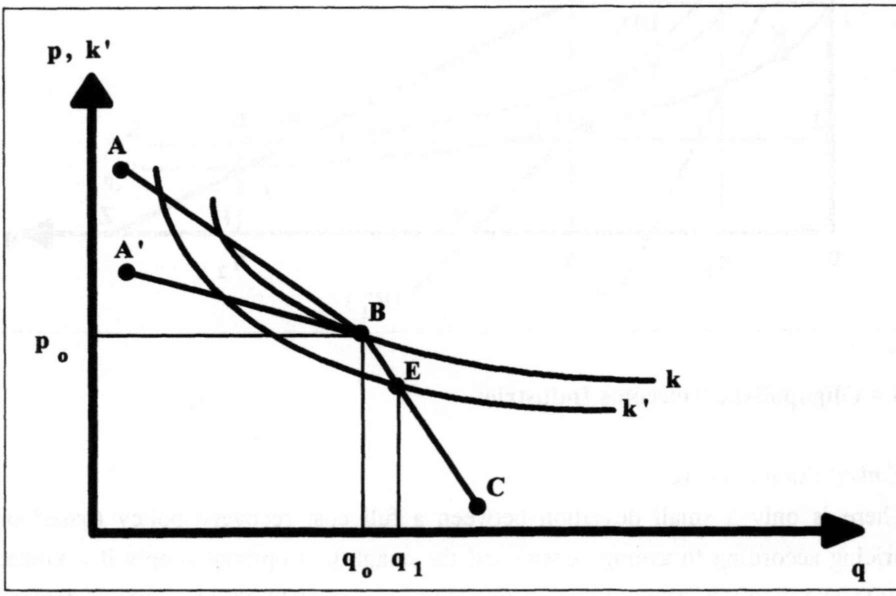

Creating Oligopolies

The main message of the above considerations is that price regulation is not necessary if one could establish a set of oligopolistic networks with at least overlapping market areas. This requires that countries are relatively big, that some duplication of networks is actively encouraged and that all kinds of telephone rerouting be accepted (within reasonable quality standards). After a one year grace period every subscriber should be allowed to switch to any other network company offering services. The oligopoly approach could work for quite some time in some countries; however, it is uncertain whether the oligopoly will be stable. Endogenous monopolization could occur after decades. This risk should nevertheless not be an argument to maintain the established state trunk monopolist.

Oligopoly is likely to be the dominant market structure of telecoms in Europe in the future. Given the risk of cartelization and re-monopolization, it would be quite important to have an efficient competition policy organized early on in transforming economies. There could be strong incentives to collude, but these incentives would in effect be weakened by the existence of a functional competition policy body and a clear pro-competitive policy stance.

3.5 Regulation of the Telecoms Industry

Competition has been an exception in the telecoms business of industrialized economies in the 1980s. It has been argued that the telecom network represents a natural monopoly in the sense that there are declining marginal costs which would lead to unstable competition and, finally, to a monopoly. An efficient economic outcome under a monopoly cannot be expected unless adequate regulation is introduced or unless the firm is organized as a public utility. For decades this argument seemed rather valid as network operation was characterized by high fixed costs and declining marginal costs. Sub-additivity of costs might have prevailed because of economies of scale or economies of scope over some relevant range. Moreover, most fixed costs were irreversible sunk costs so that incumbents faced high barriers to entry. With the emergence of new technologies - especially cellular networks and satellites - sunk costs are becoming less important. This, then, is a strong reason explicitly to abandon the public utility status and to promote competition. With competition it should be feasible to deregulate subscriber prices altogether, thus avoiding the distorting effects of regulation. Regulation can impose high costs on society to the extent that firms are induced to behave strategically and invest resources to improve their respective position vis-à-vis the regulatory agency - while market competition is neglected.

Through the privatization of the economy two competing networks could be more efficient than a single operator, who in principle could better exploit economies of scale, but who is likely to set monopolist prices. Fig. B12 shows the advantages of a joint provision of telecoms services in two regions (or countries) A and B under the assumption of initially identical demand; the more efficient B-operator will service the whole market, where the price might be higher than world market price p*.

Fig. B12: Monopoly versus Duopoly

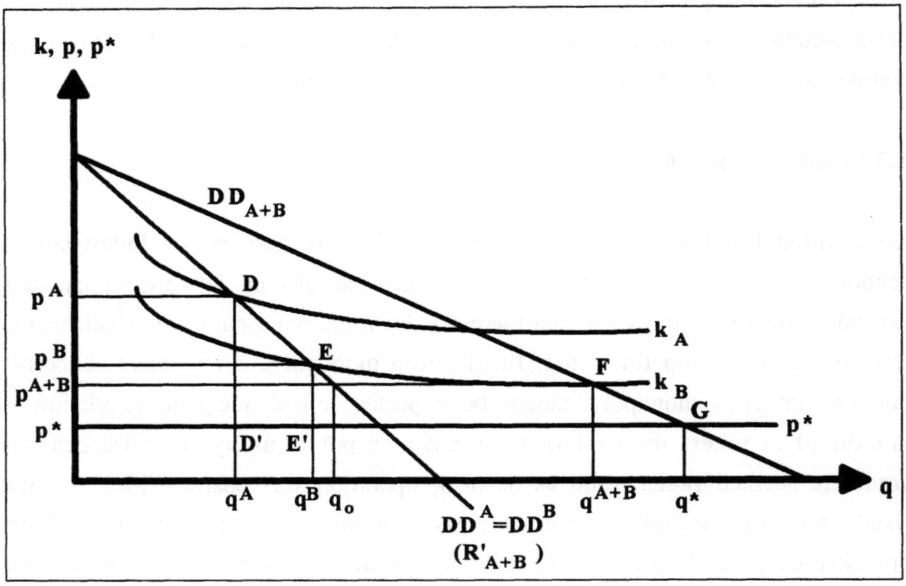

Panel a) in Fig. B13 shows the country's demand DD_0 for telecom network access resulting under a monopoly in output q^M at price p_0. If a second network operator is introduced competition could emerge; one possible duopoly outcome could be identical market shares of the two firms (firms cover average production costs k and each firm (A,B) produces the same output ($q_{1A}=q_{1B}$)). Clearly $2q_{1A}$ is bigger than q^M, and the increase in consumer surplus P_0P_1EF plus the triangle EFD (social net gain EFD) more than offsets the reduction of monopoly profit. A single producer could supply $2q_{1A}=q_2$ at average costs OH which, of course, are lower than the average costs in the duopoly case. If the increase of costs under duopoly (compared to a hypothetical regulated monopoly price p_1 at output q_2), namely the area P_1DCH, is greater than the surface of the triangle EFD, price regulation cum monopoly might be considered politically useful. However, three counter-arguments would even then support establishment of a duopoly or higher forms of competition:

(1) Regulation costs will be particularly high in the monopoly case, and these costs could more than offset society's benefit from the exploitation of static economies of scale.

(2) A monopoly is likely to cause static inefficiencies and to be less responsive to market demand than a duopoly. Launching product innovations and product differentiation is also more likely in a duopoly than in a monopoly; innovation and product differentiation are major benefits of a market economy under competition.

(3) Cost asymmetries in the duopoly will promote technical progress over time. This argument is elaborated in the sequel to this paper.

Without dismemberment of the dominant operator there will be weak competition in a (asymmetric) duopoly. Access to local networks - including innovative mobile DECT or CDMA options - is most crucial for competition in telephony.

Asymmetric costs will stimulate competition in a duopoly. Here, it is assumed that there is at least a long term flexibility of subscribers so that potential competition among the network operators exists. This potential competition will result in a uniform price (Fig. B13). Whenever one of the two firms has achieved process innovations, the competitor will come under pressure to follow price reductions by the innovative firm: Firm A has to follow firm B as it reduces its marginal costs to k'_{B2}. With monopolistic competition in a duopoly, the firm which is faster in reducing marginal costs will increase its market share transitorily; rising profits will improve future access to capital and hence opportunities for future network enlargement. If the duopoly is characterized by strongly divergent market shares a duopoly strategy is not sensible because there is the risk of tacit collusion and of "technical dependency"; the newcomer will have to rely on lines leased from the dominant operator.

Fig. B13: Duopoly: Asymmetric Costs and Competition

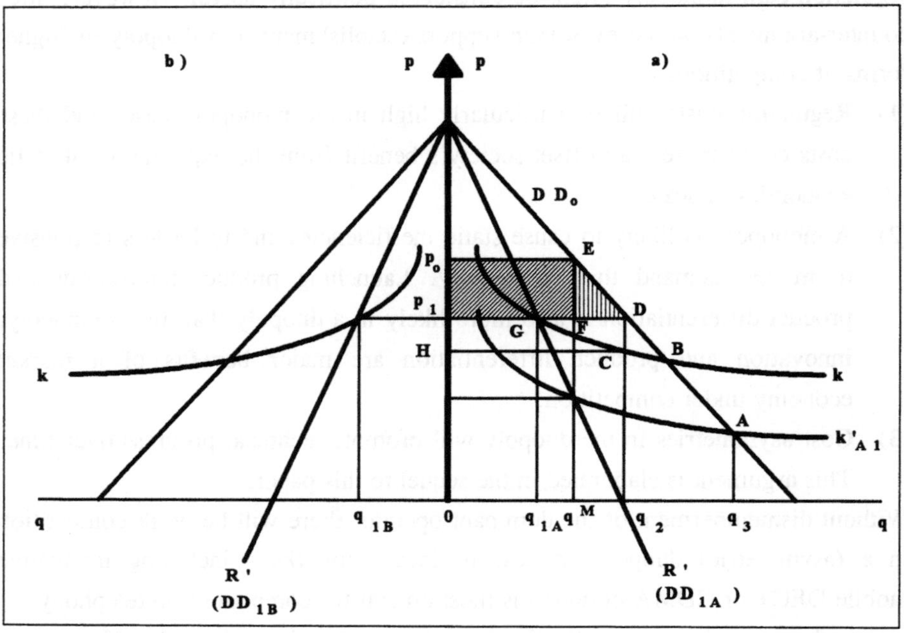

Standardization

Market entry, pricing, technical standards and market exit are the principal areas of regulation in the telecoms industry. Most importantly, all networks must form a seamless integrated network. This requires that certain common standards (technical protocols) be generally applied by all network operators; herein lies the task of the regulator. Furthermore, the use of a telecoms system requires directory information, and network operators might specify telephone numbers as proprietary know-how, while in effect the main purpose would be to exploit a monopoly rent; wrong and delayed directory information could be a tool of incumbent firms to keep new entrants outside the market. Efficient management of telecoms networks has a clear requirement: Directory information needs to be standardized and must be accessible at uniform charges for network insiders and network outsiders. Here the regulator must impose adequate rules.

Ensuring open market access to the network is also important for an efficient seamless telecoms web. Access charges are a double problem: (1) Which access charge would a local network impose on the trunk operator wanting a connection to a resident or firm; (2) Which access charge will the trunk operators charge to the local networks wanting to transmit voice/data/video out of the local area? In the US, AT&T resents high charges for the last few miles of wire to customers and is therefore eager to break the local network monopolies (of baby bell companies and other independent regional operators). If both the trunk operator and the local network operator charge monopolistic access charges, prices of telephone services will be higher than they should be - with respect to the true costs. In the US the solution could be to adopt full network competition. On the one hand, the existence of cable (TV) companies which have their own network means opportunities for local networks to expand into medium and long distance transmission; on the other hand, the existence of cable networks and cellular telephone networks offers an opportunity for long distance operators to establish at least a wider local presence. In Europe, however, separate (private) cable network companies do not exist; instead, cable networks are part of the state-owned network operator's activity in most countries. For Eastern Europe it could be useful to support the expansion of private cable network operators in all major regions. Each region would have a cable TV network and at least two cellular telephone network operators. Cable TV companies could offer private communication services which - in digitized form - would ride piggy-back on the cable network transmission. Given the huge investment costs in the telecoms industry it would, however, be quite useful to have a telecoms development framework plan accepted by a very broad political majority. Political uncertainties could sharply reduce domestic and foreign investors' interest in the telecoms sector.

Rate of Return Regulation versus Price Regulation

Since the regulator has less information on costs and innovation options, there is always the problem of achieving efficient regulation (EINHORN, 1991; WINSTON, 1993). Rate of return regulation is obviously a rather ambiguous approach to regulation because the regulator cannot induce firms fully to exploit the range of possible process innovations. Especially in periods of high potential process innovation dynamics, the regulated firm will be hesitant to implement quickly technical progress because profits would become too high; if the firm fully realizes innovations but looks for ways to spend some extra money on lavish management benefits and fringe benefits for the employees or to invest in new industries with low returns on investment, resources would not be optimally allocated (US baby bells are subject to a rate of return regulation). Price regulation seems to be a superior approach. However, price regulation is not without problems since the regulator still has poorer information on cost developments and technological dynamics than the firm. Price regulation might be necessary in the domestic network for national and local phone calls if there is a network monopoly. Price regulation is not necessary and indeed is counterproductive with respect to international telephone calls. Outgoing telephone calls should be priced at the same price as incoming phone calls, and international agreements should be changed in a way so that the lowest rate will always apply. In order to prevent US or EU telephone network operators from cross-subsidizing outgoing international phone calls, separate and standardized accounting systems for international telephone calls should be established. International long distance competition should realize and implement the law of one price for telephone calls, data transmission, etc.

A monopoly with price regulation always means insufficient incentives for technical progress and quality improvement. In a duopoly there will be at least some competition and hence incentives for innovation. At the same time there is the risk of collusion. In Central and Eastern Europe such a risk could be reduced if rival foreign investors could be attracted for each network operator. In big countries - such as Russia and the Ukraine (and possibly Poland) - more than two network operators could be profitable and efficient.

Given the long-established special links between firms and governments in formerly socialist economies and the intense political competition among the many new parties in Central and Eastern Europe, the ex-CMEA countries face particularly high risks that monopoly operators would engage in sustained rent-seeking activities in order to defend the monopoly. Scarce resources would be diverted to lobbying instead of flowing into productive use. This clearly suggests that competition should be favored over regulated monopolies. Finally, while the creation of a "new" telecoms sector requires many standards to be set by the regulator, "lean regulation" is required in the long term. As soon as competitive internationalized markets have emerged, technical standardization (by government) will no longer be necessary.

4. Policy Options for a Modern Telecoms Policy

Policy options in Central and Eastern European economies are limited in many areas since short term labor market problems, balance of payments problems, intensified social conflicts and political instability often reduce political degrees of freedom. It is all the more important to understand that the telecoms sector is a multiple key to many long term problems and to an increase in political degrees of freedom in the medium range. Using the unique situation of systemic transformation, Central and East European economies could avoid many problems identified in telecoms policies in Western Europe and exploit the exceptional opportunities of an ongoing technological revolution in the telecoms and the computer industry. There are several potential innovative uses of a modern telecoms system which all could be exploited in systemic transformation. Some bold initiatives would be required on the side of governments, however, since the political dividend would be higher sustained economic growth and an accelerated internationalization of the economy. Although Western Europe is considering limiting the intra-European East-West income gap to a politically desirable level the EU and major international organizations should encourage the modernization of the telecoms sector and of telecoms policy in Central and Eastern Europe.

4.1 Privatization of the Telecoms Industry

Network operation involves considerable sunk costs. For the potential newcomer all costs are variable before market entry. The incumbent firm could thus adopt a limit pricing strategy if the process innovations of the incumbent are not exceptionally high; recovering marginal costs minus the sunk part of costs is the lowest offer-price the incumbent will consider. For the main network operator this implies big opportunities to defend his monopoly position. This is all the more important if there are additional political barriers to exit (e.g. employees cannot be sacked); barriers to exit are, of course, barriers to entry. Given the fact that demand for telecoms services will rapidly rise in transforming economies and that there was a huge excess demand for telephone services in socialist countries the political exit barriers do not seem to be very important in the long term. In the short term, however, with real output sharply falling in the first transformation stage, the resistance of employees in the government-owned network operator could be an important problem for privatization. It would be desirable to have competition in network operation in the long term. Private competition could be phased over several years.

Both the equipment industry and the networks should be privatized in the long term. It is doubtful whether the incumbent state or private monopoly should keep the whole existing network. Rather, a second network operator might be allowed to swap some of his assets (e.g. rights of way in the case of a grid or railway company) with a strategic part of the existing network (e.g. one major city). The only condition attached to a transition period would be the guarantee by both network operators of access to each other's initially exclusive network area for a certain price.

Privatization should involve a strategic foreign investor to make sure that modern technologies are rapidly introduced and that entrepreneurial market expansion can take place. Foreign investors will bring know-how as well as an active interest in linking existing foreign networks to the newly acquired network stake.

Fiscal Aspects of Telecoms Privatization

Direct fiscal revenues from privatization of the telecoms industry are of minor importance in comparison to the indirect revenue effects, which occur in two ways: profitable private network operators and service providers will generate increased future tax receipts. Secondly, higher output resulting from cheaper and better communication input will generate jobs and thus reduce the burden of unemployment benefits on the one hand, while on the other hand there will be additional tax receipts from these prospering industries. Since future profits are uncertain and long term, external encouragement to adopt a long term fiscal view would be useful. Here the EBRD and the World Bank as well as the PHARE program could play a valuable role.

Internationalization of Telecommunications in Visegrad Countries

In the Visegrad countries a consortium of the Swiss PTT and the Dutch KPN has acquired 27% of the Czech telecoms operator for more than $ 1.3 billion in 1995. This amount is considerably higher than the $875 million paid for a 30% share of the Hungarian MATAV which went to a consortium of Ameritech and Deutsche Telecom in 1994, even though Hungary actually represents the bigger market potential. Poland was hesitant to embrace competition and privatization in the telecoms sector, but the government finally decided at least to offer two tenders for a GSM mobile licence in late 1995 (with a maximum of 49% for foreign partners) and to consider partial privatization of the network operator. Polish TPSA still suffers from major problems in the terrestrial network whose quality is relatively poor as the high failure rate for Warsaw in 1990 showed (KUBASIK, 1994): The call completion rate was 40% in the whole country, but in Warsaw only 25%. A high failure rate effectively reduces telephone density. The Czech Republic and Poland have awarded several licences to foreign companies that are willing to develop the network in rural areas.

General Network Liberalization Required

One may emphasize that modernization of the telecommunication system is at least as important as raising telephone density. Only with modernization will there be an improvement in quality, reliability and sophistication of services, which then will contribute to economic growth and higher future tax revenues. This long term tax revenue aspect is much neglected in transforming economies where governments typically focus on short-term revenue maximization. Thus they would rather privatize a monopoly than introduce competition. Moreover, governments are slow to recognize that achieving competition in the telecoms sector ultimately requires looking into liberalization strategies for all network industries: namely, electricity, gas and railways which all are potential market entrants in the telecoms business. If free market entry in the telecoms sector is allowed while these network industries maintain a monopoly position and invest the monopoly profits from their shelterd core activity in telecoms activities, there will be a biased competition in telecommunications. A general network liberalization approach has, however, not been adopted in the EU either, even though on theoretical grounds such an approach would clearly be desirable for stimulating competition in telecommunications (WELFENS, 1995).

4.2 How to Accelerate Telephone Penetration

Telephone penetration can be accelerated by various means. A primary step is to encourage market-orientation; this requires the reduction of state ownership. Listing the company on major international stock markets could be a medium measure; the critical eye of international analysts and the international listing would drive the management towards an exploition of market opportunities.

One interesting strategy could be to follow Telefonica de Espana (capitalized at end-1993 at $ 13 bill.) and its subsidiary Telefonica Internacional. The Spanish government has a stake of only 20% (1995). Telefonica International has equity stakes in many Latin American countries and also in Portugal and Romania. Within a few years Telefonica sharply reduced the waiting lists in Spain and had modernized Spain's telephone system. However, national and international telephone calls are relatively expensive in Spain. This points to the problem that network competition might bring down prices for telephone subscribers, but that modernization might take much longer than under a monopoly network operator with extra profits. The international orientation of Telefonica Internacional might partly be a remedy for lack of competition at home. Since national network operators in Central and Eastern Europe do not enjoy special advantages in internationalizing their network operations (as in Spain with its cultural, economic and languistic ties to Latin America) one might, have to resort to the competition strategy. Even for Russia combining a domestic monopoly with internationalization is not an efficient option.

Network expansion requires a growing demand for network services and growing regional and international communication needs. No network operator could earn much from the pure transmission of voice or data. A standard phone call between two US states costs about 17 cents per minute. The operator has costs of some 16 cents, whereas 7 cents go to the local network operator; about 4 cents are for the physical trunk network, and transmission itself costs about 1 cent (TYLER, 1994). With the emergence of a market economy, which typically has a wider spatial production network than a command economy, the need for regional and international communication is rising strongly. Network expansion and network competition will be facilitated by the unfolding of market economy dynamics.

Social Aspects

Poor households and people in remote areas could obtain government transfers to cope with the potential rise in the price of telecoms services. The alternative would be to auction off whole regions in a way which would give the licence to that operator requiring the lowest government subsidy for a full-area coverage and universal services. Mobile network operation could be regulated in a way so that at least 2/3 of the country's surface has to be covered. With continuing technological progress, less attractive areas will ultimately be covered as well. At the same time, areas with poor coverage will suffer from emigration. This in effect reduces part of the economic problem, yet there might still be a political problem in the sense that the government would like to stabilize population density in border regions at some critical level. This problem could, however, be solved by regional income transfers.

Some people in Central and Eastern Europe are very poor - some are simply the losers in the transformation stage, others are poor retired people. Old people especially need a telephone, an important link to the outside world (to friends, the physician, the library, etc.). One might also consider a pricing regulation which would force regional telecoms monopolies to offer telecoms services at a discount to the elderly.

4.3 Special Aspects of Mobile Telecommunications

Scarce frequencies in cities make franchising a necessary element of telecoms policy in the cellular telecoms business. Should licences be awarded in a lottery or auction or should licences be awarded to an operator in a screening evaluation in which technical and economic experts play a role? These questions were analyzed in recent economic literature (KRUSE, 1993). A lottery would attract "gamblers" and would not - despite being fair - award the licence to the most efficient operator. If a government sells a licence to a private operator while a subsidiary of the state-owned network gets its licence for free, it would be discrimination against private competitors. Only in the case of a dominant foreign investor would a licence seem to be appropriate as a means of sharing monopolistic rents. It seems obvious that the best way for awarding licences in the mobile telecoms business is to rely on standardized expert evaluation; this could include international experts. If an auction is held one could use the price of the second-highest bidder; this would reflect a policy shadowing the opportunity costs of network building and therefore be efficient.

Franchises should be given for a limited period of several years. An incentive regulation could be that companies with above-average price reductions over a five year period will automatically get a five year extension. The discretionary power of poor bureaucratic decision-making could thus be minimized. The required area coverage should be chosen in a pragmatic way: A covering of 2/3 of the country would be sufficient in most Central and Eastern European countries, which are often not evenly populated. Western experience shows that about five years are necessary to reach the minimum coverage. Radio communications are an alternative way of connecting up rural areas (which is politically attractive), and in many cases are now the lower-cost option. Since economies of density are lower than for land-links, there is a prospect of spreading telecoms connections into less densely populated areas without any need for heavy cross-subsidies.

While mobile telephony is strongly dominated (more than 90% of subscribers) by corporate users in Western Europe, residential users in the US account for a high share (about 40%) of the mobile market. The US experience suggests that residential users could play an increasing role in the long term in Central and Eastern Europe. However, the poor terrestrial network in some transforming countries suggests that at least in some regions mobile telephones would be important for residential users early on. In general, it may be assumed that corporate users will dominate the mobile market for at least a decade, since only a small share of the population can afford high mobile rates in a period of declining real income. Only in the course of a long term economic expansion process could one expect that residential demand for mobile phones will quickly increase, certainly stimulated by falling call rates and subscriber rates.

4.4 Innovative Use of Telecoms

With a rising share of firms and households connected to the telecom network(s), new opportunities for governments to rationalize other industries arise. One interesting and novel application of telecom and computer networks could be to organize a remote-control system for assessing the amount of water and energy (electricity or gas) used, while automatically collecting the charges from the client's bank account. On the customer's receipt additional information could be provided, which would encourage the saving of energy and natural resources. For example if energy prices were differentiated according to three time zones, the exact share of energy used during each zone could be shown so that incentives for a better use of the cheaper time zones would be given. Households and firms would become more cost-sensitive and would increasingly strive for energy conservation measures. This will reduce the use of natural resources, help to minimize emissions and furthermore save scarce capital since energy producers will face a reduced need to add new capacities for power generation.

Private citizens in big cities could especially organize special interest information groups. The government might subsidize networks for new entrepreneurial activities. Moreover, TV teaching courses could be increasingly used as a tool to update human capital resources.

4.5 Capital Market Aspects

Among the remarkable developments in network operation in the past two decades is the sweeping change of US network operation after the break-up of AT&T. Of the seven baby bells, Pactel is quite interesting: Pactel decided to spin-off AirTouch Communications as an independent company comprised of Pactel's cellular, international (e.g. German/Mannesmann co-activity) and paging operations. Divesting cellular business allows Pactel to bid in its own right for California franchises for personal communication services. Regulation and valuation were core considerations for spinning-off. It was believed that the wireless business - buried in the conglomerated activities of Pactel - was undervalued, causing problems with respect to access to capital for the expansive cellular expansion plans (given valuation and dividend requirement). This suggests that spinning-off should at least not be prevented in Central and Eastern Europe, where building and expanding mobile networks will be quite crucial.

The creation of functional capital markets is itself important for the efficient financing of major infrastructure projects on the one hand, and, on the other hand, for long term economic growth in private industry. Stable monetary policy and privatization of banks and insurance companies as well as a macroeconomic policy geared towards high savings rates are important ingredients for a successful modernization of the telecoms sector. A modern telecoms sector in turn will facilitate the expansion of financial market transactions.

4.6 Social Problems

In systemic transformation certain groups will suffer from transitory income losses and unemployment. There will naturally be pressure on the government not to contribute to real income losses through the pricing policies of public utilities and policy principles of privatization. Thus short term social problems could prevent long term telecoms regulation from being organized in an efficient way, which certainly requires not to focus attention on short term problems. Long term efficiency gains from the establishment of efficient regulation and opening up telecoms to competition are more important than short term problems caused by the switch to a new regulatory regime. However, social problems must and can be taken into account. Social policies could be designed in such a way as to give a once-and-for-all transfer to poor households receiving telephone access for the first time. If rebalancing subscriber charges and call rates leads to an increase of subscriber charges in the new competitive regime, the government's transfers could effectively reduce the price pitch for poor income groups while leaving higher access prices as valuable signals for investment and modernization of the telecoms sector. Competition will help to reduce real prices for all users of telecoms services in the long term, as is clearly demonstrated by experiences in the United Kingdom between 1984-91, when telecoms was opened up to competition and new price regulations were adopted (NATIONAL AUDIT OFFICE, 1993). The real cumulative price change in the period 1984-91 for exchange line rentals was 2.6 % for residential line rentals and 3.0 % for business rentals. While rental rates increased the wholesale price index rose faster, the real price of telephone calls was reduced. The change in the effective price of peak rates was -14% for peak rates and -19% for standard rates; charges for peak long distance calls - typically made by business - fell by 61 % in real terms, while cheap long distance calls made mostly by residents fell by between 23% and 28% in real terms. Residential users thus benefitted less than businesses, but private households also benefitted. Moreover, prices of information intensive goods also will fall. In post-socialist countries artificially low prices will have to be raised first, later competition will help to reduce telecoms prices and to improve the quality of services.

In a politico-economic perspective, a blockade of reforms by the short-term interests of rural and poor groups - facing rising subscriber charges while discounting uncertain falls in the real call rate - in combination with short term revenue interests with a focus on extracting monopoly profits from long distance services and international telecoms services could prevent competition.

4.7 Principles of Efficient Regulatory Policy

Competition in the telecoms sector is vital for all transforming economies which suffer from bottlenecks in telecommunication services. Efficient regulatory policies would stimulate competition and thus help to remove an important bottleneck for economic modernization, increasing internationalization of business and sustained growth in transforming economies. Achieving overall growth will require that some activities shrink, while others are allowed to expand. The number of PTT employees per 1000 direct exchange lines (DEL) was much higher in ex-CMEA countries than in OECD countries. Hungary had 23 per 1000 DELs in the late 1980s (India topped the global list with 98), while the US New England Telephone network had 0.21, Spain's Telefonica 0.14 and Japan's NTT 0.16 (AMBROSE/HENNEMEYER/CHAPON, 1990, p.13). While competition in telecoms network operation could mean slow growth of jobs in network operation, falling network user rates will bring about high growth potentials in the value-added services sector.

Through competition in the telecoms sector service quality will improve and prices fall; thus information will be more widely used and applied, particularly benefitting the least-advantaged in society. Hence long term social aspects indeed reinforce competition in telecoms services. This is not only shown by the example of UK deregulation in the 1980s but also by Sweden, which had no telecoms law prior to July 1993. Yet Sweden enjoyed the world's highest telephone penetration rate and the highest mobile penetration rate in the world - and cheap rates as well. Thanks to its competitive network environment the Swedish telecoms hardware industry is a leading player in world markets. For small open economies, a demanding home market (plus focus on global market) clearly is conducive not only to the improvement and the establishment of cheap services at home, but to the development of international competitiveness as well. One can clearly learn from Swedish, British and US telecoms competition.

Competition in telecoms services is indirectly a locational advantage; it improves international competitiveness, helps to create jobs, improves the use of knowledge in society and thus raises productivity, while helping to raise tax revenues as economic activities expand and raising consumer welfare as waiting lists are cut and service quality and variety improved.

Fig. B14: Effects of Competition in Telecoms Services

	Improves Quality of Services	*Locational Advantage *Improves Internat. Competitiveness
COMPETITION IN TELE-COMS SER-VICES	Accelerates Tele-phone Penetration	*Creates Jobs *Improves the Use of Knowledge
	Stimulates Techni-cal Progress and Falls in Real Price of Phone Calls	*Raises Tax Re-venues *Improves Econo-mic Welfare

Regulatory policy should promote competition in the telecoms industry while safeguarding a universal service and social obligations. The organization of the regulatory agency should aim to establish a politically independent body which, however, is accountable. An advisory board in which users, technical and economic expertise as well as regional and international interests are represented could support strategic decisions in regulatory policy. Innovative regulation can be promoted by creating political incentives through new constituencies: e.g. competitors, consumer groups and business user groups which all stand to benefit from prudent regulation. Moreover, - as suggested by WELLENIUS ET AL. (1993, p.10) in a World Bank analysis - major regulatory decisions could be built into sales contracts and concessions or licenses granted to the operators rather than left to the incipient regulatory process (however, in some ex-CMEA countries contracts are not at all effective).

International interests could be represented on the advisory board so that important economic partner countries (including the EU) could be involved in order to aim at integrating regional and international markets more effectively and to coordinate regulatory approaches in a consistent manner. One might create a special transition department which could deal with special problems of systemic transition; this department could deal with some aspects of privatization, with social transition aspects and with the opportunities to recollect some frequencies formerly used by the military. The main departments of the regulatory agency could be the following: a) legal, b) administration, c) engineering, d) frequency management, and d) telecoms business (see Fig. B15).

Fig. B15: Organization of Regulatory Agency

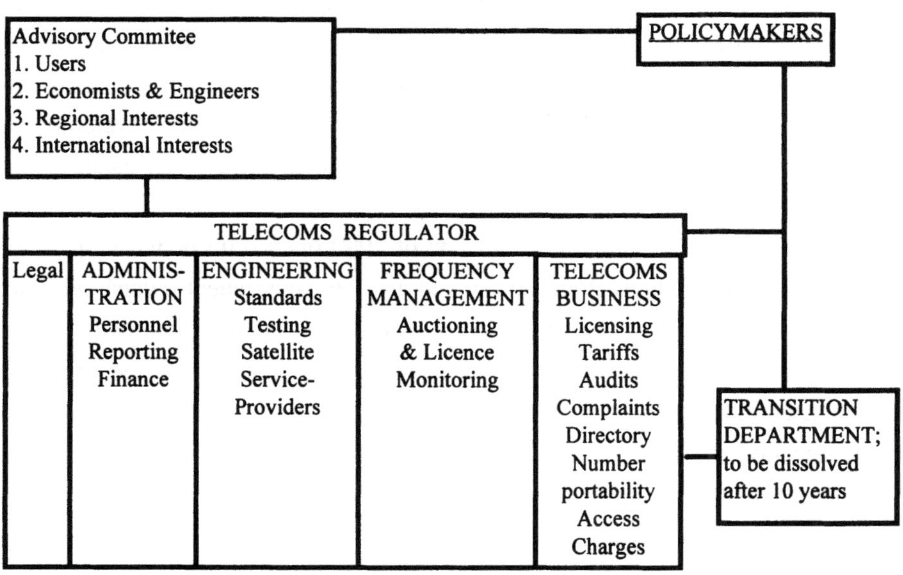

4.8 Perspectives for Poland

Poland should establish competition in its network operation and privatize its trunk operations; a minority equity stake of the state could be useful to capture part of the capital gains expected in the course of network expansion in an increasingly market-oriented and productive environment. Mobile telecoms should be strongly encouraged. Building oligopolistic networks while encouraging the internationalization of network operators seems to be most appropriate for rapid network expansion and for supporting the needs of the transformation process.

Network Expansion: Government Monopoly Versus Internationalized Competition
Poland, as a leading transforming economy, had a difficult starting point in the telecoms network sector because productivity - measured by main line per employees - was only about 1/3 of that of Spain. Since revenue per main line in Spain was about four times as high as in Poland, the value productivity of labor in Poland was less than 1/10 of the Spanish figure. Adjusting to Western European productivity levels does not require that the number of employees at TPSA be reduced by 60 % in Poland, but rather that network line capacity be quickly expanded on the one hand; on the other hand, a considerable number of employees could be transferred to a new subsidiary in mobile telecoms and new firms in the private sector. However, one cannot overlook that developing an innovative service sector is crucial for raising the demand for line capacity and raising the revenue per main line.

The Polish telecommunications law of 1990 opened up network operation to private operators in regions not served by the government-owned telecoms provider TPSA. About 90 licences were issued, but no real competition was introduced. According to the law, independents must negotiate the terms of interconnection charges with TPSA. The lack of a standardized agreement in interconnection discouraged most of the independents from further pursuing the development of their local network; without fair rules for interconnection, outside investors are reluctant to commit major resources (WORLD BANK, 1994).

There are basically two alternative approaches to network expansion in transforming economies: (I) The state monopoly operator maintains its position and assigns licenses to private operators for underdeveloped areas thereby creating a patchwork in which newcomers have no real opportunities for full network competition (not least because the national network operator can always cross-subsidize activities and impose excessive access charge). As a consequence, subscriber prices will remain high, the scope and quality of services poor, government tax revenues lower than otherwise and the telephone penetration rate much behind the most advanced countries. Such a situation indeed described the US in the early 20th century, when inefficiencies as well as long and inconclusive regulatory reviews by the state body FCC (created in 1934, but already advocated in 1907 by Bell's chief executive) characterized the telecoms sector. This finally led to the 1974 charges of the Department of Justice, which after ten years settled a compromise with AT&T, namely its divestiture (IRWIN, 1990). Central and Eastern European national telecoms networks would gradually modernize their telecoms networks and imitate most or all the liberalization moves undertaken in Western Europe. Given the large East European backlog with respect to telephone penetration such a gradual strategy is unlikely to close the East-West-gap in Europe, and even more so to stimulate a rapid expansion of the services industry.

A minimum step towards more competition would be the privatization of the government-owned network operator and network competition in international services (e.g. as in Korea). A standardized contract for interconnection charges is needed to encourage network investment from independent operators; additionally, some of the major cities should be opened up to independent operators so that some yardstick competition could occur. The comparison of call rates, speed in the reduction of waiting lists, customer service and technical reliability (fault rates) would be a useful strategy to encourage modernization and restructuring at TPSA. Simply to rely on internal TPSA dynamics of change is insufficient as one cannot expect a huge and inflexible network operator to react without external market pressure. Allowing management buy-outs for parts of TPSA's network could also be an interesting option to increase the efficiency of network operation. Unbundling of network operation and network services should be encouraged. This is the best way of privatization.

Creating Full Network Competition

Scenario II assumes the existence of at least two competing network operators, each operator possibly including one strategic foreign investor or several such investors. A foreign investor would facilitate technology transfer and access to international business - especially multinational companies. The national railways and electricity grid company are natural candidates for forming a second and a third network operator, which would in the beginning also have to rely strongly on leased lines. Government regulation could stipulate that at least 1/3 of line capacities has to be made available for leasing; this should include the option for resale of capacity. If the state network operator would sell a major part of its network the required capacity open for leasing could be reduced. An organized spot market as well as a wholesale market for telecoms network capacity could be created. If a government wanted to retain a controlling stake in network operators, a modest controlling strategy could be recommended - say a government stake of 25 % - as long as sufficient competition itself was created; and competition plus open markets is the best control against misuse of potential market power a society can get. Even more important would be for a new government regulatory body to obtain all major information on prices and investment of each company, so that the unfolding of a competition could be monitored by the government. Indeed, a government could use its equity stake in the telecoms operators less as a device to control firms, but more as a means to appropriate some of the future capital gains expected from telecoms firms in the process of systemic transition.

Standardized interconnection rules would be important. They could be reviewed by the regulator every two or three years in order to make sure that interconnection charges are not fixed in a way to prevent potential new entrants from challenging the incumbent operator(s). Some rules for cost-sharing in universal service obligations are also important. The rebalancing of tariffs along marginal costs should be encouraged, while cross-subsidization should be discouraged. To the extent that economic policymakers want to ensure that rural areas and/or poor people get access to the telephone network(s) at low rates, income transfers to the respective users are useful since they will not distort the competition process as such.

A strategy of actively nurturing competition in the telecoms sector seems to be unusual only at first sight - possibly because in Western Europe the persistence of state network monopoly is still prevalent. However, the creation of competing networks in the fastest growing industry of the late 1990s is just what leading transformation countries should look for. Telecoms network operators are increasingly forging international alliances in OECD countries, and early movements towards competition in the telecoms network operation can help ex-CMEA countries to enter several international alliances in parallel. Relying on several partners from Western Europe, North America and Japan could indeed be a strategy to prevent becoming dependent on one major international/national operator. One might even consider alliances including cooperation among ex-CMEA countries (e.g. Visegrad countries); Central and Eastern European countries with competitive networks could indeed become important source countries of foreign direct investment in the future telecoms networks sector. It is unclear whether governments in transforming economies could be bold enough to introduce network competition prior to Western Europe, even though Central and Eastern Europe stand to benefit even more from competition here than the EU. International organizations - such as the EBRD, the European Investment Bank and the World Bank Group - could encourage such innovative developments. If there were competition in network operation plus import competition, international organizations (e.g. the World Bank) which extend loans for privatization of the hardware industry could be less strict in imposing requirements for privatizing state firms in the telecoms hardware sector. With network competition market forces will finally force the hardware sector to restructure along the principles of competitive private firms.

As regards regional coverage with telecoms services, uniform prices are not optimal (KNIEPS, 1993). An infrastructure fund - suggested by KNIEPS for Western European areas - might be established which could pay subsidies to operators in peripheral regions, on the basis of the lowest subsidy required. However, in Central and Eastern Europe, subsidies to producers could be counterproductive since they would stimulate the call for subsidies by many other industries. Given large deficit problems in transformation and the backlog in privatization in most transforming countries, it is not advisable for any country to envisage strategies for infrastructure policies which will subsidize the producers. Rather, the users of telecoms services should receive tax credits. This would also be useful for firms because their drive toward profitability would thereby be encouraged; a tax credit is useful only for profitable firms. Non-profitable state firms might, however, be allowed to accumulate part of past tax credits, provided that the company is privatized.

Competition in telecoms sectors is of key importance for long term growth in Central and Eastern Europe. Poland could reinforce its position both in the region and with respect to future EU membership if privatization and competition were guiding principles of the telecoms sector. Poland has a big domestic market which could only benefit from reduced information costs and improved communication. The market is big enough to be covered by at least two competitors in long-distance and local area telephone business. Through the elimination of international communication bottlenecks, Poland would increase its attractiveness as an investment location. Moreover, international trade would be stimulated.

At least two mobile telephone licences should be awarded to each region. Licences should be awarded on the basis of an expert evaluation; auctioning off licences could be adequate if foreign investors play a major role in telephone licences. The national network operator (or competing future trunk network operators) would have to guarantee unrestricted access to the network under standardized conditions. Separate accounting for access charges and publication of financial results should also be mandatory.

Regulatory policies should concentrate on technical standards, accounting separation between network operation, leased lines and value-added services. Firms should be encouraged to create subsidiaries alongside lines of business so as to create clear market signals for financial markets as well as for the users. Regulators would find their task much easier if the government was not to be a shareholder of major telecoms firms. Many conflicts of interests will arise if government ownership plays a strong role; it should at least be reduced in a progressive way over a period of several years. Opening up the telecoms and the service market is urgently needed in Poland.

The national network operator has a natural tendency to defend his turf, but policy could redirect part of the potential of the national network operator towards international options, which could be quite interesting in the ex-CMEA area. Creating a competitive domestic environment for network operation will reinforce the long term ability to export turnkey projects in telecoms industry and to become an international player in the region; Telefonica is an encouraging example with respect to Latin America. Poland has been an early leader in systemic transformation, and by adopting an efficient regulatory approach in telecommunications its network operators could finally enjoy a first mover advantage in international network operations. Poland's skilled labor force and strong technological potential imply a favorable position for exporting value-added services, for becoming active in network operation in developed and developing countries (after privatization) and for creating a modern tele-informatics society.

Given the experience of OECD countries (WELFENS/GRAACK, 1996) and taking into account the side-constraint of systemic transformation, the following conclusions for developing a modern telecommunications sector can be drawn:

(1) a massive increase of telecoms penetration within a decade or so is necessary if the large East-West gap in Europe is to be overcome. Many new private regional telecommunication companies have to be launched to quickly raise the penetration rate. Network interconnection is important and hence regulation has to set technical standards that allow the establishment of a seamingless network. Politically independent regulation also has to encourage procedures for determining reasonable interconnection charges.

(2) Competition in the telecoms network is crucial in order to expand telecommunications services and to improve the locational quality for domestic and foreign investors. As regards the incumbent telecommunication monopolist, competition ideally requires the separation of the (terrestrial) network business while encouragement of modern bypass-technology at the local level: e.g. wireless DECT can establish a second, cable TV a third and satellite transmission a fourth direct access to households and firms.

(3) Moving towards competition implies higher profit tax revenues for government and a higher tax burden for consumers. One should recall that standard tax theory shows that a monopolist not only produces less than firms in a competitive regime but also produces on the elastic segment of the (linear) demand curve implying a tax incidence much to the disadvantage of the monopolist. Thus, under competition the demand side - basically meaning private households - will bear a larger burden of taxation in the telecommunication business.

(4) Number portability should be quickly established in order to encourage true competition. Right of ways should be clearly defined, and indeed central government (or private operators of highways) could generate new revenues from charges for using highway lanes to carry telecoms lines. Local government in turn could charge for the local terrestrial networks which would encourage the application of wireless technologies.

(5) A duopoly strategy clearly is non-optimal as a way to introduce competition in the telecoms network sector. Several long distance operators - no less than three if one follows the example of Finland - can profitably operate. In the presence of foreign investors (say x_1, x_2, x_3) there is a risk of "secondary monopolization" which occurs if market consolidation abroad brings a merger of foreign investors (X created by a merger of x_1, x_2 and x_3). Such a situation can be anticipated in the privatization process and the sales contracts, respectively. Government could retain a golden share for the sole sake of preventing secondary monopolization.

(6) If prices in local telephony initially are below costs rebalancing (higher prices) are necessary in order to stimulate market entry and to support investment financing requirements.

(7) Rapid privatization via the stock market is urgently needed in order to generate pressure from the capital market for the rationalization of the incumbent telecoms operator.

Endnotes:

1 CoCom restrictions which applied until early 1994 partly explain low technical standards and insufficient digitization of the telecoms industry in former CMEA countries.

2 The US market became truly competitive only after the dismemberment of AT&T. With the emergence of regional baby bells and two new long distance networks (AT&T, which remained in the long distance business), the US telecoms market became increasingly competitive; innovative services and falling prices become characteristic for the telecoms business.

3 See Beckmann and Künzi (1994), Mathematik für Ökonomen I, Heidelberg: Springer, p.188; with a linear demand schedule p=a-bq, the consumer surplus $S= (a-p_E)2/2b$, where p_E is the equilibrium price.

Appendix B1: Main Lines per Employee, 1982 and 1991

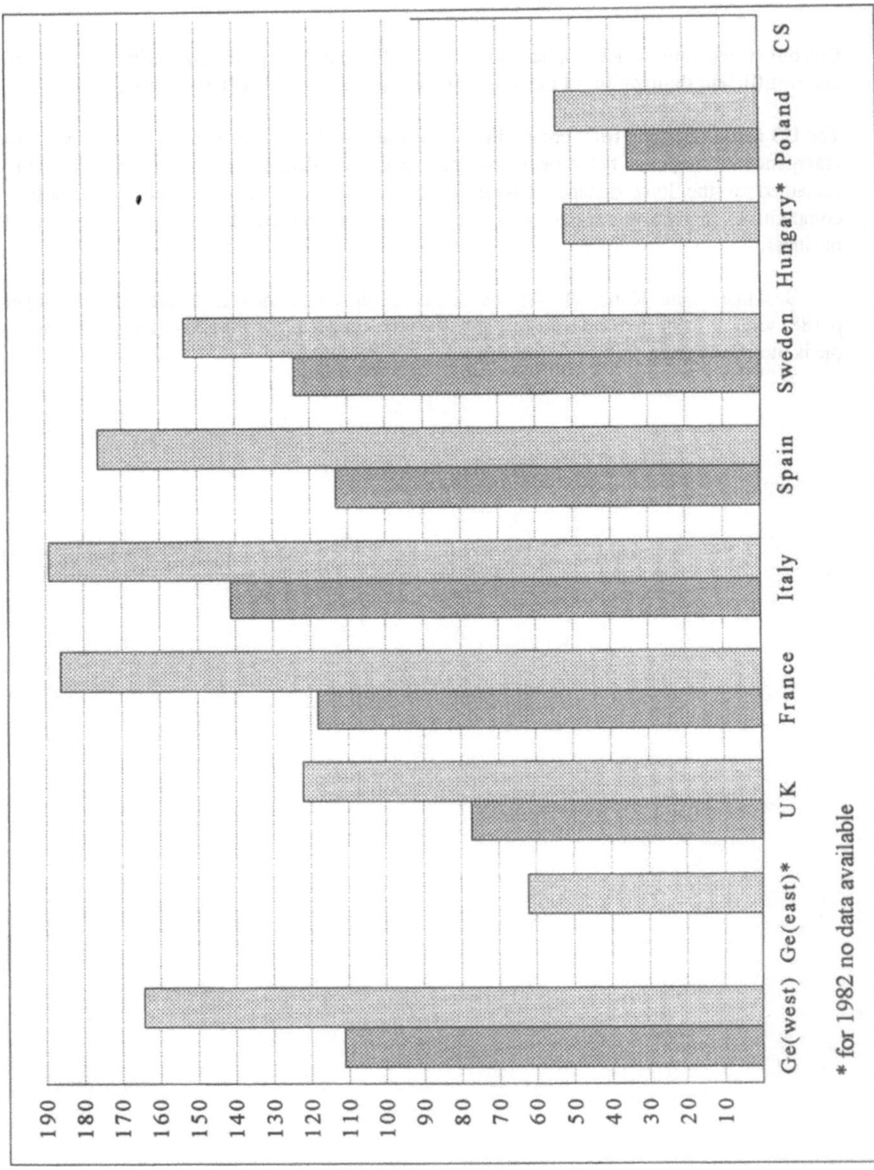

Source: ITU (1992), European Telecommunications Indicators, p.24.

Appendix B2: Waiting List for Telephone Lines

Western Europe (in thousands)

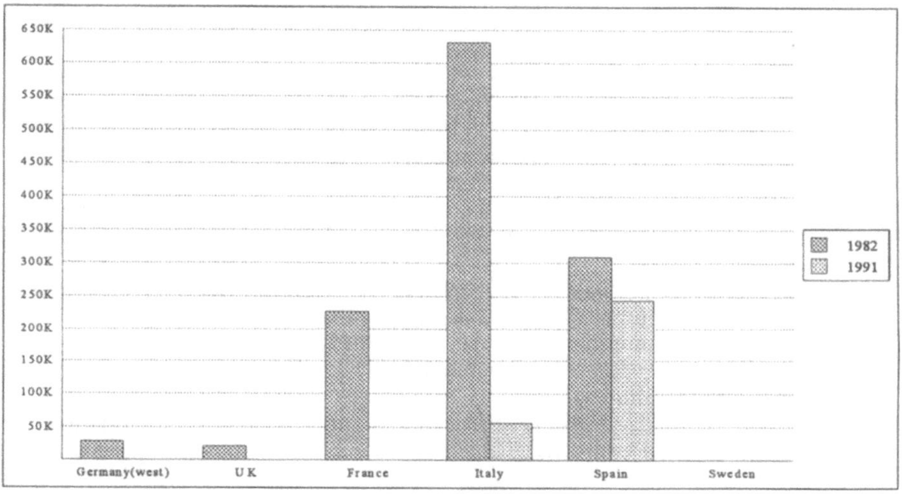

Central and Eastern Europe (in millions)

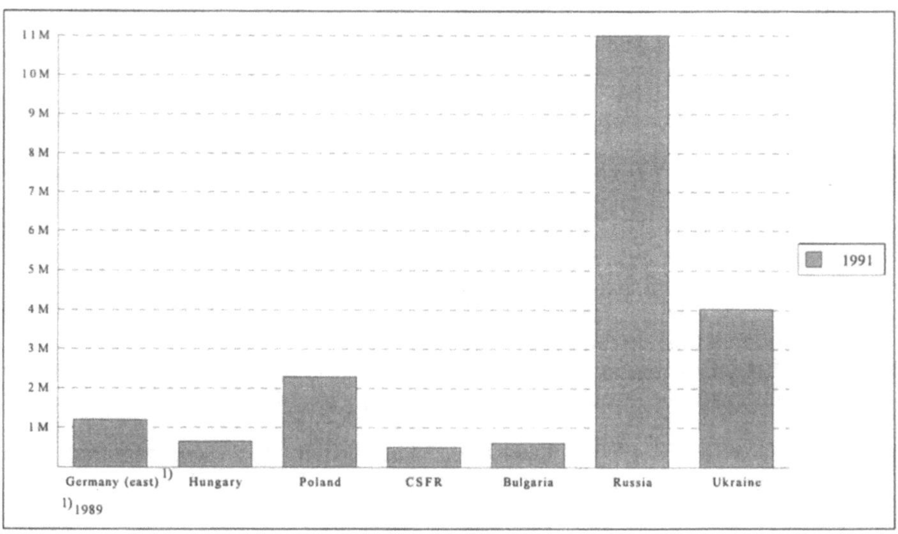

Source: ITU (1992), European Telecommunications Indicators, p.7.

References

AMBROSE, W.W.; HENNEMEYER, P.R. and CHAPON, J.-P. (1990), Privatizing Telecommunications Systems, IFC Discussion Paper No. 10, Washington DC: The World Bank.
BAR, F. and BORRUS, M. (1989), From Public Access to Private Connections: Network Strategies and Competitive Advantage in US Telecommunications, in: OECD and CEC/DGXIII, Information Networks and Competitive Advantage, Vol. 1, Berkeley: BRIE
BECKMANN, M. and H. KÜNZI (1994), Mathematik für Ökonomen I., Heidelberg: Springer.
BERLAGE, M. and SCHNÖRING, T. (1992), Telekommunikation in Osteuropa. Bestandsaufnahme und Entwicklungstrends, Wissenschaftliches Institut für Kommunikationsdienste, Discussion Paper No. 93, Bad Honnef.
BLANKART, C.B. and KNIEPS, G. (1989), What Can We Learn from Comparative Institutional Analysis? The Case of Telecommunications, Kyklos, 42, 579-598.
BLANKART, C.B. and KNIEPS, G. (1993), Das Konzept der Universaldienste im Bereich der Telekommunikation, Freiburg: mimeo.
CSIS (1992), Telecommunications in the Post-Soviet Society: The U.S. Challenge, Summary Report, Washington DC: Center for Strategic and International Studies, mimeo.
EINHORN, M.A., ed. (1991), Price Caps and Incentive Regulation in Telecommunications, Dordrecht: Kluwer.
HABUDA, J. and STOCK, W.G. (1994), Telekommunikations- und Informationsstruktur in Ungarn, Ifo Schnelldienst, 7/1994, 28-38.
HUNYA, G. et al. (1994), Central and Eastern Europe: Uneven Recovery, Vienna Institute for Comparative Economic Studies, Reseach Report No. 204, Vienna.
IRWIN, M.R. (1990), The Telecommunications Industry, in: ADAMS, W. (ed.), The Structure of American Industry, 8th edition, New York.
ITU (1992), European Telecommunications Indicators, Geneva.
KLEIN, M.W. and WELFENS, P.J.J. (1992), Multinationals in the New Europe and Globale Trade, Heidelberg: Springer.
KNIEPS, G. (1993), Europäische Telekommunikationspolitik, Freiburg i.Br., mimeo.
KRUSE, J. (1993), Lizenzierung und Wettbewerb im Mobilfunk, Heidelberg: Springer.
KUBASIK, J. (1994), Development of Telecommunications Markets and Regulation in the Central and Eastern European Countries, Country Study - Poland, draft version, November 1994)
MINISTRY OF TRANSPORT AND COMMUNICATIONS, Telecommunications in Sweden, Stockholm 1994.
MITCHELL, B.M. and VOGELSANG, I. (1991), U.S. Practice of Telecommunications Pricing, Discussion Paper No. 66, Bad Honnef: WIK and RAND.
NATIONAL AUDIT OFFICE (1993), The Office of Telecommunications: Licence Compliance and Consumer Protection, London: HMSO.
SIEMENS (1993), International Telecom Statistics, Munich.
TYLER, M. (1994), Technische und organisatorische Voraussetzungen für die Konstruktion von alternativen Netzen in Deutschland: Was kann man aus der internationalen Erfahrung lernen? Vortrag für Euroforum, 18. Januar 1994: Putnam Hayes &Bartlett Ltd.
UNCE (1994), Economic Survey of Europe 1993-94, Geneva.

WELFENS, P.J.J. (1992), Market-oriented Systemic Transformations in Eastern Europe, New York: Springer.
WELFENS, P.J.J. (1995), Achieving Competition in Europe's Telecommunications Sector, discussion paper no. 3, European Institute for International Economic Relations (EIIW), Potsdam.
WELFENS, P.J.J. and GRAACK, C. (1994), Telecommunications in Western Europe: Liberalization, Technological Dynamics and Regulatory Developments, Volkswirtschaftliche Diskussionsbeiträge der Westfälischen Wilhelms-Universität Münster, Discussion Paper No. 200, Münster.
WELFENS, P.J.J. and GRAACK, C. (1996), Telekommunikationswirtschaft. Deregulierung, Privatisierung und Internationalisierung, Heidelberg: Springer.
WELLENIUS, B. ET AL. (1993), Telecommunications. World Bank Experience and Strategy, World Bank Discussion Papers, No. 192, Washington DC.
WINSTON, C. (1993), Economic Deregulation: Days of Reckoning for Microeconomists, Journal of Economic Literature, Vol. 31, 1263-1289.
WORLD BANK (1994), World Development Report, Washington.
YPSILANTI, D. and T. KELLY (1993), Fostering Telecommunications Development. The Role of the OECD, in: STEINFIELD, C.; BAUER, J.M. and L. CABY, (eds.), Telecommunications in Transition, London: Sage.

C. Modernization of Telecommunications in the Former GDR

Stephen Economides

1. Former GDR Telecommunications System

The *Deutsche Post* was the ministry in the German Democratic Republic (GDR) responsible for the operation of the telephone system. Its other duties included regular postal services and those banking facilities typical for European PTTs (Post, Telephone and Telegraph Offices) and was approximately equivalent to the *Deutsche Bundespost* of the Federal Republic of Germany in its organizational structure and the services it provided. The legacy of the Deutsche Post in providing telecommunications services, however, is a poor one. Dilatory, lackadaisical and inefficient service was trademark of a technically antiquated and totally inadequate system which was never able to meet the demands of a modern industrialized society. This was certainly not due to the technical incompetence of the engineers and staff. The inefficient allocation of financial and material resources did indeed play a role, but this was a problem notoriously inherent in every socialist economy.

The main problem was the ideological dilemma in which the GDR's gerontocracy found itself. While realizing the need for a telephone system in order to be able to run the country, they simultaneously feared the free flow of information which all unmonitored channels of communications can provide. Almost from its inception the GDR controlled all forms of media and practiced a particularly restrictive policy regarding the access to and availability of the technical means of communication such as telephones; in effect a willful policy of "non-communication". This paranoia went so far as to ban the personal possession of photocopiers. Even university professors had to request permission to make photocopies, thereby having to state in writing the reason for the copies and the numbers involved (WALTER, 1993).

Consequently, telephone policy, i.e. who received a phone line and who did not, was an ideological and not a technical matter. Needless to say, well-known political opponents had no chance of having a phone installed. And in the event that a telephone had already been installed in the apartment of someone who may have in the meantime turned into a so-called "unreliable social element", such as in the case of the dissident Robert Havemann, the line was simply cut. Surveillance, supervision and wire-tapping became standard features in the lives of GDR telephone users. The Ministry for State Security, more commonly referred to as the *Stasi*, spied on and wire-tapped its own fellow citizens to such an Orwellian extent that it has already become legendary.

The fear that the general availability of the technical means of communication could lead to widespread abuse, especially the supposedly subversive threat posed by the establishing or maintaining of contacts with the "class enemy" in the West, resulted in the practical elimination of telephone service between East and West Germany as early as the 1950's. War-damaged lines were not repaired or replaced and those still in operation were not adequately maintained, or occasionally simply cut. As was the case with personal travel, West Berlin remained the last gap to be closed. The telephone system between the two halves of the city had been completely severed long before the Berlin Wall went up on August 13, 1961.

In a clandestine night operation on May 27, 1952 all 3,910 telephone lines along the demarcation line between East and West Berlin were sliced, ripped out, sawed through, or chopped off. In many instances this happened so quickly and improperly that the ensuing short circuits and breakdowns resulted in widespread damage to many telephone exchanges. Until the West German *Ostpolitik* first bore fruit in the early 1970's it was not possible to phone from Köpenick to Kreuzberg or from Pankow to Charlottenburg. Negotiations on governmental, not local levels, begun in 1969, eventually led to the establishment of five telephone lines between East and West Berlin in January 1971. By 1978, this number had increased to a couple of hundred but the lines were constantly busy, waiting times were long and East Berliners calling West Berlin had to pay a surcharge for "international calls" (DER TAGESSPIEGEL, 1992).

Although service was erratic and extremely time-consuming it had at least been possible during this period to make calls between East and West Germany. For example, in the case of two neighboring German villages separated by the border, one in Hessen, the other in Thüringen, the procedure was as follows: the call from the West German party was routed to the Frankfurt exchange which then contacted the Leipzig exchange via a trunk line - direct dialling was not possible. The Leipzig exchange would then relay the call to the Thuringian village. If one was lucky, it would take only half an hour to get the call through. In most instances, it took several hours. Despite technical deficiencies, for example, the calls were manually switched, the delays were for the most part intentional. The East German authorities wanted to unnerve and demoralize people and thus to discourage use of the telephone. Expansion of the existing telephone system was purposely slow for the same reasons. On the average only approximately 50,000 subscriber lines were installed annually in the GDR. Telekom put more lines into operation in 1991 alone (550,000) than the GDR had in the previous ten years (TELEKOM, 1992a; TELEKOM, 1992b, p. 3; TELEKOM, 1993, pp. 1-3; TELEKOM, 1994a, pp. 3, 8; TELEKOM, 1994b, pp. 14-19).

When the borders came tumbling down at the end of 1989, the GDR telephone system can be described as third-rate at best. Although telecommunications between the Federal Republic and the GDR were conducted over 1,461 lines (TELEKOM, 1994a, p. 4), no more than 600 were ever used at one time because the Honecker regime did not have sufficient wire-tapping personnel to simultaneously listen in on more conversations. This is based on Stasi documents made public in January 1992 (RHEINISCHE POST, 1992; TELEKOM, 1994a, p. 8). The situation within the GDR, however, was even more depressing.

With a population of about 18 million only slightly over 1,800,000 (1,826,190) telephone subscriber lines existed in the entire country, which corresponded to about 10 phones per 100 inhabitants, whereas West Germany had approximately 49 per 100 and the United States about 54 telephones per 100 inhabitants respectively. Density varied considerably from region to region. In East Berlin there were 28 subscriber lines per 100 inhabitants, whereas in Dresden there were just 9. In an international comparison the GDR even lagged behind the Soviet Union and just slightly ahead of Turkey. Government offices and party institutions as well as important areas of the economy, e.g. plants and factories, were, relatively speaking, sufficiently provided for. Forty percent of all subscriber lines were to be found in these areas (TELEKOM, 1994a, p. 3).

It was the private citizen who suffered the most from the lack of telephone service. Those lucky enough to own a phone were connected to a multi-party line.

This meant just about everybody since approximately 98% of all subscriber lines to apartment buildings and houses were multi-party, most of these being two-party lines. Applications for a phone connection increased steadily throughout the 1970's and 80's and by December 1989 1.3 million applications for a telephone had not yet been processed. In all fairness to the GDR, however, It must be pointed out that there has often been a general tendency to overestimate the value of such statistics. The figure 1.3 million definitely does not mean 1.3 million individuals living in separate households. We have now found out that in many instances every member of a family - father, mother, son and daughter - all applied for a telephone for the same apartment, evidently hoping that one of the four might be the lucky one to get one. In addition, no figures are available as to the number of people who, in view of the hopelessness of the undertaking, did not bother to apply at all. Years went by between the filing of an application and the actual installation, in some cases more than ten years (TELEKOM, 1994a, p. 3; DER TAGESSPIEGEL, 1992; WAS, 1992). Public telephone booths did not offer much of an alternative as there were only about 22,000 in the entire GDR.

The telephone exchange offices in operation did not cast a ray of light in the darkness, either. Approximately 62% were over 40 years old and of these 18% were built between 1922 and 1934. Only 1% of all exchanges were digital (TELEKOM, 1994a, p. 4).

The situation was even worse for non-voice telecommunication services. The 18,000 telex lines did manage to cover the bare needs of business and industry, but other services were either not available or, as mentioned previously, severely restricted. There were only 3,500 non-voice data lines, many incredibly still switched manually. The facsimile service, with only 3,000 fax lines and hampered by extremely limited performance characteristics, was practically meaningless as a telecommunication component. Modern data services did not exist; nor did any kind of mobile telecommunications systems. This was then the state of the GDR's telecommunications services.

2. German Unification and Telecoms Modernization

The first talks between officials of the Telekom Division of the *Deutsche Bundespost* and the *Deutsche Post* took place in December 1989 and January

1990. (The Bundespost had been restructured as of July 1, 1989 resulting in three separate divisions: *Postdienst* for regular Mail services, *Postbank* for banking services, and *Telekom* for all telecommunications services.) It was thus the task of Telekom to conduct the negotiations. Initially, the talks served little more of a purpose than to provide an opportunity to become better acquainted with each other and to take stock of the respective telecommunications systems. The ensuing shock was mutual. Telekom representatives were horrified at the dilapidated, technically obsolete telephone system and GDR officials equally so because they were not familiar at all with new digital systems already in use in the West. In the land of perpetual shortages the GDR's 43,000 telecommunications employees were more concerned with patching things up than in introducing new technologies.

Initially, it was slow going because all decisions had to be made on cabinet level between postal ministers, even though the talks, as in Berlin, sometimes dealt only with local developments. Despite the existing technological chasm both sides cooperated as equal partners. After all, the GDR still formally existed as a separate country and the GDR representatives placed a great deal of value on their independence. Nevertheless, preliminary plans on modernizing the telecommunications infrastructure were agreed upon in June 1990. Called "Telekom 2000", this was the first complex reconstruction program in Eastern Germany aimed at creating a telecommunications system on the level of the Federal Republic by 1997. 60 billion German marks were earmarked for investment. Telekom was then, and still is, the largest single investor in the new German states. It must be kept in mind that this program was initiated before reunification even took place. There are no equivalent historical parallels.

Between the opening of the borders in November 1989 and German reunification on October 3, 1990 a contagious spirit of euphoria and enthusiasm gripped the Germans. Considering these momentous events, many people understandably failed to see the actual conditions in the new German states. Statements by politicians, such as Chancellor Kohl's famous, "Nobody will be worse off than before.", or that the citizens in the new German states would be living in "flowering landscapes" were taken literally by many East Germans. They believed in the politicians, who in turn relied on the strength and efficiency of German business and industry to revitalize the former GDR. The prerequisite for the German economic juggernaut was reliance on an indispensable telecommunications system in order to do the job. Modern business and industrial operations cannot function without fast, efficient and reliable telecommunications - something which simply did not exist in the new states.

Illusions about a short period of economic and political adjustment in order to achieve complete unification were quickly dispelled.

Almost everyone underestimated the problems - and they were enormous.

The first eye-opener for the inexperienced came during the Bundestag election campaigns in the summer and fall of 1990. People, suddenly realizing how difficult, if not impossible, it was to phone from East Berlin to Bonn, or from Leipzig to Hamburg, or Dresden to Munich, expressed their displeasure very vocally. Despite Herculean efforts Telekom was subject to tremendous political pressure to speed things up. The first crescendo of criticism abated somewhat in July when Telekom started installing the first segments of its overlay network. New fiber optic trunk lines were laid, radio relay links established, and digital techniques which could make use of the predominantly analog GDR system were introduced. In addition, Telekom resorted to the use of every existing GDR network, 15 in all, to alleviate the situation. The first to be requisitioned was the GDR government network which did not prove to be of much help since it was not possible to place a call outside the system. It was a completely autonomous system as was one Stasi system, which even had wire-taps on the government network! (ERBE, 1994; TELEKOM, 1994a, pp. 5-8; WAS, 1992).

By November 1990 Telekom officials realized that they were not going to be able to do everything alone in order to reach the goal set for 1997. Consequently, four large general contractors Siemens, SEL, DeTeWe and Bosch were commissioned by Telekom to help accelerate the pace of restructuring and expansion. This is referred to as the "turnkey" project because these companies had to complete all aspects of their respective tasks, whether for a city, a county or an entire region and then just "hand over the key" to Telekom. The contractors together with numerous sub-contractors had to take care of all aspects of installing new telephone lines and exchanges: planning the networks, hiring excavation firms to dig up the ground for cables and construction firms for erecting new buildings, ordering, setting up and putting into operation all the necessary hardware, obtaining electric power from a power company (digital exchanges require constant power), and testing and connecting their networks within the national system.

The difficulties encountered can only be hinted at here. Thousands of miles of cables had to be laid. Thousands of buildings had to be rewired for telephones. Many buildings had cables that ended in a distribution box in the cellar and no one knew were the lines went from there. If cables were found that went out to the street, could they be used? And where did they go? If cables were located in distribution cabinets on the street, to what node were they connected? In many cases no plans were available. These search and grope missions were very time-consuming. A great deal of improvisation replaced systematic planning. Engineers used to three years for planning and setting up a digital telephone exchange were told that they had to do it within a year.

Disappointed at not receiving a telephone immediately, hundreds of businessmen attempted to circumvent the problem by using cellular phones as a permanent fixture in their offices. This overloaded the circuits to such an extent that it was nearly impossible to reach anyone. In Berlin, the busy signal was probably the most common message heard on both the analog and digital cellular systems.

And when a subscriber finally had a phone, what number did you give him or her? The danger of duplication was great. In Berlin alone there were around 150,000 duplicate numbers. And what did you do with two subscribers who already had the same number? Up until the mid-1970's the GDR had tacitly agreed to continue the pre-war system of allotting telephone numbers. This stopped with the construction of the East Berlin satellite cities in Marzahn, Hohenschönhausen and Hellersdorf (TELEKOM, 1993; OBER-POSTDIREKTION, 1992; DER TAGESSPIEGEL, 1992).

Cables that were newly laid were often destroyed soon thereafter by overzealous building contractors. Hoping to cash in quickly on the building boom many dug up the ground for new buildings without bothering to inquire at Telekom offices as to where new cables had been laid. In 1993 in Leipzig alone such damage amounted to over half a million German Marks. That sum was reached in one single incident in Berlin three days before Christmas 1994. An excavation firm, which was surprisingly in possession of the cable plans, nevertheless drilled through 12 fiber glass cables while doing work near the

Heerstrasse S-Bahn station. The cable television and telephone services of 160,000 and 10,000 households respectively were affected (MONITOR, 1995, pp.1-2). Telekom has subsequently filed a suit for damages.

Finally, none of the myriad tasks could have been undertaken without an organizational structure. Telekom with its 230,000 employees was in the midst of its own internal corporate restructuring and reorganization when reunification came and management had to struggle to incorporate over 43,000 East Germans into its ranks (RICKE, 1992).

The first phase of the Telekom 2000 project, also referred to as Aufbau-Ost, was concluded on July 31, 1994. At the half-way mark in the restructuring of the telecommunications system in the east all goals have either been met or surpassed. Of the 5.7 million subscriber lines targeted for 1997 over 4.1 million had already been installed by the end of 1993. All the main and junction telephone exchanges have been joined by fiber glass cables. 92% of the area (97% of the population) is roughly covered by the D1 cellular phone network. Duplication of all numbers has been eliminated and area codes for all of Germany have been introduced (TELEKOM, 1994a; TELEKOM, 1994b; HANDELSBLATT, 1994).

The Berlin telephone system was unified as early as June 1992 and the last remaining applications in the city for a phone will have been processed by the end of 1995. The 200,000 multi-party lines which still exist in the Eastern part of Berlin will have been converted to single party lines by 1997 at the latest. By 1998 all the telephone switching exchanges in the city will be digital. By the year 2000 Telekom will have invested DM 13 billion in the Berlin telecommunications network (DIE WELT, 1995; FAZ, 1995, p. 15).

Despite the myriad problems mentioned, Telekom and the German telecommunications industry have already made a great deal of progress and it is most likely that Telekom will reach its ambitious restructuring goals by 1997. Hopefully so for Telekom because it will then have to face the competitive onslaught of rival telecommunications companies when its national monopoly ceases in 1998.

References:

Various sources of information are available on Telekom's rebuilding efforts in the new German states. The numerous Telekom-Pressemitteilungen (see Supplement A), are released according to developments and not at any fixed date, but do provide the most pertinent information in short form, generally two to three pages. Of greater statistical value and with more in-depth information are the interim report of the Aufbau-Ost-Programm (Supplement B) and, of course, the final report (Supplement C). Another Telekom source very useful in locating relevant newspaper and magazine articles are the Telekom-Pressespiegel (Supplement D) which appear regularly, but also not at any fixed times. These articles are short and give the interested researcher a quick overview. Long articles are not included or mentioned. The Pressespiegel makes no claim to providing complete press coverage.

In addition to these Telekom sources, numerous articles have appeared in the German press concerning developments in the five new German states.

The following bibliography represents only some of the material presently available on the subject.

DER TAGESSPIEGEL (1992), "Das Problem mit den paar zerrisenen Strippen", Berlin, May 16, 1992.

DIE WELT (1995), "13 Milliarden für das Berliner Telefonnetz", January 4, 1995.

ERBE, W. (1994), is the head of the Telephone Exchange Office, No. 4 in Berlin, Conversation on January 17, 1994.

FAZ (1995), "Telekom beschleunigt den Netzausbau in Ostdeutschland", Frankfurter Allgemeine Zeitung, February 10, 1995

HANDELSBLATT (1994), "Kein Anschluß unter vielen Nummern in Ostdeutschland", July 5, 1994, Despite the negative sounding title a generally positive account of Telekom's achievements in Eastern Germany.

MONITOR (1995), "Zitterpartie am Heiligabend 24 Stunden harter Einsatz", Monitor - Die Mitarbeiterzeitung der Deutschen Telekom, Ausgabe 1/1995, p. 1-2.

OBERPOSTDIREKTION (1992), Die Telekom in Berlin und Brandenburg, Herausgegeben von der Oberpostdirektion Berlin, 1992.

RHEINISCHE POST (1992), "Telekom schafft in zwei Jahren soviel, wie die DDR-Post in 40", January 24, 1992, According to statements made by Frerich Görts, then a State Secretary in the Postal Ministry and now member of the Telekom board of directors.

RICKE, H. (1992), "Die Reorganisation der Telekom: Strategie, Stand, Perspektiven", in Telekommunikationspolitik in Deutschland - Perspektiven für die Zukunft -, Ausgewählte Beiträge der WIK-Konferenz "Drei Jahre nach der Postreform", Bonn, 23.-24. Juni 1992, Herausgegeben von Brigitte Bauer, Wissenschaftliches Institut für Kommunikationsdienste, Bad Honnef, 1992. Helmut Ricke bacame the first chairman of the Telekom board of directors after the passage of the 1989 Postal Reform Act. He assumed his duties in January 1990 and resigned unexpectedly at the end of December 1994.

TELEKOM (1992a), Das Aufbauprogramm der Telekom in den fünf neuen Bundesländern: eine erste Bilanz, Herausgegeben von Telekom, Februar 1992.

TELEKOM (1992b), Telekom-Pressemitteilung, Bonn, March 27, 1992, August 13, 1993, February 17, 1994.

TELEKOM (1993), Telekom 2000, Die Entwicklung der Telekommunikation in Berlin, Herausgegeben von der Telekom Direktion Berlin, 1993.

TELEKOM (1994a), TELEKOMchlußbericht des Vorstandsprojektes "Aufbau-Ost", Zeitraum: 03.10.1990 bis 31.07. 1994. Herausgegeben von Telekom, The 1,461 lines were comprised of 111 lines from East to West and 690 from West to East in addition to the 95 lines which went to and the 565 lines which came from West Berlin.

TELEKOM (1994b), Auf dem Weg in die Zukunft, Vier Jahre Telekom, Eine Bilanz mit guten Perspektiven, Herausgegeben von Telekom, 1994.

WALTER, H.-P. (1993) was a former professor at the East Berlin Humboldt-Universität, conversations during April-May 1993.

WAS (1992), "Wir nutzen alles -auch die alten Telefonleitungen der Stasi", Die Welt am Sonntag, May 3, 1992.

D. Recent Evolution of Telecommunications in the Region of Central Europe

Hanna Kontkiewicz-Chachulska

1. Introduction

The aim of this article is to present selected telecommunications development issues from the Polish perspective. The achievements of local reform are compared with solutions implemented in the Czech Republic, Hungary and the Slovak Republic.

At the end of the 80s, the gap between the levels of standard telecommunications services and networks in the European Union countries and the Central European Countries (CEC) was estimated at 20 years (based on a comparison of the penetration, automation and digitalization rates, availability and quality of services). Essentially the past totalitarian political systems favored centrally-planned command systems and did not consider modern telecommunications as a development priority. In these circumstances the bureaucratic and non-effective monopolistic Post, Telegraph and Telephone (PTT) were neither motivated nor able to sort out the permanent under-development of the sector (for years the COCOM embargo deprived CEC of access to state-of-the-art technology).

From the early start of the system reform in 1989, the CEC's governments were forced to re-consider the role of telecommunication, which was quickly promoted to the role of the general development enabler. Moreover, the strong influence of international organizations, the desire to join the European Union (EU) and the institutional imitation of West European solutions, together with general globalization trends have accelerated the transformation (KONTKIEWICZ-CHACHULSKA/PHAN, 1995). The telecommunication sector reform launched in the former socialist countries was an integrated part of the general economic transformation, which lead the previously centrally planned systems towards market schemes. The "standard" reform kit included the introduction of programs designed to achieve macroeconomics stability (the removal of price interventions and restrictions against foreign investment and trade and the introduction of fiscal and monetary policies), together with the reduction of bureaucratic control over production, distribution and consumption, followed by institutional changes. Each country adopted its own way of transforming ownership rights, a complex two-dimensional process consisting of the privatization of state enterprises and the creation of a private sector. In the case of telecommunications, the route from the old PTT to a competitive market with private providers, required the development of new and creative policies for changing the monopolistic *status quo*.

According to the European Union's experts, at the beginning of the alignment process, there were two major issues for the telecommunications sector in all the CEC: to extend the traditional services country-wide and to propose up-to-date advanced services to satisfy the needs of modern sectors of the economy and of foreign investors. The issue of telecommunications development has recently been given an additional stimulus, following the *European Council in Essen* which called for reinforced international co-operation with Central Europe in the context of the emerging "Information Society" (Bangemann Report, Delors White Paper on European Competitiveness). The conclusion of the summit was that countries in transition must be provided with the chance to fully participate in the development of new "techno-economic paradigms" (FREEMAN-PEREZ, 1988). Furthermore, the *Forum on the Information Society in Brussels* concluded that liberalization and harmonization of information infrastructure will boost competitiveness of the CEC.

The White Paper on Preparation of the Associated Countries of Central and Eastern Europe for Integration into the Internal Market of the Union underlined that basic policy principles and development targets were similar for all countries in the region. EU specialists observed that cumulative know-how in telecommunications has been rapidly developing in the region. They stressed, however, that very important elements were still missing, especially sound competition regulation, the separation of regulatory and operator functions, the elaboration of cost-accounting and cost-related tariff structure.

In this paper we will present a few aspects of the reform in Polish telecommunications as compared to the other specific solutions implemented in the region.

2. Telecommunication Administration

As early as in 1987, the General Directorate of Polish Post, Telegraph and Telephone was established as a self-financing entity within the Ministry of Transport, Navigation and Communications. In December 1989, the regulatory authority over posts and telecommunications was transferred to the newly created Ministry of Post and Communications. In 1991, two additional regulatory bodies were established: the State Radio-communications Agency (in charge of radio regulatory matters, the utilization of the frequency spectrum and the allocation of the radio spectrum), and the State Telecommunications Inspectorate (a watchdog for the wired communications industry, responsible for the regulation and the inspection of the licensed telecom operators and in charge of implementing regulatory policy).

The most important institutions responsible for telecommunications regulation in the CEC are presented below.

Tab. D1: Telecommunications Administration in CEC

COUNTRY	Ministries responsible for telecommunications	New Regulatory Bodies
Czech Republic	Ministry of Economy (sector policy and strategy); Ministry of Finance (price regulation)	Czech Telecommunication Office (majority of the regulatory duties) - section 6 and Department of Telecommunication Policy in section 4 of the Ministry of Economy
Hungary	Ministry of Transport, Communications and Water Management	Chief Communications Inspectorate
Poland	Ministry of Post and Communications	State Telecommunications Inspectorate, State Radio-communications Agency
Slovakia	Ministry of Transport, Post and Telecommunications; Ministry of Finance (price regulation)	Telecommunications Section of the Ministry (regulatory body for the Unified Telecommunications Network) and independent Telecommunications Office (issuing licenses for services outside the Unified Network)

3. Legislative Framework

Until 1989 the telecommunications in Poland was a state monopoly. According to the Post and Telecommunications Act of November 1990 a former PTT was split into two state-owned companies, Poczta Polska (Polish Post), a public utility company, and Telekomunikacja Polska SA (Polish Telecommunications S.A. TPSA), a joint stock company owned by the State Treasury (TPSA operates at its own risk and for its own account and the financial ties with the State Budget have been reduced to paying taxes). This decision could mark the first step toward privatization of the incumbent, yet private ownership seems a fairly distant prospect. The Ministry has not ruled out the partial privatization scenario, but it would not be a fast process because of its size and complexity.

TPSA started its operations on January 1, 1992. The State Treasury is represented as a shareholder in the company by the Minister of Communications, who nominates the President of the Executive Board. The company headquarters, national units and regional headquarters report directly to the Company Executive Board.

The Telecommunications Act of 1990 ended the state monopoly on the domestic services, although retained the state monopoly on the international services and state control on the domestic long-distance services. According to the Act and to its amendments of 1995, activities in the telecommunication field can be performed by the incumbent (which does not require any concession or permit - its activities, rights and obligations are formulated in the Act and in the provisions of the related documents) and entities which have been issued telecommunications concessions or permits. Permits are awarded for the installation and use of telecommunications facilities, lines and networks, and concessions for the provision of services. From 1995, new concessionaires, with the requisite capital and know-how, should be chosen in a tender process, organized by the Ministry, in order to limit the number of new entrants and to make the decision process more transparent.

The Telecommunications Act states that no concession can be issued for the provision of the public international telecommunications services. The amendments of 1995 have opened however the possibility to the non-public service providers, except when the candidate is a foreign entity or if more than 49% of the shares in the Polish company are held by a foreign entity. The basic long-distance telephone network also remains currently in the hands of TPSA. The law foresees the possibility of participation of foreign capital up to 49%, in the undertaking of inter-urban telecommunications lines, networks and services. In practice concessions for international or long-distance telecommunications will not be issued. The same limit of foreign capital (49%) is applied to the cellular telephony services providers (currently there is one provider of NMT 450 system; in January 1996 two GMS providers were chosen in a competitive bidding process).

TPSA therefore enjoys a *de jure* and *de facto* monopoly in the most profitable market segments in order to develop the national network with cross-subsidies within the sector and to recoup the huge investments made in the telecommunications infrastructure. The costs of the rural network development are met by cross-subsidization from more profitable long-distance and international calls. That, together with the costs of further extending basic voice service and the current government warranties for loans involved in funding investments and development of long distance network, is a strong argument for the government to maintain the monopolistic status in the "upper levels" of the network. The Ministry claims that an additional reason to grant TPSA a *de facto* monopoly in providing international and long-distance services (long-distance at least till 1999) is to enable it to become strong enough to face the international competition and to make its future privatization profitable.

As the last amendments of the Telecommunication Act left many issues unclear, in 1995 the Polish Parliament passed a resolution *"On the development of telecommunication"* which obliges the government to create a new legal environment for the telecommunication market no later than in 1997. The regulation should reflect the sector development strategy and attract serious and long-term investments.

In Hungary the new Telecommunications Act of 1992 liberalized the market, made possible the privatization of the incumbent operator - MATAV - and enabled the establishment of several local telephone companies (LTC), in order to attract private investment in infrastructure. The Act divided services into two categories: concessionary and competitive services. Public and mobile telephone services as well as paging belong to the first category. All other services fall in the second category. The Ministry of Transport, Communications and Water Management decided to grant the concession for international and national long-distance communication to MATAV (exclusive rights for 8 years and non-exclusive concession for a further 17 years), while the part of territorial concessions has been opened through bidding procedures to the new PSTN service providers. The concession agreements define the operators' obligations and transfer exclusive monopoly rights for a limited time period. The Ministry granted the national concessions to two GSM cellular phone providers (Pannon and Westel), and to two personal paging providers covering the whole country.

In the Czech Republic in 1992 the federal Parliament passed an amendment to the 1964 Telecommunications Act. This amendment put an end to the state monopoly over the provision of telecommunications services. The operation of basic voice services was retained by the state-owned operator (SPT Telecom), but the data communication, value-added services and mobile communications were opened up to competition (licensing agreements formula). According to the law, each organization (including the incumbent) which operates any part of the public network must obtain a license from the Telecommunications Office. For the operations that are reserved to the state, time-limited licenses were issued, which specified the rights and duties of the operators. For the other operations, time-limited, non-exclusive licenses were issued. The extent of the regulation and rules involved in the licensing procedure depend on the complexity of the service involved.

The Slovak Republic operates its telecommunications sector under an April 1993 amendment to the Federal Act. This amendment established a Telecommunications Office, with the goal of increasing the independence of the regulatory body from the political process. Long-distance and local-loop basic service provision is currently a monopoly of ST. A proposal of principles for a new Telecommunications Law is being discussed, based on the legislation in the EU. It is anticipated that the new law may open the way for possible privatization of the incumbent and the increased liberalization of service provision.

Tab. D2: Legislative Frameworks

COUNTRY	TELECOM ACT and Related Documents	IMPORTANT REGULATORY ISSUES
Czech Republic	Telecommunications Act of 1964, amended in 1992; Resolution No 428 (of 1994)	**monopoly**: in the public network construction and services (before 1995 the state share in the public operator cannot fall below 51%), some time-limited **concessions** have been granted in the monopoly area; **competition**: other networks; terminal equipment, closed systems; non-voice services time-limited, non-exclusive concessions; **local-loop competition:** rules released for 16 designated service areas (rural) + two pilot projects in Praha and Liberec
Hungary	Telecommunications Act (LXXII) of 1992, Concession Act, Frequency Management Act; State Telecommunications Policy (of 1993)	**monopoly** (service concessions are granted for a certain period of time), activities in the area in which the state has an exclusive monopoly may be carried out by state-owned companies or by concessionary companies (the development and operation of the basic network, the operation of international and long-distance services, public telephone services through regional concessions); **regulated/limited competition** (requires a license) - national mobile radio telephone, national paging; from April 30,1994 public telecommunication services, regardless of the operator's ownership, had to be offered based on a concession agreement; **competition** - mobile services (other than public), satellite services, data transmission networks, leased line resale, value-added services, special purpose and closed-loop network operations, terminal equipment; MATAV has a monopoly in international and long-distance services till 2002 and a concession for 25 years (under the terms of a long-term Concession Agreement); regulation based upon the **contractual relationships** between parties (network contracts, subscribers contracts and concession contracts, the development and quality targets are specified for each concessionaire)
Poland	Telecommunications Act of 1990, amended in 1995	**monopoly** in public international services for the national operator; **concessions** for long-distance services (for operators with minority foreign participation (<49%), national operator enjoys *de facto* monopoly) and for operators of regional networks; permits for network construction; **liberalization**: data transmission, value-added services, terminal equipment
Slovakia	Telecommunications Act of 1964, amended in 1992 and in 1993	**monopoly** in basic services: local, long-distance, international and in network construction; **open** terminal equipment market, policy allowing the construction of Ministry - approved telecom networks

The CEC had to decide what kind of competition was to be allowed and how it should be regulated. Each possible scenario brings definite opportunities for network and service development but also certain challenges to the traditional domestic operator. Whichever option is chosen, competition should not be taken as a substitute for regulation, but rather as one tool to help achieve national objectives. In each CEC the process has started with the opening of the value-added services market as well as the terminal equipment market. Subsequently the liberalization of those services which are complementary to the services of public network was begun (cellular telephony). These two steps were taken by all of the analyzed countries. The next stage of evolution leads to the opening of the public network. This decision is linked to the problem of how to fund local network development. The alternative to the traditional model of massive funding by the national monopoly is to open the local networks to other market players in order to attract private investment in infrastructure. The results of such a policy are evident, especially in the case of Hungary.

4. Development Strategy

Since 1989 telecommunications in Poland became a national priority and the government has been instigating programs and projects to achieve its goals of upgrading, modernizing and rebuilding the PTSN (public telephone switched network). Top priority was attributed to the following tasks: the development of international and long-distance networks, improvement of the quality of basic services through the automation and digitalization of the PTSN, reduction in waiting time for the installation of telephones, as well as the introduction of widespread data transmission facilities. The 1990 Act provided the legal basis for demonopolization and commercialization of the telecommunications sector by restructuring the PTT and stimulating private-sector participation through competitive market entry.

In October 1993, the Council of Ministers accepted a detailed plan for *"The Development Strategy of Telecommunications in the Republic of Poland until the Year 2000"*, guiding the modernization of the network. Major goals included improving and developing business services, continuing the modernization of international and long-distance networks, increasing the penetration rate to 25 - 27 (25% of new lines should be installed by local operators) and extending service to rural areas. Next strategic decisions were made in August 1995 when the Council of Ministers approved *"The Trends of the Structural Changes in Polish Telecommunications"*. That document, while still aiming to accelerate sector development, shifts the policy approach from promoting demonopolization and competition to strengthening the position of TPSA. From a technical point of view, by the end of this century, digital telecommunications systems and data transmission services should meet European standards. The Ministry confirmed that the State should keep strategic control over changes in the sector. Distinct programs and action plans would be formulated for the following organizational segments: international network, long-distance network, local loop and telecommunications services (VAS included). Temporary restrictions on the competition and foreign participation were established until the year 2000. The state monopoly would be maintained practically in all of the most profitable areas (a monopoly in international network, a dominant position in long-distance network up to 1999). In regional networks however, it would be possible to set up independent operating companies (the participation of TPSA in those enterprises is not excluded) - with both domestic and foreign shareholders. That diversity should encourage capital spending and improved management efficiency. A presence of the state in local segment would be confined to passing legal regulations and supervision over their observance, as well as to control the share of foreign capital. The competition in local networks should be limited to two operators.

The Ministry wants to switch the operator from a centralized management philosophy to a decentralized one, that pays close attention to market conditions. TPSA will be transformed into a corporate structure. The planned measures aim at streamlining the operation of the incumbent, increasing investment capability and strengthening TPSA's position vis-à-vis competitors. The most critical issue will be the re-orientation of staff from administering networks and services towards entrepreneurial accountability (a clear establishment of the cost and profit centers of the network and services is also intended). These changes should enhance flexibility, as staff working close to its market is able to evaluate customer needs and make on-the-spot decisions.

At the moment the discussions on the up-dated version of the policy for developing telecommunications (up to the year 2010), prepared by the Ministry, are at the final stage. The document specifies the forecasted demand for telecommunications services and defines development targets to be reached as well as the sector's strategy. As an element of the currently planned government reform, the state ownership rights of TPSA will be transferred to the newly created Ministry of the National Treasury. The Minister of Communications will still exercise the regulatory rights, monitor the marketplace to guarantee the effective performance of telecommunications, enforce standards and control universal services access. The obligations of the incumbent should be defined on a contractual basis.

5. Network Development

At the beginning of the 90s, the Polish government planned a crash-type program for the development of telecommunications. The major task was to achieve a significant improvement in the spread and penetration of the network (with the aim to balance the supply and demand for telephone services by the year 2000) and to introduce a new digital technology. An Order of the Ministry, issued in

1991, specified the technical characteristics of switching and transmission equipment which could be installed in Polish telecommunications networks (digital and upgradable to ISDN standard). Moreover, the "local content" of equipment installed and used in telecommunications systems should represent at least 50%. When in 1993, the Polish government decided to restructure the telecommunications industry through privatization, the ownership changes of the five largest telecommunication companies (Telfa, Teletra, Elwro, PZT and ZWUT) served also the additional purpose of securing three long-term suppliers of communications systems for the public network. Selection of strategic investors for these companies, based on results of competitive tenders, allowed three systems to gain legitimity in the network. Alcatel, AT&T and Siemens were selected in that way as providers of telecommunications systems to supply switches for TPSA.

Tab. D3: Evolution of the Number of Subscribers Connected 1980-1995 in Poland

Year	1980	1985		1989	1990	1991	1992	1993	1994	1995
Total number of connected subscribers ('000)	1,943	2,480		3,121	3,293	3,565	3,938	4,416	5,006	5,727
Number of new subscribers ('000)		130		169	172	272	379	478	590	721
Growth rate %		5.5		5.7	5.5	8.3	10.6	12.1	13.4	14.4
Penetration rate %	5.4	6.7		8.2	8.6	9.33	10.2	11.5	13.0	14.83
Capital expenditures (US$ million)				106	168	264	439	594	598	778.6

Source: TPSA

The development strategy consisted of measures to decongest the international and long-distance networks, to install international exchanges and fibre optic cables connecting Poland with international networks and satellite stations. A digital overlay network was planned to connect the major urban centers. The efforts which have been undertaken since 1989 to improve the state of telecommunications resulted in a level of teledensity of 14.82% by the end of 1995, in the installation of new digital exchanges and the construction of optical fibre lines, which are still under the way. The number of villages lacking access to the network was reduced from 10,000 in 1989 to 2,250 in 1995.

In last two years, TPSA invested US$1.4 billion in modernization and expansion of the network and it plans to install 4.95 million main lines in the next 5 years. According to forecasted demand, TPSA claims that the penetration ratio of 30 by the end of 1999 will satisfy customers' needs. Under these circumstances, the national operator believes that it is able to achieve this goal alone, by implementing an investment program resulting in the annual rate of growth of main lines varying between 13 - 14 %.

Tab. D4: Forecasted Growth of Subscribers Number in 1996-2000 in Poland

Year	1996	1997	1998	1999	2000
Total number of connected subscribers ('000)	6,527	7,427	8,427	9,527	10,677
Forecasted number of new subscribers ('000)	800	900	1,000	1,100	1,150
Growth rate %	14.0	13.8	13.5	13.1	12.1
Penetration rate %	16.9	19.1	21.7	24.4	27.3

Source: TPSA

The second task of the telecommunications program, accepted at the beginning of the decade, was to modernize the offer. The data network (packet switched data network POLPAK X.25 with 1,700 subscribers by the end of 1994) satellite connections (systems Inmarsat, Intersat, Intersputnik, Eutelsat, services VSAT) are being developed, and new services such as ISDN, audiotex, videophone, mobile communications (75,200 subscribers by the end of 1995), paging (65,000 subscribers), freephone services, electronic mail are being introduced. The investment program is based on the available technological innovations.

The development programs adopted in the CEC as well as the network expansion are presented in the following tables.

Tab. D5: Development Programs in CEC

COUNTRY	MAIN LINES DENSITY	TELECOMS PROGRAMS	DEVELOPMENT GOALS
Czech Republic	1990: 15.8 Goal 2000: 39.5	Telecom Project I accepted in 1991, amended in 1994	the number of lines doubled from 1995 up 2000, modernisation completed (digital backbone network built), 100% of telephone applications satisfied in 1998, investment: US$4-5 billion
Hungary	1990: 9.6 Goal 2000: 40	The Ten Year Development Program 1990-2000 with I Project 1991-1993 and II Project 1994-1997	3 million fully automated and digital lines built, national digital backbone network completed in 1993, density 30/100 in 1997, SDH transmission, 90% of telephone applications satisfied within 6 months by 1997, investment US$6 billion
Poland	1990: 8.6 Goal 2000: 27	The Development Strategy of Telecommunications until the 2000	digital backbone network completed, 10 million subscribers, new telematic services, improved quality, investment US$6-9 billion
Slovakia	1990: 13.4 Goal 2000: 35	Telecom Project I 1993-1996, Telecom Project II 1997-2000	digital overlay network, density 23/100 in 1996, investment US$530 million by the end of 1996; in 2000: quality level of services as in the EU and 70% of digital network, waiting time: 1 month, investment US$1.7 billion

The development programs for all CEC are based on huge investments. Obviously, the self-financing and the domestic sources are not sufficient, and in all CEC the option of foreign investment was taken into consideration. At the beginning of the decade, funding from such institutions as the World Bank, EIB and EBRD and foreign capital investment were the basis of development programs.

Everywhere in CEC, targets and progressive actions for the sector have started from what existed and the precise objectives were elaborated in the form of telecommunications strategies. Two basic tasks were defined: to achieve a significant improvement in the spread and penetration of the network (with the aim to balance supply and demand for telephone services at defined delay) and to introduce new technology and new services. In the first years investments were directed mainly at the development of the digital overlay network and raising the number of main lines. Digital transit exchanges were put into operation and high capacity backbone networks were completed. The new digital capacity was used to increase the number of main telephone lines as well as to accommodate large customers transferring from the analogue network, giving them better service and relieving congestion on the analogue network. The transmission systems employ state-of-the-art technology SDH. New networks are designed for the needs of national and international traffic and are an important part of the trans-European systems. Later, the development strategy should also be defined to create a subscriber connection network, taking into consideration the shortfall in connection capacity, the existing wired network and the future needs of subscribers in given localities.

Tab. D6a: Telecommunications Developments in the Czech Republic

SPT Telecom	1990	1991	1992	1993	1994
Main lines	1,623,600	1,706,500	1,818,600	1,961,100	2,150,600
Business lines	503,300	512,200	552,700	572,700	723,600
Penetration rate	15.8	16.7	17.65	19.0	20.7
Number of lines/employee		59.1	71.9	79.3	84.2
Public payphones	14,700	15,308	15,844	16,292	17,576
Cardphones			1,700	4,646	5,241
Waiting list	309,500	381,100	490,300	572,800	634,600

Tab. D6b: Telecommunications Developments in Hungary

MATAV	1990	1991	1992	1993	1994
Main lines (operated by MATAV)	995,800	1,128,800	1,291,100	1,497,600	1,731,500
Business lines	241,800	282,500	311,400	332,100	343,100
Penetration rate	9.6	10.9	12.5	14.6	17.7
Number of lines/employee	45.2	51.6	63.1	77.9	89.8
Public payphones	25,683	26,752	28,321	30,631	32,952
Cardphones			1 218	6 846	10 528
Waiting list	607,000	657,000	753,000	771,900	690,900

Tab. D6c: Telecommunications Developments in Poland

TPSA	1990	1991	1992	1993	1994
Main lines	3,293,000	3,565,000	3,938,000	4,416,000	5,006,000
Business lines	724,460	770,000	833,000	867,000	924,000
Penetration rate	8.6	9.33	10.2	11.5	13
Number of lines/employee	51.1	53.6	55.9	61.7	69.4
Public payphones	28,282	28,630	29,198	30,018	31,717
Cardphones		500	1,265	4,683	7,820
Waiting list	2,412,000	2,334,000	2,268,500	2,268,500	2,352,500

Tab. D6d: Telecommunications Developments in the Slovak Republic

ST	1990	1991	1992	1993	1994
Main lines	711,000	757,000	821,000	892,766	1,003,829
Business lines	185,900	194,000	209,600	230,600	256,300
Penetration rate	13.4	14.2	15.5	16.9	18.8
Number of lines/employee			55.6	56.9	66.9
Public payphones	6,400	6,600	6,727	6,774	6,929
Waiting list	109,400	124,900	160,259	166,498	184,656

Source: Annual Reports of the operators

6. Private Local Operators

In Poland the 1990 Act empowered the Ministry of Communications to issue local network operating licenses to independent companies.

The first few licensed local networks have been put into operation, although all of them are relatively small - like in rural communities of Laka, Tyczyn or Swidnik. Some of them are organized following the model implemented in the US and in Finland, where the development of telephone networks in rural areas is based mainly on a co-operative structure. Tyczyn Telephone Co-operative and Wist Telephone Co-operative, set up by the US National Telephone Co-operative Association can serve as an example. Another possibility is to secure foreign capital resources for that sort of projects. This is the case of RP Telekom holding, which was built around joint ventures: Netia (RP Telekom-75% and Telia-25%) and SRP Telekom (RP Telekom-41.7%, Sprint International-25%, replaced in February 1996 by Telia, IFC-11.7% and other investors). Netia controls and operates 11 project companies with licenses to construct and operate regional networks, while SRP Telecom controls and operates a project company in Silesia. In this way RP Telekom (shareholders: Danker Investment Ltd, Trefoil Capital Investors, Goldman Sachs Group) is engaged in the development of local telecommunications networks in ten voivodships (together with municipalities), providing financing and supervision of each project. By the end of 1995, three exchanges had been put into operation with the capacity of 24,000 lines. Some of new customers have been on TPSA waiting lists for years, and others are switching over. Investments of the holding, in years 1995-1997, will be over US$500 million, with credit lines provided by IFC and EBRD. In 1996 alone RP Telekom group is planning to connect 100,000-150,000 main lines, depending on the state of negotiations with TPSA. Most new potential operators plan to install up-to-the-date cost-effective technologies such as wireless communications in the local loop or combined cable TV and telephone networks.

Larger players, including Polish companies quoted on the Warsaw Stock Exchange, like *Optimus* (computer industry) and *Elektrim* (trading), are also preparing their entry into the telecommunications market (*Elektrim* moreover is the shareholder of consortium *Polska Telefonia Cyfrowa SA* - one of the winners of GSM concessions). *Optimus* wanted to obtain for its subsidiary *Polish Telemedia* the concession for a network covering 30% of Polish territory - regions where the company is present on the computer market. The company plans, calling for the investment of US$2 billion to provide 3 million telephone lines, had to be reduced to 450,000 new subscribers, when in July 1995 it turned out that the concession was granted only for 4 voivodships, without the municipalities.

Today, the Ministry forecasts, that 20 - 30% of new lines should be built by local, private operators. From 1996 the Ministry is going to announce subsequent invitations to tenders in chosen voivodships. The first two procedures were opened on 27 May, 1996, for voivodships of Bydgoszcz (1.1 million of inhabitants, teledensity 17%, minimal concession fee ECU747,000) and Lublin (1.02 million inhabitants, teledensity 13.7%, minimal fee ECU538,000). Winners will be awarded the concession to provide telephone and ISDN services and to lease lines. The permits to construct alternative network will be granted at the same time. License holders are normally obliged to commit to the construction of a specified number of lines within a given period. The bidders offering the fastest network growth will be preferred by the Ministry.

Opening the local networks to competitive provision in order to attract private investment in infrastructure (no limitations on participation of private domestic or foreign capital were specified), is the alternative to the traditional model of massive funding by the national monopoly. Unfortunately, the 1990 Act missed several legislative elements which are vital to support and control the introduction of competition into a former monopolistic market-place. The major missing elements were: regulatory instruments to prevent anti-competitive behavior of the dominant operator, independent control and supervision of the sector and provisions to accompany and control market entry.

The law states that the operator of the public telecommunications network cannot refuse to connect another telecommunications network to its network but the conditions and charges for interconnections are to be specified in a contract between the owners of the two networks. The absence of precise interconnection rules and of an independent body having the right to mediate in case of a potential conflict created a major obstacle in the development of real competition. The predatory behavior of TPSA is often criticized by its potential rivals. In particular, in its dealing with small rural operators, TPSA prolonged the negotiation process, frequently in order to use this time to develop its own network in the given area and to improve its competitive position against the new entrants (JASINSKI, 1995). For instance, one of the private operators, Telecom Silesia SA has been negotiating the co-operation rules with the regional management of TPSA for 8 months, from July 1994 - till February 1995. Then the Directory Board of TPSA refused to sign the agreement, and the re-negotiations lasted till September 1995 (the internal procedure requires the conditions to be discussed at local level and the agreement to be signed by the

Directory Board of TPSA). Similar procedural delays were experienced by Telecom Torun, Tel-Star, Telekom Modlin and Telekom Lublin, operators from RP Telekom holding. In 1995, the Anti-Monopoly Office took the unprecedented decision of fining TPSA US$200,000 for hindering the business of local operator Lublin Telekom and of ordering it to abstain from using monopolistic practices, including the unjustified delay in connecting the new operator's network to the national telecommunications network.

Despite the fact that more than 80 permits have been issued (varying in size from local community, telephone districts, and numbering zones, up to voivodship level), only a limited number of operators started providing services. In September 1995, the Ministry decided to examine the results achieved by the private operators and to withdraw the licenses from those entities whose performance had not been satisfactory or were still non-existent. Till February 1996, 27 licenses were withdrawn, and as of February 1996 there are 68 licenses for telephone services in public network, 70 for closed networks and 21 for non-public services (data transmission, e-mail etc.). The problem of interconnection has not yet been solved since the Communications Act was passed in 1990. In February 1994 the Ministry tried to find a remedy (under the pressure of the *Association of Independent Operators*) and issued a provisional document called *Settlement Principles between Telecommunications Operators*. The amendments of 1995 have given the Ministry the authority to impose, on a party's motion, the terms of agreement if two operators fail to sign the contract within 3 months. On April, 1996, the Ministry issued an additional regulation specifying the general rules for economically based and non-discriminatory cost reflective interconnection agreements. Precise terms and conditions for interconnection are to be reached amongst the parties, and only in the case of lack of agreement, the Minister will administratively decide the terms for settlement.

There is also a risk that the current situation, due to cross-subsidies profiting TPSA, can have far reaching negative repercussions for the development of competition in Polish telecommunications. The tariff structure, despite the Antimonopoly Office's efforts to enforce its rebalancing, still discourages investors from entering the local loop as the pay-back period is long and the potential for profit is low. While a clear and unambiguous interconnection policy mandated by the Ministry is still missing, it is very difficult for potential new entrants to compile a strong business case and initiate investment.

In Hungary, at the beginning of the decade MATAV prepared a first version of Ten-Years Development. This policy was supported by a Three-Year Program, in which an overlay network was built to decongest the traffic and to satisfy the needs of the business community, especially with respect to international links. To overcome the associated infrastructure expansion problems, this policy was complemented by regional liberalization. As one way of stimulating rapid additional investment and management input to the telecoms sector, Hungary has tendered the exclusive provision of local basic telecommunications services in some franchise areas to separate companies. This solution was considered to allow rapid entry of capital, technology and management skills. The rule is that the incumbent operator continues to be responsible for the long-distance and international services, and to provide local service in the major part of the country, while the responsibility for developing services within some particular areas falls to the local licensees (which can apply for licenses to operate cable television in parallel). The main operator provides also the interconnection services to the local companies. A government decree (158/93) made the network contracts obligatory and determined how different networks should be interconnected, regulating the obligations and rights of the dominant service provider as well.

40% of the country area was reserved in 1993 to be tendered out in 25 regional licenses. Bids for local telephone licenses were received from a large number of consortia including foreign telecommunication operators. The result was that several partly foreign-owned consortia received concession rights for local telecommunications services in 18 telephone districts, which cover approximately 15% of the potential domestic market. In this way Hungary attempted to handle the trade-off between the rapid increase in network capacity and the introduction of competition. Among the participants of foreign consortia which have acquired shares in regional telephone companies are Telecom Denmark, the Hungarian Telephone & Cable Corporation (USA), Bezeq Corporation (Israel), Compagnie Generale des Eaux (France), Swiss Telecom PTT, Telecom PTT Netherlands and United Telecom Investment (USA). Licenses issued for regional operators are exclusive for a limited time period, and they stipulate development targets (increased capacity and quality of service). The most important obligation imposed on the concession holders (regional as well as MATAV) is to raise the number of main lines by a yearly minimum of 15.5% in the first six years. From 1997 MATAV should meet 90% of all demands in six month.

Hungarian experience with the LTCs has shown, however, the difficult problem concerning the assets transfer agreements. The practice was that the new concession companies prolonged negotiations with MATAV about the value of these assets, often arguing that the obsolete infrastructure they stood to inherit would have to be completely replaced and should be priced at between 20 to 60% of the book value MATAV was asking for. The situation was further complicated by the fact that the MATAV's 1994 business plan was prepared on the basis of the assumption that the incumbent would offer services in all 54 primary districts only till end of June 1994, after which the hand-over of the services assigned to the LTCs should have taken place. Unfortunately, until the beginning of March 1995, MATAV could not transfer the local public telephone service to 10 local telephone companies in 13 districts out of 18 districts. After a substantial delay, two external arbiters (*Magyar Paribas* was one of them) were called to solve the problem in all litigious matters, evaluating the transferable assets to the average of 67% of the book value.

Hungarian experiences confirm, that opening of the market requires careful preparation. Not only is there an apparent risk of the wrong evaluation of new entrants, but once the choice has been made, the multi-player market brings up many potential conflicts. The existence of several participants and their opposed interests needs complicated regulation. The most controversial issue in the Hungarian approach was whether granting parallel concessions in local areas (i.e. competition in regions) would not have been a better solution than awarding exclusive concessions. There are, however, also some positive outcomes of the process. The creation of telecommunications market with several participants brings an indirect (yardstick) competition and allows the regulator to compare the level and the quality of services (the best performance can provide a benchmark to measure results achieved by other players). The sector has the opportunity to be enriched by additional money and intellectual capital and to gain experience before the European market liberalization.

The entry of new, local operators into selected local telecommunications networks of the Czech Republic (16 local areas to be opened, which represent 5% of the market) was carefully prepared and planned. In November 1994 the Ministry of Economy stipulated the basic conditions (two possible models of regional operators were proposed: joint ventures established together with SPT and independent companies competing with SPT, with the participation of foreign partners limited to 33%), the principles for commissioning, preparing and evaluating projects in a tender, the principles of network interconnection agreements, the content of the authorization (MINISTRY OF ECONOMY, 1994). As local operators were excluded from the most profitable services - long-distance and international communications - through the sharing of revenues, their financial position in comparison to SPT was forecasted to be relatively weak. The tender did not generate many offers, the majority of propositions came from companies which were going to build their own networks - mostly by cable TV operators. Potential entrants have established teledensity goals which are higher than SPT and declared they would reduce significantly the waiting list. The Ministry published the list of winning bidders at the end of 1995.

There are also two pilot projects under preparation: in Prague by Dattel and in Liberec by Kabel Plus - the Czech cable television company, claiming to be the largest cable television company in the CEC and the first cable-TV operator in the Czech Republic to enter the telecommunications market. In 1995 US West International acquired a 28.6% stake in Kabel Plus. In a joint announcement, the two companies stated that their aim is to build a multifunctional information highway.

It seems that in the long term, also in Slovakia, the share of private ownership will come from the involvement of CATV in local voice services. Current financial position of CATV operators does not allow them to enter the local telephone market but new joint ventures of foreign investors together with domestic CATV operators and local municipalities have already appeared.

Tab. D7: Private Operators in the CEC (Part I)

COUNTRY	NEW PRIVATE OPERATORS	OWNERS	ACTIVITY
Czech Republic	Radio Kontakt;	Radio-communications Praha and Telediffusion de France;	paging;
	Eurotel Praha; Eurotel 900;	SPT 51%, Atlantic West BV (Bell Atlantic 24.5% and US West 24.5%);	cellular NMT450; cellular GSM;
	Radiomobil;	RP 51% + T Mobil 49% (Skoda Mannesmann Mobilcom, DeTeMobil, Stet, Telekomunikacni Montaze Praha, Sporitelni kapitalova spolecnost - unit of Czech savings bank)	
	Kabel Plus;	US West International 28.7%, Ceska Pojistowna 34%	licences for 3 local networks;
	Dattel;		licence in Prague (*Metronet*);
	Alias, KT Jesenik, Opatel, FDP Telecom, Telecom 21;		new local operators chosen in December 1995;

Tab. D7: Private Operators in CEC (Part II)

Hungary	Westel 450;	Matav 51%, US West 49%;	cellular NMT450;
	Westel 900;	as above;	cellular GMS;
	PannonGSM;	Finnland PTT 14%, TeleDanmark 16%, Telia 14%, Telenor 14%, PTT Netherlands 16%, others 26%	
	BankNet, Satnet, Digitel, Montana;		data;
	Operator MMV, Eurohivo, Euro Call Hungary;	TDF(40%) + Magyar Musorszo Vallalat (60%);	paging;
	Deltav	Companie Génerale des Eaux (CGE)	local loop-several operators concessioned in February 1994
	Digital 2002,	CGE, Tal Giltek	
	Bakonytel, Dunatel, EgoCom,	United Telecom Investment	
	KisdunaCom, Raba-Com, Kelet-Nograd-Com,	HTCC, Telecom Denmark	
	MTT Monor Telephone Company, Jaszcom, Emitel	MATAV, Aphrodite BV (majority owned by Bezeq)	
Poland	PTK Centertel	TPSA: 51%, Ameritech: 24.5%, France Telecom: 24.5%);	cellular NMT450;
	Polska Telefonia Cyfrowa SA;	major partners - Elektrim: 33%, US West: 23%, DeTeMobil: 23%;	GMS - concessions awarded in January 1996;
	Polkomtel SA;	major partners: Plock Refinery, Polish Coppermine, TeleDanmark and AirTouch - each with 19%, Polish Power Grid Co.: 12%	
	Polpak, Telbank, Kolpak;		data
	Polpaget, Metrobip, Belpagete;		paging
	RPTelekom holding, Telekom Lublin SA (7,000 subscribers); Cuprum 2000 (7,700); OSR Tyczyn (4,200); Telefony Opalenickie (3,170); Telefony Brzeskie (2,520); Wist (2,350)		local telephony
Slovakia	RadioContakt Operator;	ST, Telediffusuion de France;	paging;
	Eurotel Bratislava	ST 51%, Bell Atlantic 24.5%, US West 24.5%	cellular NMT450, data

7. Privatization of National Operators

"The Development Strategy of Telecommunications in the Republic of Poland until the Year 2000", accepted in 1993, laid down a general scheme under which TPSA was to be gradually privatized. Very soon, at the beginning of 1994, the strategy was updated and the paragraph announcing privatization was deleted. There are a few official explanations for such decision. It is argued that the government guaranteed loans which should be paid off. Further, new investments to TPSA will increase the value of the firm which will lead to a higher selling price afterwards. As regards further ownership changes, the rules and the scope of privatization should be prepared by the Ministries of Privatization and of Telecommunications no earlier than by the end of 1996. Three independent consultants are to be chosen to propose the privatization strategy. The most probable scenarios include the flotation of the shares on the Warsaw Stock Exchange and the choice of a strategic or passive investor.

Hungarian leaders have chosen an active foreign involvement to achieve the development objectives, as it was rather unlikely that MATAV would carry the ambitious program alone. Already in August 1993 the government issued a formal public invitation to interested parties to participate in an open tender for 30% of MATAV shares. The deal was concluded in December 1993 with the sale of the stake to a consortium of Deutsche Telekom and Ameritech (MagyarCom). The Deutsche Telekom's recent experience in Eastern Germany as well as its involvement in a great number of telecommunication infrastructure projects in Central and Eastern Europe, was an important contributor to its success with MATAV. The German carrier's role in these projects enabled MagyarCom to mount a strong case for, among other things, its ability to establish Budapest as a telecommunications hub for the region, the strategic issue for Hungarians. MagyarCom acquiring 30% equity interest in MATAV was able to secure *de facto* control over MATAV by obtaining veto rights over major decisions taken by the Board of Directors and by creating an operating Committee responsible for day-to-day operations of the incumbent.

The Hungarian authorities claim that the confidence of the international financial institutions gained after a start of co-operation with foreign partner and the accessibility of financial means necessary for rapid development of Telecom network are the key outcomes of the privatization process. MATAV has no problems in meeting the concession requirements (concession fee, development targets, quality parameters). A new company structure is emerging as the result of a continuing process of adapting new working methods. The consortium got to work quickly on MATAV's out-dated systems. In 1994 it ploughed about US$500 million into the company and has started to introduce new technology and know-how, as well as attempting to change an outdated customer service attitude (BOLCSKEI, 1995).

Nevertheless, there are some objectives which have not been met. MATAV has not managed to elaborate a long-term development strategy and it has not been transformed into a real business-oriented company (the new management of the company is about to be formed), operational efficiency has only increased slowly. The harmonization of different cultures (and overcoming the language barrier) seems to be a very difficult issue. The partners are not of the same size. It is sufficient to look at the MATAV's 1993 revenue of US$800 million - minuscule compared with its foreign partner's combined revenues of US$48 billion. The adaptation of new owners to the Hungarian environment - local institutional and legal framework - is not an easy task. Deutsche Telekom and Ameritech executive experienced a corporate culture shock when they arrived at MATAV. Profit orientation was either non-existent or underdeveloped. The company closed its financial books once a year and reported results up to five month later. MATAV's foreign partners are introducing monthly profit and loss statements and new accounting methods. MATAV was overstaffed by international standards. The company had just 78 lines per employee, compared with 331 lines par employee at Ameritech and 164 lines per employee at Deutsche Telekom at the end of 1993.

The ever-changing economic environment in Hungary makes the financing process very complex for the foreign partners. They have nevertheless raised MATAV's borrowings, pushing them to 45% from 31% of equity, and capital expenditures jumped 50% in 1994 to US$645 million. In contrast, the Czech national company planned to invest only US$300 million in 1994; Poland's national operator, serving a population nearly four times that of Hungary or the Czech Republic, wanted to spend only US$650 million on capital expenditures, according to EBRD sources.

The MATAV privatization raises the important question whether the host government needs to hold a majority ownership interest in order to ensure that the telecommunications company remains responsive to national interests (GLUCK, 1995). At the time when MATAV was privatized, the government was required by law to retain 50% plus one vote - under Hungarian corporate law a stake of 50% plus one vote gave effective control over the company. In May 1995, legislation was passed reducing the minimum government stake in MATAV to 25% plus one vote. For the next stage of privatization, which was planned to be launched in 1995, the Hungarian Government chose a consortium consisting of CS First Boston and Deutsche Bank to act as advisors. The sale of up to 40% of MATAV (reducing the Hungarian privatization agency's APV Rt holding, 64% at the moment, to 25% plus one) was anticipated to generate about US$1 billion for the Hungarian government, which was under pressure to sell its remaining stake to help meet a target of privatization revenues in the 1995 State Budget. MagyarCom might have increased its stake when the government approved a further sale, while an international public offering and private placement were also likely to be undertaken to sell a stake to global investors. The company, slated for sale, was telling potential investors to ignore its US$26 million consolidated loss for 1994. Nevertheless, the 1994 results was a factor investors took into consideration.

As in 1994 the consortium blocked a planned public offering of MATAV's shares (because the price per share would have been lower than it had paid), the government was expected to prefer passive investors to take up the new offer, preventing MagyarCom from gaining a majority. The final decision, agreed at the end of December 1995, was taken in favor of MagyarCom which bought about 35% of the stake, paying US$852 million. The consortium argued that the additional investment would enable even more rapid creation of value, with accelerated application of operational and marketing skills, critical for steering MATAV in the right direction for the liberalization and competition which are certain to increase after the year 2002, when MATAV's monopoly in international and trunk calls is going to terminate. The group intends to focus on increasing MATAV's efficiency and improve business services. MagyarCom's future plans call for the flotation of MATAV's shares on foreign exchanges, including American stock exchanges, but not before end of 1997. By that time, MagyarCom anticipates that its investment plan will significantly improve the operator's performance guaranteeing a greater return on the investment made. MagyarCom agreed not to sell more that 10% of its stake within two years without the Hungarian government's approval and not to sell more than 24% of its share before MATAV was introduced on the stock exchange. MagyarCom, holding a 67% stake in MATAV has control of seven out of eleven seats on the board of directors.

In 1994 and 1995 MATAV was in the process of restructuring and received large capital inflows. It seems that between 1996 and 1997 it will break even and by 1997 it will become profitable. At the end of 1994 the design of new 3-year plan (1995-1997) was being finalized. Total investment was expected to reach US$2 billion during this period. The basic issue in future perspective is how to reconcile the public service with the commercial activities and profitability as the strategic investors expect a return on the investment. The Hungarian authorities think that an adequate tariff structure and favorable conditions for investment would provide a moderate return on investment at the first stage of development with a level corresponding to international standards to be reached at the next stages.

MATAV has a large backlog of connections to clear and is required by the government to increase the number of access lines by 15.5% per year. It does not expect to be able to meet this demand by installing fixed lines alone. The radio communications are taking the chance to leapfrog over wire or cable. Connecting remote areas via radio, in association with all techniques being developed for mobile connectivity should be both cheaper and quicker than copper. A contract, signed in 1995 with Motorola, will involve the installation of 200,000 lines across the country between 1995 and 1997.

The major trends for the telecommunication market are defined till the end of 1997. After that limit, the further development will be stimulated by the solvent demand for new high quality services. It is expected that after 1997 the market will be characterized by increasing demand for mobile telecommunication and multimedia services.

The second telecommunication company in the region of CEC opened for bidding was the Czech state operator SPT Telecom (transformed into a joint stock company on January 1, 1994, with the National Property Fund as the sole owner of the company). As in Hungary, the Czech government has decided to retain a majority share of the incumbent in the first phase of transformation. The former Federal Ministry of PTT has prepared the Amended Telecommunications Act which was approved in April 1992, enabling demonopolization and liberalization of telecommunications, separation of the state administration from economic activities and the start of the privatization of telecoms.

The First Telecommunication Project was one of the key investment projects of SPT. It covered the development of the Czech telecommunications network from 1992 to 2000 with the aim to balance supply and demand by the year 2000. The basis of the digital overlay network was completed in the first half of 1994. In August 1994 the Czech government approved "*The main principles of the State telecommunication policy*" (Resolution No. 428). The most important ideas of the document were as follows: (1) to increase by more than twice the number of main telephone lines up to the year 2000, (2) to maintain the integrity of the incumbent (the question was whether SPT should be privatized as an integrated whole or should be broken up into a number of regional telephone companies) and (3) to allow a strategic partner to acquire a stake in the joint-stock company SPT Telecom under the condition that the State share would be at least 51% (at the end of 1994, the accelerated version of the plan was elaborated according to which the goals were to be reached two years sooner than originally proposed, i.e. in 1998). The tender for the choice of strategic partner was planned to be decided on by the beginning of March 1995 (CZECH TELECOMMUNICATION OFFICE, 1995). The tariff policy, with a formula based on the inflation rate and the correction coefficient, was to be completed by the end of November 1994 and the state regulatory framework by the end of June 1995. SPT Telecom obtained the exclusive right for providing long distance and international telephone services up to the year 2000 (the exclusivity period and new tariff regime were intended to allow SPT to generate the necessary cash flow to fund accelerated infrastructure investment program). However for the local networks the participation of other market participants was considered. The Ministry of Economy decided to issue two licenses for mobile telephone services.

As part of the on-going privatization program, certain former commercial subsidiaries, construction, assembly and some production activities were separated off and were already privatized. To privatize the main activities, the Telecommunication Privatization Project was prepared. The privatization of SPT Telecom began with the creation of the state joint-stock company at the beginning of 1994 and the participation in the second wave of the voucher privatization of the Czech economy. As a result, 70% of stocks remained in state ownership, 26% were in the hands of individual shareholders, 3% were left for restitution and 1% for the National Foundation. The next step of the privatization process was the selection of a strategic foreign partner who was to take a 27% equity stake in SPT Telecom. The announcement of a short-listed group of five bidders, originally due in December 1994, was delayed by the unresolved issues of the tariff policy reform and the changes on the company's board.

The bidding process required the candidates to compete in terms of network development as well as price. The year-long contest ended in June 1995, when the second - highest offer of TelSource, the Dutch-Swiss consortium was selected as the most suitable. The transaction documents contain detailed provisions concerning the services the strategic partner is obliged to provide. TelSource is committed to providing US$130 million in contributed services in addition to the US$1.32 billion bid price. Contributed services consist of seconded employees, training programs and intellectual property. They are designed to facilitate the transfer of knowledge and expertise to enhance the ability of SPT Telecom to compete effectively in a liberalized market. TelSource is backed by AT&T, which will bring its knowledge concerning customer care. TelSource has said it will add two million phone lines by 2000 and bring about a "substantial decrease" in the waiting list for new lines. Another advantage of TelSource lies in its participation in Uniworld, together with Swedish and Spanish operators and AT&T. This was considered as proof of its "openness", a critical factor for the Czech government. It also promised to turn SPT into an international hub for Unisource NV, a joint venture of the Dutch and Swiss phone companies along with Sweden's Telia AB, which provides pan-European telecommunications services for big businesses.

TelSource had submitted the best plan for modernizing the telecommunications network and was considered by the government as being capable of transforming SPT into a commercial organization, able to face competition. TelSource represents to the Czechs a friendly, modern, commercial alternative to the big phone companies. The same importance was put on commercial factors as on participation in development plans, the contribution to commercial and management skills, the ability to accelerate the cultural transformation into a commercially and customer-driven entity and the international openness. The bid commission has preferred to choose the small, partially privatized operators, integrated in the international telecommunications network, as they will not threaten SPT's own future as a regional force. Of the two winning partners, the Dutch operator KPN is seen by industry analysts as the more aggressive and, unlike state-owned Swiss Telecom, it is also a publicly traded company. By standard industry measures, KPN's Telecom subsidiary is one of the world's most efficient phone companies. Czech and Dutch operators have already been working together in the software development company they established together a few years ago.

The transaction documents contain precise annual development/service quality targets and obligations as well as details concerning the services the partner is obliged to provide (SURREY, 1995). To enable the strategic partner to meet contractual requirements, it has been given a substantial measure of influence over the affairs of the company (major role in the development of annual business plans and majority in an operating committee managing the company's day-to-day affairs). To counterbalance the position of a strategic partner and to secure the influence of the Czech government regardless of the future evolution of events, one of the shares owned by the National Property Fund was changed to a Golden Share.

The negative scenario was also taken into consideration. The decision was that if the strategic partner fails to meet its development commitments, it will be subject to an array of sanctions, the severity of which depends on the extent of the non-performance. These penalties range from monetary fines to loss of management participation by the strategic partner.

The Minister of Economy explained that domestic sources were sufficient to finance network expansion but he stressed that due to the global trend towards multinational alliances in anticipation of market liberalization, internal changes inside SPT Telecom were urgently needed. Top management should quickly transform a giant with partially socialist working morale into an efficient and productive company. It was the task of the selected partner, who would have relevant experience in management, marketing and strategic planning, to face the issue of company restructuring and customer orientation. In the new techno-economic paradigm, the objective of satisfying customers requires a cultural change in the operator, calling for modern managerial skills of innovation, productivity, service and marketing. Developing customer services and an adequate marketing policy should be a major point of attention. The main objective of the privatization program (in the Czech Republic as well as in Hungary) can be summed up as the metamorphosis of the bureaucratic PTT into a modern company with innovative, flexible and efficient management styles based on competitive marketing providing the customers with a wide choice of services (KONTKIEWICZ-CHACHULSKA/PHAN, 1996).

The Slovak network is operated by Slovak Telecommunications (ST), a state-owned company established at the beginning of 1993. The pending transformation of ST to a joint-stock company depends, according to Slovak authorities, on several factors where the interests of the state, the banking and entrepreneurial sectors and of society must be harmonized with the objective of ensuring satisfactory services. The new principles of regulation, capital structure and competition have been carefully prepared starting in 1994.

Tab. D8: Privatization Programs

COUNTRY	INDEPENDENT NATIONAL OPERATOR	PRIVATIZATION PROGRAM	OWNERS
Czech Republic	SPT Telecom 1993, from 1994 joint stock company, 95% of the market	from 1993 as part of the country's second wave of voucher redemption, tender for the choice of strategic partner who bought 27% stake, previously planned for 1994, delayed till 1995	from September 1995: National Property Fund: 51.1%; TelSource N.V.: 27% (PTT Nederland: 13.77%, Swiss PTT: 13.23%), Investment Funds: 16.06%, Individual Owners: 2.92%
Hungary	HTC Matav (established in 1990, joint-stock company from 1991), 85% of the market (39 of 54 primary regions);	1993: MATAV has been partially privatized, further sale was decided in 1995, the basic network is operated and developed by concessionary companies chosen in 1994 through competitive tendering	MagarCom (Deutsche Telekom and Ameritech): 67%: {30.2% (1993; $875m) + 37% (1995; $852m)}, EBRD: 1.97%, IFC: 0.99%, State Privatization and Holding Company
Poland	TPSA 1992	purely speculative, decisions to be taken in 1996	100% State
Slovakia	ST Telecom 1993, to be transformed to joint-stock company in 1996, 100% of the market	delayed schedule, no earlier than in 1997	100% State

The government of prime minister Mr. Meciar had decided in 1995 to abandon the sale of state companies to citizens through the voucher program started in the former Czechoslovakia. Three-and-a-half million of Slovak voucher holders were promised government bonds in compensation. Privatization has continued through management buy-outs financed by state bank loans - the preferred mean of selling off enterprises. While the government underlined its strong will to modernize the telecommunications network, it was not likely to seek a foreign partner in the foreseeable future. Long discussions on privatization of ST so far have not considered the possibility of selling a majority stake of the enterprise. In July 1995 the Slovak Parliament adopted the Law on Securing State Interests During the Privatization of Strategic State Enterprises and Share holding Companies. According to this Act ST need not be privatized at all (JURZYCA, 1995). Nevertheless, at the beginning of 1996, the prime minister promised new opportunities for foreign investment and said that Slovak turbulent politics did not threaten its economy. Among the new opportunities that would be open to foreign investment, he mentioned possible future foreign investment in the Slovak telephone company and the opening of a tender to award a GSM license. A few months later, the government chose N.M. Rothschild to provide advice on the privatization strategy. The Slovak government has decided to transform ST into a joint stock company in the first half of 1996. Initially, 100% of the shares will be held by the government, but in 1997 an international tender will be launched to seek a strategic foreign partner, who is expected to participate in the development plans in return for an initial 20% stake in the company.

8. Tariffs

In Poland, prior to the reforms, prices for telecommunication services were fixed by the Ministry of Finance according to specific social objectives. From 1991, according to the Telecommunications Act TPSA is solely responsible for local tariffs, but the Ministry of Communications needs to approve international tariffs before they come into force and has the right to impose a maximum price for domestic voice services.

Currently TPSA maintains high rates for long-distance services compared to the local rates and so the rate structure does not reflect the relative costs. Without the pressure caused by real competition or a legal obligation, TPSA has no incentives to determine the true costs of providing each service and to alter its rate structure.

Similar tariff policy was carried out in the past in Hungary. Until the beginning of the 80s tariff decisions strictly followed political considerations, resulting in low tariff levels which could not generate sufficient funds for development. Later, under the pressure of extreme budgetary problems and increased inflation, the tariff structure was modified, but was still unable to generate the proper level of investment funds. The new plans discussed at the beginning of 90s incorporated a more sophisticated approach to telecommunications pricing. The regulatory framework was redefined at the end of 1993 (decree 30/93), with caps on tariffs directly linked to the evolution of the domestic producer price index.

One of the basic reforms undertaken in Czech economic transformation from central planning to market oriented mechanisms, was the price liberalization proclaimed at the beginning of 1992. It did not concern, however, certain price groups which were state regulated - such as rents and prices for basic public telecommunication services. A new version of tariffs was therefore to be prepared by the operator and discussed with the Czech Ministry of Economy and Ministry of Finance. A new tariff structure has been implemented but is still far from being a cost-based structure. The elaboration of tariff policy was one of the key aspects of the regulatory framework decided on prior to privatization. The decision taken was to adopt a form of price cap regulation (based on a formula PPI minus 1 or 2% - to be determined on an annual basis), to take effect in 1997, prior to which the first significant steps towards rate rebalancing will have been taken (SEDA/HRUBY, 1995). Tariff policy was designed to provide a high degree of predictability until the year 2000.

When in 1992, in order to accelerate the development, the Slovak government offered guarantees for foreign bank loans, it had to approve at the same time new tariffs rules (as in Poland, the tariff structure is partly conditioned by a need to maintain the World Bank recommended annual cost-revenue ratio). Prior to the changes in July 1995 the prices reached only 30% of the OECD level and 50% of the Central and Easter European Countries level. Pricing policy goals for 1996 cover the following: introducing a price system based on an inflation basket, introducing off-peak call charges and introducing cost-oriented prices for services which are not regulated.

Tab. D9: Annual Fees Paid by Subscribers

	1993	1995 (business customers)	1995 (residential customers)
Czech Republic	156	197	185
Hungary	227	407	333
Poland	365	320	320

(10% of connection fee + annual subscription fee + 700 local calls + 200 long distance calls-up to 100km), in US$

Source: Siemens, TPSA

9. Cellular Telephony

There are some other common approaches to the development of telecommunications in the CEC. A new strategy is based on mobile networks. Joint-ventures have been created to offer mobile services. Each country in the region has licensed NMT 450 mobile services via a joint venture between a private company and the state operator (51% of shares). Such enterprises allow the introduction of external capital and expertise, and are a way of providing a rapid service. Mobile communications was limited initially to the capital cities and their surroundings and were installed relatively quickly as overlay networks. At the beginning, business executives and government agencies were the customers of the system, i.e. groups which could afford to pay relatively high rates. Now the digital mobile services are making an entrance, opting for the GSM standard preferred by West European countries.

In Poland the tender for GSM concessions was launched in 1995. Two equal GSM concessions were to be granted to two operators (independent of each other), for identical and extendible periods, of initially 15 years. One of the main conditions for taking part in the bidding was the creation of a partnership in which the majority of capital would come from Polish investors, but in which contributions from foreign firms with suitable experience would be necessary. Each license required a fee of US$130 million to be paid up-front, with additional charges to be paid over the next 5 years. TPSA refused to bid, blocking its partners from the joint venture operating analog network from participating (because of a contractual agreement, which states that all PTK Centertel partners must participate together). Ameritech and France Telecom insisted however that the Ministry had already promised PTK Centertel a GSM license. In their opinion, when the Polish government awarded the NMT license, conditions of the agreement stated that the operator would automatically receive a GMS license, when frequencies became available. The Polish authorities claimed that changes to the telecommunications law in 1995, requiring a bid for each license, made void the agreement. Arbitration proceedings were therefore initiated.

That situation, together with the high license fees and huge investments to finance the network development (estimated at more than US$1 billion) and the lack of interconnection rules (to be negotiated with TPSA in the future) have dampened enthusiasm in the Poland GSM market.

In Hungary, as in the Czech Republic, one of two GSM concessions were attributed to NMT 450 operators. In Hungary an initial concession was offered for the period of 15 years (with possible extension to a further 7.5 years). The concession could not go to the same bidder or involve overlapping ownership and a minimum of 25% + one share of the equity was to be owned either by a Hungarian resident or by the state. Westel, launched in October 1990, became a GSM operator in March 1994. The second license went to Pannon (Telecom Denmark 16%, PTT Telecom Netherlands 16%, Telia International 14%,

Telecom Finland 14%, Norwegian Telecom 14%, Antenna Hungaria 5%, MOL Rt 15%, Videoton Holdings 1%, Wallis Holdings 5%).

The licenses were granted for exclusive country-wide GSM coverage including international roaming for a 15-year period. According to the terms of the licenses signed in October 1993, both companies are required to provide national coverage within two years, with a more stringent one-year requirement for Budapest, the main roads and the Balaton area. The service was launched in March 1994 and was the first digital cellular mobile telephone in the CEC.

In the Czech Republic the existing NMT-450 operator EuroTel Praha, a joint venture between Atlantic West BV and SPT Praha, automatically received one of the two GSM licenses. The other licensee was awarded to a joint venture consisting of Czech Radiocommunication (51%) and a partner with the remaining 49% chosen through a competitive tender.

In Slovakia a GSM tender is to be announced in 1996. The operators of cellular telephony in the CEC are presented in Tab. D10.

Tab. D10: Cellular Telephony Subscribers

	1990	1991	1992	1993	1994
Czech Republic		348	4,651	11,151	27,433
Hungary	2,650	8,477	23,292	45,712	160,000
Poland			3,365	15,699	38,942
Slovak Republic		119	1,537	3,125	5,946

Source: ITU, Annual Reports of operators

10. Conclusions

As candidates for EU-membership, the CEC prepare for the gradual rapprochement bringing their national legislation in line with EU policy. Different steps were also taken aiming at revitalizing the telecommunications sector with foreign equipment, investment, expertise, know-how and competence.

In the above article the stress was put on the different approaches to the privatization of the incumbents and the creation of independent operators. We observed three options:

- development is accomplished and funded mainly by the incumbent,
- strategic partner is chosen,
- independent local networks are established (with exclusive or non-exclusive rights).

The progressive shift from Ministerial control to the monopoly, followed by limited competition towards open competition in the future, implies a new focus for the regulatory decisions on interconnect issues and market entry. Obviously there are a few key elements necessary to attract investors: a reasonable potential for profit, a stable and clear policy as well as legislative basis.

Hungary was the first to decide to sort out its obsolete telecommunications by selling the incumbent to foreigners, who would inject the necessary capital and know-how to make it competitive. In 1993, the first step of MATAV's privatization was the largest transaction of that type not only in Hungary, but in the whole Central and Eastern Europe region. Three years later, after the second transaction, the strategic investor is controlling a majority stake in the company. In 1994 the Czechs started with the voucher privatization of SPT Telecom which was an initial step to market control of the operator. They continued the privatization process with the selection of a strategic partner for the incumbent. The procedure was considered to be even more relevant than in Hungarian case. The tender was a logical consequence of a well-designed policy towards the sector and regulatory development: the bidding documents effectively balanced pricing considerations against other competitive criteria; the tender procedure and criteria were transparent. The monopolist position of SPT Telecom which secures the exclusive rights on 95% of the market resulted in a price which was higher than the price paid for the MATAV shares. In the Czech Republic both the chosen framework and structure provide the conditions for a gradual introduction of competition into the sector while control over the development of the sector and the position of the market players is retained. In Hungary the MATAV monopoly was significantly reduced by the entrance of new private operators into primary areas.

In practice the price which can be obtained depends mainly on the outcome of the restructuring of the operator and on adopted legal regulations. The interests of overall national and telecom sector development conflict with the interests of any particular player in the game. The government itself has conflicting interests: for example the Ministry of Finance is usually interested in maximizing the revenue from the sale of the operator's shares, which implies maximizing the incumbent's monopoly power. In the national interest, on the other hand, is the reduction of costs and expansion of services - which requires some degree of competition. In each case, the government must weigh these trade-offs to set clear policy guidelines and elaborate a strategy before it enters into the privatization process.

The exclusive rights granted to Hungarian regional operators was a conscious choice to attract serious investment engagement in order to balance the urban and rural telephone penetration. While independent operators have taken over regional networks in Hungary, Czech alternative operators will build parallel networks. In Poland the major part of the telecommunications market is still in the hands of TPSA and only 32,000 main lines were installed by a few from more then 60 licensees. The sector's structure is conceived here in favor of limiting competitive entry into the market-place until the dominant operator has an opportunity to complete its investment program, internal restructuring and consolidate its position.

Poland and Slovakia are implementing investments at their own pace. Given the need to service World bank and EIB loans they are focusing on increasing the penetration in more dense urban areas in an effort to obtain the fastest return on investments. The governments' policy aims at strengthening the position of the incumbent before the future privatization scenario will be chosen. As it was argued there has been inadequate progress in some crucial areas of Polish structural reform of the sector: the separation of regulatory responsibilities from the ownership of TPSA, the definition of pertinent tariff policy and clear conditions for effective entry of other operators.

Generally it can be said that the evolution of the sector is a response to the institutional set-up, which defines the criteria of selection of market players, chooses operators and steers their actions. It seems therefore that the efficient reform of telecommunications sector should provide a space for dynamic organizations able to adjust and evolve within the regulatory framework. The success of the sector's transformation depends undoubtedly on the state's strategy and overall institutional context within which the operators (incumbent and newly created organizations) need to undertake economic activities.

Bibliography:

BOLCSKEI, I. (1995) "The Hungarian Model"; paper prepared for A. Smith Conference Warsaw, October 1995, "Comparative Models in Telecommunication in Central and Eastern Europe", paper prepared for Comnet Conference, Prague, May 1995

CZECH TELECOMMUNICATION OFFICE, (1995), "Telecommunications of the Czech Republic", Prague

FREEMAN, C.; PEREZ, C. (1988), "Structural crises of adjustment: business cycles and investment behaviour" in "Technical change and economic theory, ed. by G. Dosi (et al.), Printer Publishers Limited

GLUCK, G. (1995), "Privatisation of Telecommunications Companies: MATAV and the Hungarian Experience", paper prepared for A. Smith Conference, Warsaw, October 1995

GOVERNMENT RESOLUTION (1994), "The main principals of the state telecommunications policy", Government resolution no. 428, 1994; Prague

JASINSKI, P. (1995), "Competition rules and regulation in telecommunications: The case of Poland's intent to join the EU", paper prepared for the EAEPE Conference, Cracow, October 1995

JURZYCA, E. (1995), "Main problems of the Slovak Telecommunications", unpublished paper

KONTKIEWICZ-CHACHULSKA, H.; PHAN, D. (1995), "Institutional Change, Paradigm Shift and International Strategies in Telecommunications: Perspectives for Selected Central European Countries", paper prepared for ITS Regional Conference, Jouan les Pins, September 1995

KONTKIEWICZ-CHACHULSKA, H.; PHAN, D. (1996), "From path-dependent processes of structural change to a diversity of market models in Central European Countries' telecommunications", paper prepared for ITS Biennal Connerence, Seville, June, 1996

MINISTRY of ECONOMY of the Czech Republic, "Conditions for Entry of Additional Business Entities into Local Telecommunications Networks", 1994;

SEDA, R.; HRUBY, Z. (1995), "Czech Telecommunications: Financing of Network Expansion and the Choice of Strategic Partner", paper prepared for Berlin Workshop on Network Utilities, October 1995

SURREY, R.S. (1995), "Czech Telecom Privatisation - Fast treck to a modern telecommunications sector', paper prepared for A. Smith Conference Warsaw, October 1995

E. Modernizing Telecommunications in Central and Eastern Europe: A Business Perspective

Marec Béla Steffens

1. Structural Reasons for the Current Eastern Demand for Telecommunications

The telecommunications infrastructure of Central and Eastern Europe is underdeveloped not only in comparison with leading Western countries, but also in comparison with the networks of the dynamic threshold countries of South East Asia. This underdevelopment is dangerous, since telecommunications are a key factor for success in global economic competition - nowadays to a larger extent than ever before. Even for countries enjoying much more favourable economic circumstances, it sometimes took 10-15 years to bring the main line density from a level of 13 per 100 inhabitants - which can be taken as a starting point comparable to that of many Eastern countries around 1989-1990 - to 30 main lines per 100 inhabitants, the minimum of what is today acceptable. However, the examples of France and South Korea show that such a progress can be made within 5-6 years as well (see Tab. E1).

The reasons for this unfortunate situation can be found in the communist past (STEFFENS/SUNDRUM 1993). It was not only the Western technology ban which prevented the Eastern countries from gaining access to modern, digital telecommunications technology. The Eastern block itself, in its striving for autarchy, relied too long on hopes to be able to develop comparable technology. But the necessary resources were never devoted to this aim, as the communist planners focused on large steel works and other prestige projects and, for ideological reasons, regarded all infrastructure as non-productive. As a result, neither the industry nor the telecom operators (PTTs) received the means to develop sufficient telecommunications networks.

Tab. E1: The Process of Increasing the Main Line Density

The Process of Increasing the Main Line Density

from 13 to 30 per 100 Inhabitants

	68	69	70	71	72	73	74	75	76	77	78	79	80	81	82	83	84	85	86	87	88	89	90	91
Belgium		☎	☎	☎	☎	☎	☎	☎	☎	☎	☎	☎	☎	☎	☎	☎	☎	☎						
Germany (West)			☎	☎	☎	☎	☎	☎	☎	☎	☎	☎												
France								☎	☎	☎	☎	☎	☎											
Spain								☎	☎	☎	☎	☎	☎	☎	☎	☎	☎	☎	☎	☎	☎	☎		
Taiwan														☎	☎	☎	☎	☎	☎	☎	☎	☎		
South Korea																	☎	☎	☎	☎	☎	☎	☎	

Source: SIEMENS AG, ed., International Telecommunications Statistics, Munich, div.vol.

The consequences, however, were not as dramatic as they would have been in a market economy. A socialist command economy is based mainly on vertical flows of information. The nerves, so to say, are the red telephones, the direct links between state planning commission and branch ministry, branch ministry and combine director, further to the factory directors and so on. Information is channelled top-down and bottom-up, and because there are only few interlinks between units on the same level of hierarchy, no widespread communications network is required for the functioning of the system.

The situation is different in a market economy, the efficiency of which is based on a large number of direct, horizontal contacts between acting units who do not rely on hierarchy to make decisions. Enterprises are in direct contact with customers and suppliers and cannot be competitive on the market without efficient national and international communication services. That is felt nowadays in Central and Eastern Europe by the subsidiaries of multinationals as well as by the newly-emerged small and medium private enterprises.

In addition, free flow of information is also characteristic of democracy, and was accordingly regarded as suspect by the then communist rulers. It is symptomatic that it was one of the first measures of General Jaruzelski's military regime in Poland to switch off the public telephone network for several weeks. Even in quieter times of communist rulership, supplying all the 'ordinary' citizens with telephones was regarded as being 'not in the public interest' - as the guidelines of the Hungarian PTT officially stated (see PAP 1990). Today, both market economy and democracy as the two main goals of systemic transition require the implementation of up-to-date telecommunication networks.

2. Pumping Fresh Money into Old Structures is not Enough

Whereas the huge investment demand is undisputed, the main question still is where the necessary resources will come from. In the old times the socialist state confiscated the PTT's profit, if there was any, as well as any other state-owned company's. The PTT could not on its own decide upon investments, but could only bargain with the responsible state planning authorities for the necessary means. In this bargaining process the PTT was not in a very favourable position, as the posts and telecommunications ministry was by far not as powerful as, say, the ministry for heavy industry.

Under this framework, effective investment control could never take place. The state-set pricing for PTT services created more bias in this respect: the cross-subsidizing of parcel delivery or newspaper distribution diverted money away from the profitable telephone services.

When the World Bank, the European Investment Bank, the European Bank for Reconstruction and Development and others entered the scene and started to devote major loans to the Eastern telecommunications sector, they demanded structural changes right from the beginning. Their aim was to ensure that new money was not pumped into old structures, but that new wine was poured into new bottles.

For the Central and Eastern European countries themselves it was much easier to react to these demands than it was for Western countries, since the necessary institutional changes could be made in the framework of general systemic transition. The trend of deregulation and privatization in telecommunications which could be observed in the United States and the European Community (there, primarily in Great Britain) was considered to be in line with what had to be done in virtually every sector of the transforming Eastern economies.

Only some discontent arose among those people who regarded the telecommunications network as some kind of 'national wealth' which should remain in the hands of the state and should be protected against foreign influence. However, the weak performance of the Eastern networks, experienced by residents and visitors day in day out, clearly shows that there is no impressive 'wealth' to speak of. Therefore all possible measures should be taken to improve the situation. Every serious investor capable of contributing to such an improvement should be welcomed - that creates more benefit to the economy and society than direct state control.

3. How to Attract Investors' Interest

For a supplier like Siemens financing has become one of the key issues in business. If operators as customers do not have sufficient means for the desired investments, how can they place orders? Suppliers can of course try to arrange or even grant loans. But since state or state bank guarantees are less and less available, this means an increased risk which often will not be covered by export credit agencies. The suppliers, however, will not be able to grant unsecured loans for every contract.

But fortunately telecommunications has a big advantage over other sectors of infrastructure in that it creates immediate and direct returns on investment. In general institutional and private investors will be eager to invest in such a lucrative sector. As the emerging local financial markets will hardly be able to raise sufficient funds, it is especially important to deal with international investors. How can their interest be attracted?

First, one should get away from purely technical demand calculations. In a business perspective, demand for a good or service is only given where the willingness and ability to pay for it exists. A calculation of how many telephone lines per 100 inhabitants exist in a given country as compared to the European Community average, and which investment is necessary to reach the same penetration in that country, is a point of departure, but not at all sufficient to convince investors that they should risk their money.

Instead, costs and revenues have to be carefully analysed. The main task is to define a certain project, to isolate its costs and revenues from those of other areas of the telecommunications sector, and to prove that this very project is a profitable business. The latter should ideally be done by an independent outside consultant showing that the project is feasible from the technical, commercial and legal point of view.

Then it must be shown that neither the invested money nor the associated revenues can 'disappear' within the general funds of such a large organization as a network operator usually is. Control over investment, that is what an investor most basically requires - not control in the sense of day-to-day interference in business, but as an instrument to ensure that the invested money is being used in the interest of realizing the defined project in a profitable way.

In many cases, a special joint venture will be founded in order to ensure a proper assignment of costs and revenues. Such a joint venture can act as an operator in a certain area or for certain services, depending on the business aims of its founding parties and on the regulations applied.

However, the joint venture is not an island independent of the outside world. It has to maintain links with other operators, because usually only a small part of its traffic remains entirely within the joint venture's own network. Therefore it is essential that regulation not only gives enough operational freedom to the joint venture, but also safeguards non-discrimination in the area of cooperation with other operators - especially with the former monopolist who can still be expected to have a dominant market position. The respective regulatory measures are a key factor for market success.

On the other hand, regulation should leave the joint venture the leeway required for running the business. A flexible tariff policy must be possible in order to react to changing market demands or to fluctuating exchange rates. The latter is especially important: without tariff adaptations, the joint venture will lose its ability to pay back the foreign currency loans usually required for equipment purchases.

In general, investors cannot reasonably be expected to risk their money in a venture where the regulatory authorities have all possibilities to determine whether the venture will be a success or not. No serious investor will appear on stage if sudden and arbitrary regulatory measures must be feared. On the other hand, if both the project and the framework are attractive, this will not fail to have the desired effect.

4. Special Networks (Overlay, Mobile, Wireless Local Loop)

As investment resources are scarce, they should be used where they meet the most urgent demand. Economically, that means: where they create the highest returns. Such investment projects are to be defined where the customers' willingness and ability to pay is above average. The returns on these investments will then allow for further investments.

The concept of overlay networks has been applied by a number of Central and Eastern European countries. It means that a special network consisting of a number of digital exchanges and transmission links is 'laid over' the existing network in order to solve the most critical bottleneck problems first. Calls generated in the standard network will be transmitted through the overlay network for part of their way, and some of the customers are directly connected to the overlay network. In the course of the years, more and more parts of the standard network will be replaced by overlay network elements, until the overlay network has become the standard countrywide.

Although overlay networks could be run by a separate operator, they have, in Central and Eastern Europe, usually been installed by the PTT as former monopolist. In contrast, the mobile networks in these countries are being operated by joint ventures founded by the PTTs together with Western operators as strategic investors.

In addition to the other advantages which made them very popular in the West, mobile networks are a good means to solve bottleneck problems. Once the basic infrastructure is in place, which is a matter of months only, new users can be connected to the network without any need to dig up streets in order to lay cables. The operator can charge significantly higher tariffs than in a fixed network. And because of the attraction of mobility and other features, it is likely that subscribers will not give up their mobile phones when the fixed network has reached a sufficient performance level.

The introduction of cellular mobile networks made it possible to develop small handheld telephones which are very convenient for the users. Now Central and Eastern Europe is on the move to introduce GSM, the digital mobile standard which has been spreading throughout Europe and far beyond. Whereas all analogue mobile networks, such as some Central and Eastern European countries' NMT networks, are national only, the main advantage of the GSM standard is its internationality: a GSM phone can be used in any GSM country without formalities for the subscriber.

In addition GSM subscribers can be reached under the same phone number, no matter whether they are in France, in Hungary, or in the Philippines. This attracts travelling business people as well as tourists and leads to a much higher rate of international calls. Accordingly, GSM networks are extremely interesting for both operators and financial investors.

The advantages of a mobile network with regard to installation can also be applied for customers who are not interested in mobility as such and not willing to pay mobile tariffs. Wireless local loop (WLL), also called radio in the loop (RITL), is an attractive solution especially for countries with a long waiting list for telephone lines. WLL means that subscribers are connected to the network with a radio link, but they are restricted in terms of mobility. The WLL terminal can be used like an extended cordless phone - in house and garden or in the pub at the corner, but it mainly serves as a substitute for a plain old telephone at home.

The operator can connect tens of thousands of subscribers within a few months, and can impose somewhat higher tariffs than in the fixed network. However, tariffs cannot be as high as in a genuine mobile network.

Overlay, mobile, and WLL networks are good means for a rapid improvement of the situation in telecommunications. Bottleneck problems can be significantly reduced, which reduces pressure from public opinion. Sometimes, however, it is feared that operators will concentrate forever on such lucrative business networks and totally neglect the demand of the general public.

But it is not very likely that operators will behave that way. Such special networks as described here are rather a preparation for, not an alternative to, a substantial improvement of the general network. In the long run, business subscribers will not be satisfied if they can talk only with each other, but cannot reach their customers. It can even be argued that the public interest is served better by investments that increase the inequality first but generate much more means for re-investment than could be generated by an investment policy directing equally low means to every sector and region (STEFFENS 1991). In addition, regulators can carefully impose some conditions to be fulfilled by operators in the interest of neglected rural areas, etc.

5. Regulators' Possibilities and Contradictions

Regulators are confronted with a number of tasks and aims - some of which are contradictory. The best known example is the conflict of interests in the privatization of a PTT as former monopolist. The government as owner is interested in a high selling price for its shares. The price is, of course, the higher the better the business opportunities are - and they are best if the monopoly is not challenged. This contradicts the usual policy of deregulation but, in reality, the conflict is usually solved in favour of higher government sales revenues. That is by no means typical for ex-socialist countries alone, as the Thatcher government did the same when privatizing British Telecom (GRANDE 1989).

A field where the government tends to develop a similar appetite is that of licence fees. Some governments are keen to licence a large number of operators and to use all means including auctioning in order to maximize licence fees. Such governments overlook the fact that ultimately it is not the operators or their (sometimes foreign) shareholders who pay the licence fees, but the ordinary subscribers - in terms of higher tariffs and worse network performance due to less investment funds available.

If licence fees are to be imposed, it is a possible solution not to ask for cash payments, but to require the special network operator to make in-kind contributions to the country's general network. That is of course not what an economist is used to suggesting from an efficiency point of view, but it is not too likely that a cash licence fee will be spent by the government in an efficient way. In addition, it may contribute to a better public acceptance for the special network operator if its activities visibly contribute to improving the general network.

Apart from the financial side, regulators have to analyse and control the impact of special networks on the development of the general network. They have to take care of compatibility questions and to make sure that, at least in the long run, cream-skimming is prevented. As already said, the former monopolist should be prevented from applying discriminatory practices. And the former monopolist should be convinced to regard special network operators not only as (unwelcome) competitors, but also as a source of additional revenue from interconnection fees, or from dividends where the former monopolist is a shareholder.

Furthermore, regulators should pay special attention to the consistency of telecommunications policy and economic policy in general. It does not help at all when, e.g., decision-makers in telecommunications strongly favour local manufacturing in Poland, whereas the customs regulations discriminate against the local factories by levying a higher duty on the import of necessary components than on the import of finished products. It was announced that this odd tariff setting will change in the near future.

In all their activities the regulators should rather set frame conditions than interfere in the operators' day-to-day business, as is consistent with the general principles of a market economy. Regulators should set up rules and enforce them, but they should do so according to criteria which are known to the public. If the regulators are following general guidelines, their decisions can be anticipated, and actual and potential players in the market know what is expected of them. For example, a licence condition that all tariff increases require the licensor's approval leaves room for widespread lobbying and even for arbitrary decisions. If conditions are given under which the operator is allowed to raise tariffs (e.g., when wages, financing costs, or the consumer price index in general are rising), then the game has clear rules.

A final appeal to regulators: whatever you do, do it fast. A perfect solution serving all the diverging interests is very often impossible, and political discussion then tends to repeat the same arguments over and over again. Great harm can be caused by delaying key decisions which make a number of subsequent decisions pending and drive the whole telecommunications sector into a stalemate at a time when rapid progress is most urgently required.

6. Siemens's Role as Supplier and Local Manufacturer

In most of the questions addressed here, Siemens is more than only an observer. Siemens has been an active player in Central and Eastern Europe for almost 150 years. The consequences of the cold war restricted Western telecommunications sales in the region to a minimum, but the political changes immediately created opportunities for resuming and considerably expanding the traditional activities.

Right from the beginning, it was Siemens's concept not simply to export products and to collect revenues, but fully to integrate countries like Poland in the company's international work-sharing. As the systemic change allowed that, modern technology was not only transferred to Central and Eastern Europe, but local specialists could participate in the adaptation of the digital technology to local requirements and in its further development.

Siemens led the field by founding joint ventures in almost each and every country in Central and Eastern Europe (see Tab. E2) - at least one could say that before USSR, CSFR and Yugoslavia were split up into their respective successor states... The aim was always the same: to combine the local partner's specific market knowledge with the technical and managerial know-how of the Western side. A typical example was the foundation of the joint venture CEWIS together with the leading Polish manufacturer ZWUT, who held 51 % in CEWIS. The manufacturing of the digital switching system EWSD began at CEWIS's premises in 1992. CEWIS then was the first manufacturer of digital switching technology in Poland.

Through the establishment of CEWIS, Siemens gained better access to the Polish market and benefitted from the local partner's market knowledge and established customer relationship. The benefits for the Polish side are obvious, too: the creation of local value-added directly contributes to the country's Gross Domestic Product. Jobs for highly qualified personnel are created and secured. The reduced demand for imports improves the country's balance of payments. In addition, the experience of Polish personnel might also be used for projects in other countries which would then create foreign exchange revenues.

In 1993, Siemens took over ZWUT and integrated CEWIS into it. In contrast to popular views on the behaviour of multinationals, it was the Polish side who initiated this acquisition. The government decided to privatize the country's telecommunications industry and made clear that only vendors who participate in this process would be eligible for future orders of switching technology.

Tab. E2: Siemens Public Communication Network Group - Joint Ventures and Subsidiaries* in Eastern Europe

Country	Partners	Name	Location	Siemens Share	Manufacturing Area** TN	NK	MN	Start of Production
Bulgaria	Incoms-Telecom	Digicom	Sofia	60 %	X			1996
Czech Republic	---	Siemens Telekomunikace (Ltd.)	Prague	100 %	X			1992
Czech Republic	Kabelovna Decin	Kabelovna Decin	Decin	30 %		X		
Hungary	---	Siemens Telefongyár****	Budapest	100 %	X			1992
Hungary	Magyar Kábelmuvek, Hungarian Telecom. Comp.	Dunakábel	Budapest	49,3 %		X		1992
Poland	Employee Shareholders	ZWUT + ELWRO****	Warsaw Wroclaw	99,21 % 91,21 %	X X			1992
Romania	Electromagnetica, Rom Telecom, EBRD	EMCOM	Bucharest	35,51 %	X			1992
Russia	P. O. Motozavod, Svyazstroy, SSSK	Izhtel	Izhevsk	40 %	X			1995
Russia	Morion	Kamatel	Perm	34 %	X			1992
Russia	DeTeMobil, EBRD, France Télécom, MGTS, ASVT etc.	Mobile Tele-Systems	Moscow	9 %			X	1994
Slovenia	Iskra Telecom	Iskratel	Ljubljana	47,7 %	X			1990
Ukraine	P. O. Korolov, Telecom***, Karat***, and ZNIIS	MKM Telekom	Kiev	30 %	X			1993

*** Not including representation offices etc.**
**** TN = Telecommunication Networks NK = Telecommunication Cables MN = Mobile Radio Networks**
***** formerly part of P. O. Korolov**
****** The former foreign partner was taken over on its demand**

Source: Siemens(1995)

This limitation of market access has sometimes been criticized. But there are important reasons for this policy: the relatively low labour costs in Poland are not a sufficient incentive to attract a substantial amount of foreign investment in technology-intensive branches like telecommunications. The Polish authorities realized that large-scale commitments can only be expected if the market is not fractionalized and if investors have a chance to achieve a sufficient market share. So it was consequent that market access was limited to those manufacturers who engaged themselves in the privatization of the Polish telecommunications industry. And they could acquire Polish firms only after a tendering process which included an evaluation of the competitors previous engagements and future commitments.

In Czech Republic and in Hungary, too, joint ventures turned into 100 % subsidiaries. The former joint venture partner in Hungary named Telefongyár, a manufacturer with a 100-year tradition, felt that a complete takeover by a potential Western partner was to be a better solution for the entire firm's future than the creation of a joint venture as an "island". In Prague the Czech partner decided to sell its joint venture shares to Siemens after a few years.

But there is no general trend against joint ventures: in other countries of the region, the Siemens joint ventures are successfully operating with unchanged shareholdings. And the Romanian joint venture Emcom even attracted the European Bank for Reconstruction and Development as an additional shareholder.

Now the introduction of the GSM digital mobile standard will bring about new possibilities for the Siemens joint ventures and subsidiaries in Central and Eastern Europe. The Siemens GSM technology, called D 900, is based to a large extent on the fixed network switching system EWSD, so that the local production lines can easily be adapted to the new requirements. As the system affinity of D900 and EWSD is not typical for the branch, Siemens hopes that its local content possibilities will contribute to the company's further success in the region.

It is part of Western investors' experience that some of the joint venture partners, the government or part of the general public have unrealistic expectations about a Western partner's possibilities. Especially in telecommunications as a technology-intensive sector, no large labour forces are required, and a manufacturer of digital equipment needs much fewer workers than one of analogue systems. If, in a region, a steel works with thousands of employees was closed down, a joint venture for telecommunications cannot improve the labour market situation very much. And especially since a Western partner is expected to transfer not only technical, but also managerial know-how, it cannot be expected that a joint venture is as overstaffed as socialist entities usually were. Another aspect is that almost every country desires to supply the home market entirely with locally produced equipment and to export to the neighbouring countries - who are not inclined to import, because they have the same intentions.

But the overall picture is clearly positive. One of the most convincing features of Siemens's experience in Central and Eastern Europe is that the company's usual quality standards could be met right from the beginning. ZWUT e.g. applied for the ISO 9000 Certificate of Quality. Auditing will take place in late 1994, and ZWUT anticipates that is will become the first Polish telecommunications manufacturer to be awarded with this Certificate. This experience in quality proves that the local employees' skills and abilities, if stimulated by the right economic system, are second to none. The task of modernizing telecommunications in Central and Eastern Europe is tremendous, but the encouraging signs cannot be overlooked.

References:

GIBBS, D. (1994), Information and Communication Technologies in Poland, Telecommunications Policy, vol. 18, no. 5, pp. 363-366.

GRANDE, E. (1989), Vom Monopol zum Wettbewerb? (From Monopoly to Competition?), Wiesbaden: DUV.

NULTY, T.E. (1993), Stimulating Vitality and Innovation. Privatization As a Development Tool in Central and Eastern European Telecommunications, Telcom Report International, vol. 16, no. 4, pp. 10-13.

PAP, Z. (1990), Telefon in Ungarn (The Telephone in Hungary), in: Forschungsgruppe Telefonkommunikation, ed., Telefon und Gesellschaft (Telephone and Society), Berlin: V. Spiess, pp. 214-224.

STEFFENS, M.B. (1991), A Principle of Justice and the Reality of Hungarian Telecommunications, paper presented at the Centennial Scientific Days of the PKI (Research Institute of the Hungarian Posts and Telecommunications), Budapest, 20-22 Nov. 1991.

STEFFENS, M.B./SUNDRUM, L. (1993), Political Change Opens Telecommunication Markets, Telcom Report International, vol. 16, no. 4, pp. 19-22.

F. Telecommunications Reform in the United States: Promises and Pitfalls*

Marius Schwartz

1. Need for Pro-competitive Regulatory Reform, and the 1996 Telecommunications Act

The United States Congress recently enacted sweeping legislation to overhaul the rules governing competition in telecommunications services. The Telecommunications Act of 1996 (see CONGRESSIONAL RECORD, 1996) is the first major rewrite of the Communications Act of 1934. It also supersedes the 1982 antitrust consent decree that broke up AT&T and barred the seven new regional Bell operating companies ("Bells") from manufacturing equipment and offering long-distance service.

The stakes are high, as telecommunications ("telecom") are a critical element of a modern economy's backbone. The U.S. telecom sector's revenue in 1994 exceeded $200 billion: $150 billion in telephone service, $42 billion in broadcasting, and $28 billion in cable television (ECONOMIC REPORT OF THE PRESIDENT, 1996, chapter 6). The importance of the regulatory reforms in the 1996 Act as perceived by those in the best position to know--market participants--is reflected in the frenetic lobbying leading up to the Act (especially on its telephone provisions), and in the recent wave of corporate restructuring and shifting alliances reportedly driven by expectations of a new competitive environment.

Regulatory reform enjoys broad support, but there is less agreement about its appropriate pace and nature. The road to reform holds both promises and pitfalls. This paper discusses the underlying economic issues, the progress made by the Act, and the challenges lying ahead as we move from regulated monopoly to competition. Although the U.S. is starting with regulated private monopolies, some of the discussion will be pertinent also for a transition to competition when starting with state monopolies.

1.1 Reassessing Government Intervention in So-called Natural Monopolies

Much of existing regulation in telecom (and other public utilities like electricity, and oil and gas pipelines) is premised on the belief that certain services such as local telephone and cable service were *natural monopolies*. A natural monopoly exists when it is less costly for a single firm than for several firms to supply any relevant quantity of the product (or products in the case of a multiproduct natural monopoly). Natural monopolies can arise especially in industries characterized by fixed costs that are large relative to the relevant market demand, e.g., the cost of running telephone or video cables to the home, the cost of railroad tracks, or the cost of electricity transmission lines.

In such circumstances competition simply may not be viable, and an unrestricted monopolist would set an inefficiently high price and therefore inefficiently low output. Even if competition were viable (if rivalry among firms would not be so intense as to reduce price to marginal cost and hence below average cost), competition in a natural-monopoly industry would prevent full exploitation of scale economies. In theory, government intervention could increase efficiency: to fully exploit scale economies the government might allow only one firm to operate, and require it to set a lower price and higher output than would obtain under no intervention.[1]

In practice, intervention has worked less well than hoped. It is by now widely recognized that government control of natural monopolies, through direct ownership (as in Europe) or through regulation of private firms (as in the U.S.), entails major inefficiencies. Incentives for cost reduction and for product innovation have suffered, and the administrative burden of regulation itself has been heavy. The burden consists of direct costs and, more importantly, of losses from rigidity and delay--in changing prices and in making investments. Growing awareness of the costs of monopoly regulation is a key factor accounting for the broad support for regulatory reform in telecom.

A second key factor is the dramatic technological changes in telecom conducive to competition. An industry's "natural monopoly" status can erode over time, due to growth of market demand or, more importantly, due to technological innovation that alters cost conditions. Technological innovation can reduce fixed costs by offering a new way to provide a given service, e.g., wireless technologies may offer a lower cost way than cables to reach a customer's premises. It also can reduce fixed costs by enabling the same facilities to supply additional services at a lower extra cost than if supplying the additional service alone, e.g., by reconfiguring phone or cable lines so each can supply both services.

Technological changes of the latter sort tend to blur industry boundaries, making competition possible between providers of traditionally different services. Such changes also introduce hybrid services that do not fall neatly into any traditional "industry," e.g., video conferencing, multimedia, and data transmission. Attempts to preserve artificial industry definitions in order to maintain regulation under traditional monopoly franchises then become arbitrary, futile, and counterproductive.

For both the above reasons--inefficiencies in government intervention and technological changes favorable to competition--the regulatory regime should not be cast in stone. Policymakers should constantly ask if the industry should continue to be treated as a natural monopoly or if conditions are ripe for replacing monopoly with competition.

Competition may be desirable even in industries that remain textbook natural monopolies if economies of scale are modest, because the "static" losses from having more than one firm can be outweighed by the many dynamic benefits from replacing regulation with competition--reduced internal slack, greater innovation, greater responsiveness to customer demands, and avoidance of regulatory costs. Moreover, the textbook "natural monopoly" rarely applies, because firms typically offer differentiated products rather than perfect substitutes, and the variety gains that competition may provide can outweigh the static losses from failing to exploit fully economies of scale. (This is the fundamental criticism of Chamberlain's famous proposition that monopolistic competition is wasteful because it leads to "excess capacity.") The important point is that competition need not be perfect in order to be superior to regulated monopoly.

Nowhere is a reexamination of regulatory policy more appropriate than in telecom, with its dizzying pace of technological change. There is so much uncertainty about the shape of future communications networks and sufficient promise of successful competition that the best course is to give market forces freer rein. Policymakers should not prejudge the outcome by assuming, for example, that local phone and cable services are natural monopolies best provided by regulated franchise monopolists.[2] Rather, regulatory regimes should adapt flexibly to changing conditions, to help shrink the boundaries of the regulated sector and rely more on competition.

1.2 Challenges to Regulatory Reform

There is broad agreement that moving to competition is desirable and that it requires relaxing government barriers to entry, as well as redundant restrictions on the range of services firms may provide. Substantial disagreement remains, however, over the proper transition process to competition and the role for government once that transition is complete. There has been much rhetoric in the U.S casting the debate as "regulation versus deregulation." This characterization is misleading. Certainly the "need for caution" can be misused as an excuse for delay and the status quo--delay that should be resisted strongly in industries where structural conditions clearly make competition viable and where only legal barriers stifle the rapid growth of competition. However, many traditional natural monopolies do not neatly fit this characterization, and pose difficult challenges for good regulatory reform. Immediate hands-off deregulation of such industries often is not the best approach.

Even if legal barriers are removed, entry into traditionally regulated industries will take time and, at least initially, the incumbent monopolist will retain significant market power. Premature deregulation without competitive safeguards can allow the exercise of market power until competition arrives. It also could delay the arrival of competition.

Deregulating a monopolist's prices before competition has emerged (e.g., in cable TV) could lead to price increases that create the standard monopoly inefficiency. Perhaps more importantly, it could unleash a political backlash against deregulation broadly, threatening the political consensus for a move to competition.

There also is a government role in protecting competition. As explained later in the paper, allowing a firm to enter related markets while it remains price-regulated in its monopoly market could enable it to circumvent regulation by (a) cross-subsidizing its unregulated operations and, more importantly, (b) discriminating against competitors in unregulated markets in the access terms it grants to its monopolized bottleneck facility (e.g., elements of a local telephone network). Such behavior leads to higher prices in the regulated markets (if cross-subsidies occur), and inefficient distortion of competition in the potentially competitive markets, harming both rivals and consumers. Competition also can be stifled if mergers are permitted between monopolists who are likely future competitors, e.g., between local telephone and local cable companies.

Finally, some worry that even successful competition will raise some prices and threaten "universal service"--widespread availability of affordable telecom services. They see a government role in guaranteeing universal service.

Disagreement in the U.S. over the proper role of government in telecom policy reflects differences over important underlying questions:

- How rapidly will competition develop? This affects for example one's attitudes towards the appropriate timetable for decontrolling cable rates.
- How much competition is needed to discipline prices? Is threat of entry enough, as in the theory of contestable markets, or do we need strong actual competition, as stressed by critics of contestability (e.g., SCHWARTZ, 1986a)? This affects, for example, the prospects for replacing regulation with competition in local telephone service.
- What if any are the monopoly bottlenecks, for example, the entire local telephone network or only the local loop? This affects whether structural separation between local and long-distance telephone service is appropriate.
- How effective is regulation? This colors attitudes towards continued regulation of cable rates. It also shapes attitudes towards the feasibility of preventing the Bells from discriminating against long-distance rivals in access to local networks if the Bells are allowed into long distance (assuming that a significant monopoly bottleneck remains in local-telephone service to make such discrimination a serious concern).
- How much of a need is there for a "referee" during the transition? Should Bells seeking to enter long distance be required only to meet a competitive checklist "guaranteeing" competitors' equal access to their local networks, or must regulators remain heavily involved to ensure the checklist's implementation?
- Would many low-income consumers drop their telephone subscriptions if local rates and fixed charges rose somewhat but long-distance rates fell substantially, as is likely if current cross-subsidies from long-distance to basic local service

are reduced? This affects the perceived need for a government role to promote universal service.

These are not easy questions, and they leave substantial scope for legitimate differences of opinion. This paper offers my perspectives on these issues and on the 1996 Act. Section 2 discusses the removal of artificial governmental restrictions to competition. Section 3 discusses the continuing role for government in several areas, while Section 4 discusses its role in promoting competition in the huge telephone industry. Section 5 addresses an important consideration in moving towards more competition in the telephone industry, namely, removing cross-subsidies while preserving universal service. Section 6 concludes with a brief assessment of progress to date and of what lies ahead.

2. Increasing the Reliance on Competition and Markets

2.1 Relaxing Legal Entry Barriers

For many years local telephone and cable companies were protected from competition by what was believed to be natural monopoly technology, *and* by a patchwork of legal barriers: state laws prohibited competition for local telephone companies, municipalities could grant monopoly cable franchises, and federal law prevented telephone companies from offering cable service. It is time to remove these legal entry barriers, and test whether these markets indeed are natural monopolies.

A key pro-competitive aspect of the 1996 Act is to strip away much of the authority of States and local governments to impose legal or other barriers, such as discriminatory access to rights-of-way, that may hinder competition in *telephone and other telecom services.*[3] Given the size of the local telephone market, which in 1994 had revenues of about $100 billion, opening it to competition is quite important. Some States already took steps in this direction, but the Act sets a clear nationwide standard and orders immediate removal of most barriers in all States. The Act tries to put real teeth into its requirements by authorizing the FCC to preempt State and local government barriers to competition. As I discuss in Section 5, the Act arguably does not go far enough; in the name of protecting universal service it leaves States some rights impose "competitively neutral" requirements on entrants which nevertheless could inefficiently impede entry, such as requiring entrants to serve the exact same region as an incumbent. But the Act certainly takes an important step.

The Act also aims to increase competition in *cable television*. It eases restrictions on the ability of telephone companies to offer video programming, and weakens the barriers posed by local cable franchising requirements. As regards the former, the 1984 Cable Act had codified a 1970 FCC rule prohibiting local telephone companies from offering in their service areas video programming directly to subscribers.[4] Court decisions in the last few years have found this prohibition unconstitutional, but the 1996 Act clarifies matters by unequivocally repealing the cable-telephone cross-ownership restrictions.

The 1996 Act also weakens the barriers to cable competition posed by the franchising authority of local governments. Although the 1992 Cable Act prohibited the award of exclusive (i.e., monopoly) cable franchises, enforcement of this prohibition can be problematic: the Cable Act lets plaintiffs seek redress through the courts, but this process can be slow and unreliable. The 1996 Act allows a common carrier (an entity--such as a phone company--that offers service to the public on a nondiscriminatory basis) to provide any of the following while exempt from local franchise regulations: (a) video programming using wireless technologies (a common carrier or any other entity then requires only a radio license); (b) transmission of video programming on a common carrier basis (under Title II of the Communications Act); or (c) video programming as an "open video system." (The Act also allows (d) provision of video programming as a cable system under Title VI of the Communications Act, but that is subject to local franchising.)

The open video system is a newly created regulatory category, which entails certain additional obligations compared to a regular cable system, but in return qualifies for streamlined regulation.[5] Importantly, the Act delegates certification of open video systems to the FCC and requires the FCC to process applications and resolve disputes expeditiously. Thus, open video systems may offer an important entry mode for telephone companies to compete with cable. Recognition that this entry mode is available may also sway local authorities to relax their franchise restrictions for regular cable systems.

As regards *broadcasting*, the 1996 Act removes the limit on the number of radio stations that a single entity may own nationally, and relaxes the ownership limits in local markets (based on local concentration). It also ends the limits on the number of television stations that a single entity may own nationally, and allows such an entity to reach 35% of the nation's TV audience, up from 25% today. These relaxed limits could help broadcasters realize greater economies of scale, while competitive concerns with, say, mergers could still be addressed on a case-by-case basis. But the Act retains the ban on cross-ownership of broadcasting and newspapers, a ban which may be desirable for reasons discussed in Section 3.

Finally, the Act eases restrictions imposed by the 1982 AT&T decree on the ability of the seven Bells to manufacture equipment and to provide long-distance telephone service. However, as explained in Section 4, those restrictions should not be portrayed as "artificial entry barriers" and relaxing them must be accompanied by competitive safeguards, along the lines provided in the Act.

2.2 Liberalizing Restrictions on Foreign Investment

It is important to remove barriers not only to domestic entry but also to foreign entry. U.S. law limits foreign ownership of a radio license; that is, a license to provide any radio-based services such as wireless communications as a common carrier, or radio and television broadcasts. Other countries have cited this restriction, at times overstating their impact, as a reason to deny U.S. companies access to their telecom markets. However, through a combination of administrative rulemaking by the FCC and legislative changes in the 1996 Act, the U.S. is now able to lift these restrictions.[6]

Taking advantage of this ability, the U.S. made a comprehensive offer at the World Trade Organization (WTO) negotiations on basic telecom services to open its markets to foreign trade and investment. The offer covers all telecom services (including voice, video, data), delivered through any technology (wireline, terrestrial- or space-based wireless), using facilities owned by the provider or leased from others (as with resale), and including international, long-distance and local service. (The ability to include local telephone markets was enhanced by the 1996 Act's preemption of State entry barriers.) The U.S. offer is conditional on receiving sufficiently favorable responses from enough countries. The WTO negotiations are due to end April 30, 1996, and hopefully will bear fruit. But even if they fail, the FCC's rulemaking and the provision in the 1996 Act enhance the ability of the U.S. to pursue liberalization of trade and investment in telecom services, through bilateral or multilateral agreements.

2.3 Making Better Use of the Radio Spectrum

The radio spectrum is the range of electromagnetic wave frequencies used in wireless applications, including radio and television broadcasting, paging, and mobile phones. The spectrum is an increasingly scarce resource. Assigning in an efficient and equitable manner the rights to its use has taken on increased urgency as demand for spectrum has risen with the growth of wireless technologies. The exciting potential of wireless services is illustrated by the rapid growth of cellular telephones, which in the U.S. grew from virtually no subscribers in 1984 to about 5 million in 1990 and about 30 million in 1995; and by the growth of television delivered via direct-to-home broadcast satellite service ("DBS") which, since its inception in June 1994, already has attracted over 2.5 million subscribers.

Spectrum in the U.S. is allocated largely through non-market mechanisms. The FCC, which manages spectrum used by the private sector, designates particular regions of the spectrum for particular uses (akin to land zoning).[7] It then assigns licenses to individual applicants for the designated uses. Historically, assignment was free of charge, in ways that created both inefficiencies and inequities (KWEREL/WILLIAMS, 1993).

At first, license assignment was done through administrative hearings: applicants argued the merits of their case, and FCC Commissioners judged which applicants would best serve the "public interest." Administrative hearings create obvious inefficiencies: spectrum need not go to the most productive users, and the proceedings themselves caused losses from rent seeking and especially from lengthy delays in introducing new technologies such as cellular telephones. For example, although the spectrum allocation process for cellular services began in 1968, the first commercial cellular licenses were only assigned in 1982, partly due to the delay imposed by the hearings process.

In 1981 Congress authorized the FCC to assign licenses in certain cases using lotteries. Lotteries reduced delays in awarding licenses, and subsequent resale of licenses allowed users who valued spectrum highly to try to obtain licenses in a secondary market. However, using the secondary market can entail inefficiently large transaction costs, especially in assembling suitable blocks of licenses. The lotteries also created windfall gains to lottery winners--windfalls that became transparent when certain lottery winners resold their licenses at huge profits.

These visible windfalls fueled political pressure to assign at least some spectrum through auctions: applicants bid for licenses and the proceeds from leasing this public asset go to all taxpayers. In 1993 Congress gave the FCC limited authority to use auctions for assigning licenses to provide services for which subscribers pay fees (unlike advertising-financed broadcasting), such as personal communication services or PCS--advanced mobile two-way voice and data communications services. Auctions tend to assign spectrum to the highest-value users (who are willing to bid highest), while avoiding delays and rent-seeking costs. In addition to these efficiency advantages, auctions also are more equitable: they raise revenue for all taxpayers from the disposition of rights to use a public resource, instead of conferring windfall gains on a lucky few.

Designing good auction rules for PCS and other spectrum licenses poses novel and difficult problems.[8] The FCC developed innovative auction rules to address these issues, and has conducted very successful auctions. For example, auction winners were able to assemble suitable aggregations of PCS licenses over frequency bands and regions, as needed to form efficient communications networks (CRAMTON, 1995). Auctions have been held for PCS licenses over both narrow and broad geographic regions, and for DBS licenses. They have attracted participation by established and new players, including small businesses. And since their inception in July 1994, the auctions have already raised almost $20 billion for U.S. taxpayers in lieu of granting windfalls to a lucky few.

The Senate bill preceding the 1996 Act would have continued the move in the right direction.[9] Although these provisions were dropped from the final Act, they may well resurface in other legislation. If adopted, they will help deploy spectrum in its more productive uses, raise revenue for all taxpayers, and increase the public's confidence in the political system by making the license assignment process more transparent and "fair." But more remains to be done in ensuring that spectrum goes to the right uses and the right users, and that users face correct incentives to economize on spectrum use. To achieve this, some economists advocate giving all licensees (including existing ones) flexibility, subject to technical non-interference constraints, to pursue any applications or to sell their spectrum licenses to others.

3. Role for Government: Media Cross-ownership, Cable Rate Decontrol, Telephone-cable Consolidations

3.1 Media Cross-ownership

The Administration opposed easing the limits on national radio and TV ownership and on cross-ownership (one entity owning different outlets such as broadcasting and publishing) (see the speech by Vice President GORE, 1995). It is difficult to oppose expanded national ownership limits on narrow economic grounds. Antitrust tests for market power would look not at how many stations one entity owns nationwide, but rather on its *market share* of facilities in the relevant market, which often is local not national (and may include alternative outlets to television). Therefore, the objection must be evaluated on other grounds, e.g., that moving from local stations to national networks would skew the content of information provided away from local news (protecting diversity of information indeed was a major Administration concern).

The objection to cross-ownership also is open to the criticism that the relevant market power test depends on concentration, not the absolute number of different outlets owned by one party. This narrow antitrust objection is valid. But there is a broader objection to cross-ownership. If cross-ownership becomes the norm, there may not be sufficient independent voices to speak out on issues that affect the interests of any media segment at the expense of other industries or consumers. This danger is nicely illustrated by the debate over the pending proposal to give free spectrum to TV stations for digital broadcasting, spectrum whose value market value is estimated at billions of dollars. At present, television broadcasters are noticeably mum, and the proposed giveaway is attacked in other media such as the press. But if all media are cross-owned, who will be left to blow the whistle?[10]

3.2 Cable Rate Decontrol

Critics of cable rate regulation stress the difficulties of regulating cable entities that, unlike telephone companies, are not only delivery conduits but also producers of content, namely programming. They argue, somewhat inconsistently, that regulation reduces program quality, and that regulation is cumbersome because of the scope to evade it through retiering--shifting programs from more heavily regulated tiers ("basic") to less regulated ones ("expanded basic" or "premium").[11] They further believe that regulation of the traditionally monopolized delivery function will soon be unnecessary because of growing competition to cable companies from telephone companies, DBS, or other sources.

The Administration worried that such competition will develop slower than touted, and that until it develops cable operators will retain substantial market power to raise rates. It preferred to deregulate rates only once entry has brought sufficient competition to substitute for regulation. The 1996 Act strikes a reasonable balance between these positions. It deregulates immediately rates charged by small cable companies for cable programming services and some basic services. For large systems, it deregulates cable programming services by March 31, 1999. And it lifts all rate regulation sooner if a system faces "effective competition," defined as the offering by a rival of comparable video programming in the cable system's franchise area using the wireline or wireless facilities of a telephone company (but not DBS). Of course, operationalizing the "effective competition" test is likely to be contentious.

3.3 Cable-telephone Consolidations

In most traditionally unregulated industries, the prospects for present and future competition are largely determined by actual competitors--the firms already in a market. But in traditionally regulated monopolies, future competition must largely come from outsiders--from current potential competitors. For this reason, the Administration opposed excessive loosening of restrictions on mergers and cross-ownership between cable and phone companies in the same local area. Although there are technological challenges in using phone wires to deliver video, and cable wires to deliver phone service, cable and phone companies nevertheless are likely potential competitors, as both have wires to the home. Thus, mergers or joint ventures among them could delay competition.

Antitrust enforcers could challenge such anticompetitive consolidations on a case-by-case basis, but doing so for a potentially large number of transactions in different regions would be quite costly. Maintaining clear prohibitions may be preferable, as long as such consolidations yield no significant economies, and as long as local cable and telephone companies remain among each other's most likely potential competitors.

The 1996 Act prohibits the acquisition of more than a 10% interest by a telephone company in a cable company operating in the telephone company's service area, or by a cable company in a telephone company operating in the cable company's franchise area. It also prohibits joint ventures among such entities. However, it makes an exception for rural areas with fewer than 35,000 residents (provided the cable system serves less than 10% of the households in the telephone company's service area).[12] The presumption is that small markets could not support separate phone and cable systems; but it is based on no solid evidence, and the exception could delay competition in local phone and cable service. Still, these exceptions are much narrower than in earlier drafts of the legislation.

4. Role for Government: Telephone Competition

4.1 Background

4.1.1 Local and Long-distance Service

The consent decree that broke up AT&T was filed with the court in August 1982 and went into effect January 1984. It aimed to separate (a) those segments of the business that--because of natural monopoly characteristics or for jurisdictional reasons--would remain rate-regulated (local telephone service) from (b) those segments that were viewed as potentially competitive and eventually could be unregulated (equipment manufacturing, and long-distance service). AT&T inherited the potentially competitive segments, while seven Bell companies inherited regulated regional monopolies over local service. The Bells were prohibited from re-integrating into long-distance service or equipment.

The distinction between "local" and "long-distance" service is inherently somewhat arbitrary. The decree distinguished them by dividing the part of the country served by the old AT&T into about 160 newly created Local Access Transport Areas (LATAs). The region served by any of the new seven Bells encompasses multiple LATAs, but the decree permits a Bell to use its facilities only for service within a LATA (intraLATA). To complete interLATA calls it must use facilities of a long-distance carrier, also known as Interexchange Carrier (IXC), such as AT&T, MCI or Sprint. (Thus, for purposes of the decree, long-distance service means interLATA.) The thinking was that technological conditions made interLATA service potentially competitive, but that actual competition would emerge only if the Bells, that controlled the vital local exchange networks, were barred from long distance (more on this below).

Today, Local Exchange Carriers (LECs)--the companies that provide the lines into subscribers' premises--remain virtual monopolists in their local service areas. Although there are many independent LECs especially in rural areas, LECs owned by the Bells account for about 75% of all LEC revenues nationwide. LECs' prices, for local calls and for access to IXCs, are regulated by the States and the FCC. In contrast, long-distance service is largely unregulated and relatively competitive. Four IXCs have national networks, several others have regional networks, and many offer service using leased facilities. Reflecting the change in market structure since 1984, the FCC ruled in October 1995 that AT&T should be reclassified as "non-dominant."

In 1994 consumers spent about $135 billion on wireline phone service[13] (compared with about $25 billion on cable TV). Of this, IXCs received about $63 billion, but paid about $28 billion to LECs in access fees. In addition, LECs earned approximately $72 billion from local service (about $13 billion from intraLATA toll calls, and the rest from purely local service), bringing their total wireline revenues to $100 billion.

The Act seeks to increase competition in long-distance service and introduce it to local service. As mentioned, it relaxes legal barriers to entry into local service; in addition, it requires LECs to share their networks with entrants (by providing reasonably-priced interconnection as well as access to pieces of their networks--"unbundling"). It also abandons the structural-separation policy embodied in the 1984 AT&T breakup by superseding that decree and--under specified conditions--allowing the regulated Bells to provide unregulated long-distance service and manufacture equipment.

The issue of Bell entry into long-distance was probably the most contentious issue in the legislation. Supporters argued that the Bells would add needed competition to the long-distance market. Opponents, including the Administration, argued that competition would be hampered by allowing Bell entry before they face meaningful competition in local networks and without sufficient regulatory safeguards.

Allowing Bell entry on these terms raises two dangers: (a) distorting competition in long-distance, as the Bells could abuse their monopoly control of vital local networks to gain inefficient advantages in long distance; and (b) delaying local competition, by granting Bells' the "carrot" of entry into long distance before they have seriously opened their local networks to competitors, thereby reducing their incentive to continue cooperating with entrants. The 1982 AT&T decree stresses only point (a), the threat to long-distance competition, as has the Department of Justice throughout. However, a strong economic argument can be made also for (b), that is, for conditioning the Bells' authority to enter long-distance on their cooperation with local entrants--since a major goal of the 1996 Act (a goal not addressed by the 1982 decree) is to promote local competition.[14]

The issue of Bell entry not only is controversial, but is probably the most important economic issue in the legislation, because of the large size of the telephone industry, and because Bell entry would cause a significant change in market structure that would be very difficult to reverse if competitive problems arose. It also is widely misunderstood--it is mischaracterized as "deregulation" and "promoting competition," inducing reactions such as "what is wrong with allowing entry?" We therefore begin by recalling why the AT&T breakup was sought in the first place (see also BRENNAN, 1987; BROCK, 1994).

4.1.2 Why the 1984 AT&T Breakup: Regulatory Evasion via Favoring Unregulated Affiliates Distorts Competition and is Inefficient

A price-regulated monopolist can use its affiliates in other, less regulated markets to evade price regulation, in the process inefficiently harming competitors and consumers. Before its breakup, AT&T's local rates were subject to Cost-of-Service-Regulation (COSR), but its rates in long-distance service and its equipment prices were regulated more lightly. Such asymmetric regulation creates incentives to shift profit to the less regulated markets, since the affiliates' profits go to the same stockholders. Profit shifting entails favoritism towards the firm's less regulated affiliates, through cross-subsidization or, more importantly, through access discrimination against competitors.

Cross-subsidization involves misattributing *costs* of other less regulated businesses to the regulated one, where--under COSR--higher costs can be used to "justify" requests for higher rates. The cost shifting could involve real transfers, such as inflated payments to affiliates for inputs, or accounting allocations of supposedly common costs. (Such cost shifting is akin to the use by multinationals of transfer pricing to minimize tax liabilities by shifting costs into jurisdictions where tax rates are higher.) For example, AT&T allegedly favored its equipment affiliate, Western Electric, by paying it inflated prices for possibly lower quality equipment than offered by independents.[15] In principle, cross-subsidies may pose a problem whenever a regulated firm also operates in unregulated markets; but it is more likely to escape regulatory detection when markets are closely related.

Access discrimination against competitors in the less regulated segments is likely to pose the greater threat to competition. The bottleneck monopolist may try to evade regulation by forcing customers who demand an unregulated service that hinges on access to the bottleneck service to purchase from it also the unregulated service at a higher price ("tying"). To do so, the monopolist reduces competition in the unregulated market, by using its technological control of the bottleneck facility to discriminate against competitors of its affiliate in the non-price access terms to the bottleneck, in ways that are difficult for regulators to detect or prevent. Such discrimination steers customers from competitors towards the monopolist's unregulated affiliate. Consistent with this theory, AT&T was alleged to have discriminated against long-distance rivals like MCI, e.g., by providing them poorer-quality connections to the local network, and by imposing unnecessary delays in responding to requests.[16]

The precise distortions depend on the specific method of price regulation. COSR creates incentives for both cross-subsidization and discrimination. (COSR has other bad effects that are present also for a firm operating in a single market, such as reducing incentives to contain costs; the focus here is on bad incentives specific to the multimarket case, which arise from asymmetric regulation in the different markets.)

In theory, incentives to cross-subsidize can be avoided by replacing COSR with other methods of price regulation such as *price caps*, which decouple the allowable price from those elements of cost that the firm can control. The firm's incentives to allocate unregulated costs to the regulated business are eliminated if higher reported costs do not lead to a higher allowable price. In practice, however, price caps are never pure--they are not set initially forever, but are revised in light of a firm's experience with costs. Price caps therefore amount to cost-based regulation with a longer lag, so a firm still has some, albeit weaker incentives to shift costs of unregulated businesses to the regulated one.

Moreover, price caps leave intact incentives to discriminate against competitors in the unregulated markets. Since the firm's regulated prices are capped, while its prices in unregulated markets are not (more generally, allow higher margins than in the regulated market), the firm gains from shifting business in unregulated segments from rivals towards its affiliates by reducing rivals' access to the monopoly bottleneck facilities.

Regulatory evasion through cross-subsidies or discrimination is inefficient. Cross-subsidies inflate the reported cost of regulated services, leading to monopolistically higher prices. Prices of unregulated services--whose costs are under-reported--may fall, but need not (e.g., if the cost shifting involves only allocation of fixed costs, or if rivals are driven out of the unregulated market by the cross-subsidies); even if prices fall, they would be artificially below cost, and consumption of unregulated services inefficiently high. Also, sales may be diverted away from more efficient competitors in the unregulated markets, because the regulated firm attains an artificial advantage through cross-subsidies.

Discrimination in access terms against competitors in unregulated markets creates similar distortions. It raises prices for those unregulated services, because excluded competitors might have been more efficient and because even equally efficient competitors could curb the monopolist's prices more effectively than can regulation alone. In addition, consumers lose the variety and innovation that competitors would bring. And potential competitors who may be more efficient or innovative are denied profit opportunities. Importantly, these resulting losses from discrimination can far exceed the gain to the regulated monopolist: the monopolist is willing to exclude a rival that would generate large benefits to consumers if exclusion raises its own profit even modestly.

The Department of Justice successfully sought AT&T's 1984 breakup because it felt--and the court agreed--that regulation alone could not, without imposing undue burdens, prevent the many ways in which AT&T could use its control of local telephone service to inefficiently favor its affiliates in equipment manufacturing and, especially, long-distance service. Indeed, the Justice Department and AT&T tried to negotiate a settlement without divestiture; the result was a draft consent decree which for its length and complexity became known as Quagmire II, or the Telephone Book decree...

4.1.3 Unbundling Regulated Bottleneck Monopolies from Potentially Competitive Segments: Alternative Approaches

The old AT&T case illustrates a generic question: how does one introduce competition into industry segments that are potentially competitive but that hinge on access to a bottleneck monopoly segment that is price-regulated? Another example is electricity, where generation is potentially competitive but requires access to transmission, which for the foreseeable future will remain a regulated (natural) monopoly. Similar issues are likely to arise today in telephone service, where the potentially competitive long-distance segment still requires access to the monopoly local network; even within local service, certain functions such as marketing and billing are potentially competitive, but providers would require access to the underlying monopoly physical network. How to inject competition into such markets is a difficult challenge. We are wrestling here with hard questions, to which there are no easy answers. But as we look at the 1996 Act and beyond, it is worth noting the pros and cons of three regulatory approaches: mandated equal access, structural separation, and global price caps.

- *Equal Access.* This approach regulates the monopolist's conduct, but not market structure directly. The bottleneck monopolist is allowed to operate in the potentially competitive segments, but its prices for access to the bottleneck are regulated and it is required not to discriminate against competitors' access in any way, whether through quality of connections, speed of responsiveness to requests, etc. This approach, which we call (mandated) equal access, attempts to prevent discrimination in the face of strong incentives by the firm to engage in discrimination. The next two approaches instead reduce the incentives to discriminate: structural separation bars the monopolist from unregulated markets; global price caps extend price regulation to cover the monopolist's sales in the unregulated markets.
- *Structural Separation.* This was the approach taken in the AT&T breakup--keep the regulated bottleneck monopolist out of related potentially competitive segments, to prevent favoritism through cross-subsidies or access discrimination. Criticisms of structural separation include: (a) It sacrifices economies of scope between the bottleneck and related segments. (b) The monopolist may have other unique know-how and skills that it could usefully deploy in the related markets. (c) There is uncertainty over where to draw the line between the allowed and prohibited markets--the correct line depends on knowing where the monopoly bottleneck lies, but the bottleneck can change over time with technology. For example, in local telephone networks, does the

bottleneck include regional trunk lines, only switches and local loops, or only local loops?

- *Global Price Caps.* The pros and cons of structural separation versus equal access are discussed further in section 4.2, where we see that both have serious drawbacks. Given these drawbacks, it is worth mentioning a third approach that had been noted by earlier authors but has recently been forcefully advocated by LAFFONT and TIROLE (1996).

The idea behind global price caps is based on two principles. First, price caps reduce incentives for cross-subsidization and other cost misallocations. More importantly, conventional price caps preserve incentives to discriminate against competitors only because such caps are *partial* and thus amount to asymmetric regulation: the monopolist's access prices to competitors in unregulated markets are capped, but the monopolist's own output prices in those markets are not regulated.[17]

Structural separation avoids this regulatory asymmetry by barring regulated monopolists from unregulated markets. An alternative is to regulate symmetrically, by extending regulation to cover the monopolist's prices in the potentially competitive markets (but exempt competitors who do not control any monopoly bottleneck). *Global price caps* take this approach. A global price cap involves placing a ceiling (cap) on a weighted average of the monopolist's prices in all markets. In telephones, the global cap on a monopolist LEC would include prices for local service, access fees to IXCs for long distance service, and the LEC's own retail prices for long-distance service. Under a global price cap regime, if the LEC can provide long-distance service less efficiently than an IXC, the LEC could cut its cost by switching traffic from its facilities to the IXC, and thereby raise its profit since the LEC's allowable revenue would not change. Thus, global price caps in theory eliminate incentives to discriminate.[18]

In practice, there will be various problems. As mentioned, price caps are always revised over time, so incentives remain to shift costs towards unregulated segments. Second, granting the monopolist latitude over prices may allow it to engage in predatory pricing against potential competitors. For example, if the monopolist's bottleneck is not completely secure (and it never is), and if entry into the bottleneck is made significantly more likely by the presence of competitors in complementary segments, then the monopolist may still want to stifle the growth of such competitors, possibly through predation. But a reasonable response to this is to supplement global price cap regulation with antitrust enforcement against predation.

Third, a global price cap gives the firm latitude over the price structure (subject to the overall cap), but the firm's profit-maximizing choices may differ from the regulator's. For example, in the U.S. regulators have encouraged cross-subsidies to local service from long-distance service, through setting high access prices for long distance and using the revenue to subsidize local service rates. A LEC operating under a global price cap would generally not choose this same pattern. But the correct response to this concern is to try and move away from cross-subsidies find other ways to achieve the same underlying goals, for example, by subsidizing poor people directly rather than distorting relative prices away from costs. Cross-subsidies in telecom are probably inefficient and, in any case, become less tenable as one moves to competition. As discussed further in Section 5, other ways will have to be found to meet the underlying social goals.

4.2 Bell Companies' Entry into Long-distance Service

Having discussed approaches to fostering competition in markets that rely on a key segment which remains monopolized and the rationale for the 1984 AT&T breakup, we turn to the issue of Bell entry into unregulated markets today. Almost from the outset, the Bell companies have clamored for repeal of the decree's line-of-business restrictions, especially those provisions barring them from long-distance service. Proponents of Bell entry into long-distance argue that their entry would create substantial efficiencies and that regulation could adequately prevent access discrimination. Opponents counter that the benefits from Bell entry would be modest, and that discrimination could not be adequately prevented--the same dangers and distortions discussed earlier in the context of the 1984 breakup could resurface today.

4.2.1 Arguments for Bells' Entry: Costs of Structural Separation

Proponents of Bell entry raise several points, which opponents vigorously contest:

- *"Long-distance market is not competitive."* Bell proponents argue that the long-distance market is far from perfectly competitive and that Bell entry would bring much lower prices.[19] Opponents retort that the long-distance market is very competitive, citing several indicators. There are four independent national long-distance networks and many regional ones, offering a proliferation of price and service packages. Reflecting this competitive jockeying, in 1994 about 20% of all customers changed their long-distance company. Also, customers increasingly originate international calls in the U.S. rather than abroad (through call backs and similar devices), presumably due to lower U.S. prices--that are indicative of greater competition in the U.S. than abroad since the breakup of AT&T and the emergence of competitive IXCs.[20] In addition to the fall in prices, IXCs also point to the great innovation spurred by competition, for example, Sprint's deployment of its fiber optic network prodded AT&T to accelerate deployment of its own fiber network.

  Overall it is safe to say that the long-distance market is at least fairly competitive. Moreover, the perfectly competitive benchmark may not be relevant in an industry such as this that still exhibits significant fixed costs.
- *"Significant benefits to consumers from integrated service."* Most observers see significant consumer benefits from seamless, one-stop shopping for all communications needs; for example, an integrated provider could offer

simplified calling plans. The Bells could provide such integrated service if allowed into long distance. IXCs argue, however, that delivering such consumer benefits does not require Bell entry: others could do the job if regulators forced the Bells (and other LECs) to resell their local service at appropriate discounts to IXCs, as the Act indeed requires (see the next section).[21] IXCs or other companies could resell such local service together with long-distance and other services.

- *"Technological economies of scope between local and long-distance service."* Even if the long-distance market were not competitive, a key question is why entry would be more profitable for the Bells than for other firms. This could be the case because only the Bells could use such entry to circumvent local rate regulation (a "bad reason"), or because they have special legitimate cost advantages in offering long-distance service (a "good reason"). The Bells cite the latter.

They argue with some justification that LATA-based distinctions between "local" and "long-distance" service can be quite artificial, at times amounting to arbitrary regulatory lines with little economic or technological underpinnings. For example, a Bell sometimes may use its own facilities to complete a fairly long call (intraLATA toll call), but must use an IXC's facilities to complete short calls to a neighboring LATA. Allowing Bell entry into long distance would let them tap economies of scope, since a Bell could link its existing regional networks to provide long-distance service at lower cost than would be incurred by a de novo entrant deploying new long-distance capacity.

4.2.2 Arguments against Bells' Entry: Preventing Access Discrimination

Opponents concede that there exist some technological economies of scope in providing local and long-distance service. But they note that such economies existed also when the decision was made to break up AT&T. The policy judgment then was that (a) breakup was necessary to protect competition in the potentially competitive segments--regulation alone was not enough--and (b) that the gains from competition in those segments would outweigh any losses from sacrificing economies of scope. Fundamentally, opponents of Bell entry argue that not enough has changed since 1984 to justify a policy shift.

Some forty leading economists, on behalf of three Bell companies that sought to repeal the 1982 decree and be allowed into long-distance, argued that circumstances indeed have changed. BRENNAN (1995) assesses their arguments and overall, finds them not compelling. Following is a sampling of the arguments and my assessments.

- *"Better regulation."* The move from COSR to incentive-based regulation (e.g., price caps) adopted by many States, reduces incentives to cross-subsidize. However, as noted earlier, price caps are just cost-based regulation with a longer lag, and thus reduce but do not eliminate cross-subsidization incentives. Moreover, even pure price caps (unless they are global) do not reduce incentives for access discrimination. Finally, requiring the Bells to provide unregulated activities through separate subsidiaries also does not eliminate discrimination incentives because the subsidiaries' profits flow to the same shareholders.[22]

  Proponents of Bell entry further argue, correctly, that regulators today are more attuned to the dangers of discrimination. But regulators would still find it difficult to effectively police all avenues of technological discrimination. My own instinct is that "where there is a will there is a way"--once the Bells are allowed into long-distance, they would find ways to discriminate against competitors. Discrimination may be especially difficult to prevent when IXCs require customized (rather than standard) arrangements in order to provide enhanced services that depend on the local network; and enhanced services is where many project the most rapid growth. The Bells might delay their responses, charge higher fees, etc. For recent examples of such behavior see CAULEY (1995). Moreover, even if regulators can detect discrimination, their enforcement ability and political latitude to act may be limited.

- *"Established long-distance competitors."* Major IXCs have made substantial investments that are largely sunk, so the Bells are unlikely to drive these IXCs out of the market. Nevertheless, the Bells still could use their control of key local facilities to squeeze rivals in the short run, and threaten their viability in the long run by discouraging their future investments and innovation. Thus, Bell entry into long-distance could well reduce prices in the short run, while rivals remain potent competitors; but in the long run it could stifle competition and undo many of the benefits of the original breakup.
- *"Local telephone competition is just around the corner."* A rhetorical response is that if competition really is so imminent, why don't we wait a few months until it arrives? More to the point, in evaluating the state of competition, it is important to distinguish between various local network services. Competition has been growing in limited services to certain customers: competitive access providers have been bypassing LEC trunk lines and connecting large business customers through private lines to IXCs. But the LECs still face no serious rivals for providing switched network services, with the result that smaller businesses and residential customers still largely depend on the LECs for access to IXCs, as well as for virtually all their local calls. The LECs' persistent dominance is reflected in the fact that, despite the heralded growth of competitive access providers, the LECs still command about 96% of all fees paid to access local exchange networks.[23]

In short, there are legitimate concerns that if the Bells are allowed to reenter the unregulated long-distance and equipment markets before they face local competition and thus remain price-regulated in local service and access, regulation alone could not prevent them from inefficiently excluding competitors in unregulated markets. Under such circumstances, allowing the Bells to provide unregulated long-distance service would give them incentives to discriminate against long-distance rivals. Allowing them to manufacture switches and other network equipment could enhance their *ability* to discriminate, by making it easier to retain proprietary control of important technical information needed to interface with long-distance and other unregulated services that rely on local networks.

4.2.3 Competitive Safeguards in the 1996 Act

In my view the most compelling answer to "what has changed since 1984 to justify repealing the consent decree and allowing the Bells into long distance" is this: unlike the decree, the 1996 Act makes a conscious and serious effort to promote also local competition--by repealing state regulatory barriers (as discussed in Section 2.1) and by requiring incumbent local phone companies to cooperate with entrants (see below and Section 4.3). The Act reflects an increasing belief, prompted partly by technological advances, that competition is feasible in at least some segments of local networks.

Local competition, in addition to its inherent benefits discussed in Section 4.3, can help prevent access discrimination by the Bells against IXCs. Local competitors offer IXCs alternative ways of reaching some customers. They also provide regulators a useful "yardstick" for policing discrimination: a Bell's claim that it is not feasible to provide certain services to an IXC will ring hollow if the local competitor finds no difficulty providing such services.

An approach that prevents discrimination against IXCs while minimizing the extent of hands-on regulatory involvement would be to allow the Bells into long-distance only after sufficient local competition has developed to eliminate the Bells' market power over local networks. However, that requisite degree of local competition would take quite some time to develop.[24] Such a delay in authorizing Bell entry would have been politically untenable, and may be economically inferior to the alternative discussed below.

That alternative approach is to require less local competition as a precondition for allowing the Bells into long-distance, but leave more room for regulatory discretion. This is essentially the approach taken by the 1996 Act.[25] The Act: (a) imposes regulatory safeguards against Bell discrimination and other abuses; and, importantly, (b) links the Bells' entry authority to their ongoing cooperation with entrants into local networks, giving government "referees" the discretion to make key judgments.

The Act specifies the following process for authorizing Bell entry into unregulated (what the Act calls "competitive") activities:

- *Separate affiliates*. Long-distance service, as well as the manufacturing of equipment, must be offered through separate affiliates, to help regulators detect and prevent cross-subsidization from the regulated local telephone service to the unregulated businesses. A regulated Bell company may not discriminate in any way in favor of its affiliates. Separate affiliates must be maintained for three years after a Bell company is authorized to offer long-distance service in its region (see below), unless the FCC extends the requirement.
- *Long-distance service from out-of-region States, and equipment manufacturing*. A Bell company may offer long-distance service immediately originating in any State where it does not provide local service (on the theory that it does not control local networks in such out-of-region States and hence cannot disadvantage IXCs). It may manufacture telecom equipment as soon as it receives FCC approval to offer long-distance service originating in any "in-region" State (one where it provides local service).

- *Long-distance service from in-region States.* To offer such service, a Bell must receive FCC approval on a State-by-State basis. FCC approval is granted only after the following requirements are met.
  - *Competitive checklist implemented.* Within 6 months of the Act's enactment the FCC will issue rules for interconnection and network unbundling (discussed further in Section 4.3), that all incumbent local exchange companies must follow in dealings with new local competitors. Before receiving long-distance authority in an in-region State, a Bell at a minimum must *offer* price and other terms that are consistent with these FCC rules. Consistency is to be certified by the State public utility commission (PUC). Moreover, if the applying Bell has received an interconnection request from a new local competitor, it must first have *fully implemented* a binding interconnection agreement with the competitor. That agreement must satisfy the FCC rules; the competitor must provide local exchange service to both *business and residential* customers in the State (pure access providers, for example, do not suffice); and it must use exclusively or predominantly its *own facilities.*[26] In short, the local competitor must have more than a token presence.
  - *FCC "Public Interest" determination, with Justice Department consultation.* Certification by the State PUC that the applying Bell has complied with the FCC's local-competition rules is necessary but not sufficient to authorize the Bell to provide long-distance service in that State. The FCC also must consult the Justice Department regarding the likely competitive implications and accord the Department's evaluation "substantial weight." Based on this input and its own evaluation, the FCC must determine that Bell entry would be in the public interest.

This approach makes considerable sense. Implementing the competitive checklist will help promote local competition, but does not guarantee its imminence or durability. Some uncertainty remains about the feasibility of widespread local competition, and there is the risk that the Bells would cease cooperating with local entrants and stifle competition once they are allowed into long-distance.[27] Accordingly, the Act provides the additional competitive safeguards in the FCC's public interest test. The strong consultative role for the Justice Department is important given the leading role played by the Department's Antitrust Division in bringing competition to long-distance telephone service through the AT&T breakup, and its general expertise and mandate in competition matters.

Some have criticized this discretion given to the FCC and Justice Department as overly regulatory. However, it is important to allow regulators and antitrust enforcers latitude to make subjective judgments that significant and durable local competition is developing because, by its very nature, a checklist cannot hope to capture all the relevant competitive eventualities.[28] Indeed, the success of the 1996 Act in protecting and increasing competition in long-distance service will largely hinge on the ongoing, hands-on involvement by the FCC, state regulators and antitrust enforcers to implement the Act's equally important other goal, of promoting local exchange competition.

4.3 Promoting Local Competition

4.3.1 Local Telephone Networks and Types of Entrants

To understand how the Act attempts to promote local competition it is helpful to review briefly the main elements of a local public switched telephone network. These include: (a) wireline *local loops* connecting customer premises to the LEC's switch; (b) *switches*, essentially computers that route calls through trunk lines; (c) *trunk lines*, transporting multiple calls between switches; (d) *signaling and control systems*, for example, to assist in call routing and processing; and (e) *operator* services.

There is some debate over which if any of these segments are natural monopolies today. The most likely candidates are the local loop and the signaling and network control elements. Technology, especially wireless, is moving rapidly, making it hazardous to guess beyond, say, five years. But even if network elements are no longer natural monopolies, most observers agree that entry on a large scale comparable in scope to an incumbent LEC would take a long time, given the massive investment required. An entrant would require cooperation by the LEC, and the Act requires such cooperation. In discussing the requisite cooperation, we can distinguish three types of entrants: *full facilities*, *reseller*, and *partial facilities*.

- A *full facilities* entrant--one that is vertically integrated to the same degree as the LEC--would still need *interconnections* to the LEC's network, to terminate calls to or receive calls from the incumbent's customers, to enable customers to maintain their telephone numbers if they switch carriers ("number portability"), and to access common signaling facilities and databases. Good and reasonably priced interconnection is vital to an entrant because telephone service exhibits such strong positive network externalities: the network's value to a subscriber increases with the number of subscribers reached by the network. This is a critical distinction between telephones and, say, textiles: a new textile manufacturer does not need much cooperation from incumbents; but an entrant into local telephone service needs substantial cooperation from the incumbent. Since an entrant will at least initially be much smaller, the incumbent can use interconnection terms as a strategic weapon against entrants.
- A *reseller* would lease access to all physical facilities from a LEC in bulk, but would perform all customer-related functions such as marketing, and billing ("retailer" would be a better term than the commonly-used "reseller," as the latter wrongly suggests only an arbitrage role). It would resell local service to its customers, along with any other services that would be jointly marketed, for example, it might jointly market local and interexchange service. Such joint

marketing is important because customers value one-stop-shopping. Since a reseller undertakes costly functions otherwise performed by the LEC, to succeed it must buy the LEC's network service at a sufficiently discounted wholesale price. The discount relative to retail prices should reflect the cost savings the LEC could realize from cutting back on these functions and delegating them to a reseller.

- A *partial facilities* entrant would purchase only some elements from the LEC and supply the rest itself. It would need interconnection as would a full-facilities entrant. In addition, it would need access at economical rates to facilities that it does not provide but instead wishes to lease from the LEC. For example, it might install its own switch but not wires to individual homes, and seek to use the incumbent's local loop. Such an entrant requires *network unbundling*--availability and economical pricing of a subset of network elements (as opposed to all elements as required by resellers).[29]

4.3.2 Benefits of Resale or Partial-facilities Entry

The social benefits of entry by firms that are not entirely facilities based can be quite real. These benefits can be grouped into two categories. First, there are inherent benefits from cutting prices and costs in the segments that are entered,[30] and from increasing product variety and convenience, e.g., by allowing one-stop-shopping for local and long-distance service.[31] Second, resale or partial-facilities entry can assist and accelerate the transition to full-facilities competition.

The transition issue arises because entrants rarely attain a large market share overnight, but instead grow over time. In long-distance service some entrants began mainly as resellers and added their own capacity as their name recognition and subscriber base grew. Requiring entrants to be entirely facilities-based at the outset could saddle them with unnecessarily high fixed costs and excess capacity (while subscribers are being added), making entry more risky and more costly. Granting entrants access at reasonable prices to complementary LEC facilities during the transition could permit a faster and more economical transition to full-facilities competition.

In turn, full-facilities competition could bring the above benefits of competition to a broader range of services, and an additional important benefit: it would reduce the need for regulation. For example, in long-distance service there is an active wholesale market in capacity because multiple facilities owners compete to provide capacity. In contrast, implementing resale of local service is likely to require complex regulation as long as LECs retain monopolies over local networks.

4.3.3 Requirements of the 1996 Act: Interconnection and Unbundling

The Act requires any incumbent LEC to cooperate with local entrants, by offering: *interconnection* to its network and *unbundled access* to network elements and services (including the entire network's services at discounted wholesale rates for resellers, as discussed shortly, and including databases and signaling functions). Interconnection and unbundled access must be offered "at any technically feasible point" and "on rates, terms, and conditions that are just, reasonable and non-discriminatory" (compared with those offered to any other including to the LEC's own operation). It also requires all LECs--incumbents and entrants--to offer number portability (allow a customer to retain its phone number if switching to a rival carrier), dialing parity (allow a rival's customers to dial the same number of digits as its own customers to reach other parties, rather than requiring additional digits), and reasonable access to rights-of-way and poles and conduits.[32] Inevitably, the Act provides only a blueprint, and leaves such "details" as the pricing of interconnection and of unbundled services to be determined by the FCC and the States. But these details will be crucial, and their implementation will not be easy.

For example, how large is the right wholesale discount from a LEC to resellers? This question already has generated considerable controversy, embroiling state regulators in having to define and assess the costs a LEC avoids by delegating retailing functions to resellers. To illustrate, in Rochester AT&T requested a 45% discount off the retail price, while the incumbent LEC offered 5%. And in the debate over the 1996 Act, a seemingly trivial but actually important issue concerned whether the wholesale discount should reflect the LEC's "avoided" costs or costs that are "avoidable".[33] Prospective resellers of local service, such as IXCs, worried that "avoided" leaves the LECs excessive discretion to play games: a LEC might refuse to cut its marketing staff even if resellers could take over many customer related functions, in order to argue that no costs were "avoided" and therefore that the wholesale discount to rivals should be minimal. But if the wholesale discount is too small, then even a reseller that could perform marketing and other retailing functions more efficiently than a LEC would be unprofitable.

Unbundling of marketing and billing functions from others will be hard enough, as demonstrated by squabbles over the correct wholesale discounts to resellers. Unbundling of network elements is likely to be even harder. There are many fixed and common costs; network congestion is an important element in determining efficient prices; and some unbundling may pose technological challenges (for example, regulators will have to grapple with the phrase "any technically feasible point").

The difficulties with requiring LECs to share their local networks with competitors illustrate a general problem: it is notoriously hard for regulators to force a firm to do something against its interests. Regulators are far less informed about the business. Even if informed, they often lack adequate enforcement powers. Providing a firm incentives to cooperate is in general a more effective approach than issuing specific, detailed mandates. A powerful incentive to elicit cooperation with local entrants from LECs owned by the Bells (Bell-owned LECs account for about 75% of all LEC revenues nationwide) is to condition--as the Act does--a Bell's right to offer long-distance service on its cooperation with local entrants. If properly enforced, such a linkage will accelerate the development of local competition, thereby safeguarding competition in the long-distance market by reducing the Bells' ability to discriminate against IXCs; it also would bring the benefits of competition to the larger local market.[34]

In short, introducing competition into local networks is likely to be a complex and contentious process, requiring continued active involvement by state regulators, the FCC, the Justice Department, and possibly the courts. Nevertheless, by defining the broad rules and providing for active government involvement to implement agreements and referee disputes, the 1996 Act holds the promise of stimulating ubiquitous, vigorous competition with potentially enormous benefits to businesses and consumers.

5. Reducing Cross-subsidies and Revamping Universal Service

A long-standing policy goal in the United States and many other industrialized countries has been promoting universal service--widespread access to service at a reasonable price. To some extent this goal can be defended on narrow economic grounds of positive network externalities, i.e., the benefits of subscribing to telephone service accrue not only to the subscriber but also to others who are interested in calling that person. Support for universal service, however, is based also on broader considerations--that all members of a society should be entitled to a certain level of key services.

Where attaining universal service is thought to require government intervention, because without it prices would be deemed too high in certain regions or to certain customer groups, economists generally advocate using targeted, explicit subsidies, financed through broadly based taxes. Traditional U.S. regulatory policy has *not* taken this route. Instead, regulators have pursued universal service by setting the rate structure of regulated telephone monopolists to implement a complex system of cross-subsidies.

Cross-subsidization arises when the price in one market does not cover the incremental cost of serving that market and the deficit is financed by charging a price above incremental cost in another market (more on this definition shortly).[35] The "markets" can involve different products (e.g., long-distance versus local calls) or different identifiable customer groups for a given product (e.g., residential versus business customers of local calls). Moving to competition from regulated monopoly-cum-cross-subsidies is likely to reduce some prices, but may increase others. The fear that some prices may rise could lead some groups to oppose competition. Inter alia, they might allege that entrants will engage in *cream skimming*--selectively serve only more lucrative markets and abandon the rest.

The possibility that some parties may lose from a move to competition of course is not sufficient reason to block such a move. One should consider the impact on overall welfare (in economic jargon, is the move a potential Pareto improvement?). To evaluate the overall impact, we must step beyond the rhetoric of "cross-subsidies" and "cream skimming" and ask if the current system of regulated monopoly pricing is the best way of achieving its purported goal of universal service.

Section 5.1 notes that cross-subsidies may be relatively efficient in certain circumstances, but the key question is whether these circumstances apply today. Section 5.2 reviews briefly the U.S. system of cross-subsidies, its tensions with competition, and its economic distortions. It echoes the belief of many economists that today's cross-subsidies are inefficient and incompatible with a move towards increased competition in telephone service. Section 5.3 discusses the need for revamping universal service and the relevant steps undertaken in the 1996 Act.

5.1 Cross-subsidies, Common Costs, and Economic Efficiency

The notion of cross-subsidies actually is not always well defined. The simplest case is where there are no fixed or common costs and the various markets could be served independently by different competitive industries at constant marginal costs. Cross-subsidies would involve regulators creating an artificial multimarket monopoly, and requiring the monopolist to set prices in some markets below marginal costs (these marginal costs are then the only relevant incremental costs of serving those markets) while funding the deficit from other prices that exceed marginal costs.

Defining cross-subsidies gets more complicated when serving the different markets entails *large fixed and common costs* that create a multimarket natural monopoly, as can occur in network industries. If the monopoly is to be regulated and if it is to be entirely self financing (receive no outside subsidies--an important and controversial assumption), the task of covering the fixed costs falls to regulators.[36] To cover these costs, prices in some markets will have to exceed marginal costs. And since the allocation of fixed and common costs is inherently arbitrary, the notion of "incremental cost" of serving a market is no longer clear cut, nor is the proper definition of cross-subsidies. In particular, if cross-subsidies are defined as entailing different markups of prices over marginal costs in various markets, cross-subsidies so defined may be efficient.

As stressed in the theory of Ramsey pricing, charging different margins can be efficient when demand elasticities differ across markets, because then it is least distorting to cover the fixed costs by charging relatively higher margins where demands are less elastic (see, e.g., BROWN/SIBLEY, 1989; LAFFONT/TIROLE, 1993). Doing so minimizes the inefficiencies from reduced consumption due to prices that exceed marginal costs. Even setting prices below the marginal costs of serving some market might be efficient if there are strong positive externalities between markets (rather than demand independence as assumed in the basic model of Ramsey pricing). As noted earlier, telephone ownership by different customer groups likely generates such positive network externalities, which might make it efficient to price phone service below marginal cost to certain groups in order to encourage universal service.[37]

It is important to recognize that the issue of cross-subsidies can arise even when prices in all markets exceed the marginal costs of serving those markets, because of the thorny issue of how to allocate contributions for covering the fixed and common costs. Alternative technologies may be available to serve some of those markets at lower prices than the prices charged to those markets in the common-cost monopoly regime. Those markets are then said to be paying the monopolist more than their "*stand-alone costs*" and in this sense cross-subsidizing the other markets--even though prices in the latter may exceed marginal costs. The presence of such alternative ("bypass") technologies means that a multimarket natural monopoly may not be sustainable. That is, even a monopolist that earns only zero profit while pricing efficiently[38] could be vulnerable to inefficient "cream-skimming"--selective entry that targets only a more profitable subset of the monopolist's markets (BROWN/SIBLEY, 1989; LAFFONT/TIROLE, 1993).[39]

In short, cross-subsidies can be an elusive concept and a regulated monopoly exhibiting cross-subsidies conceivably may be efficient. If this view of the industry--as a ubiquitous multimarket natural monopoly that is pricing efficiently to recover common costs but is vulnerable to cream-skimming entry--were accurate, policymakers would face a tradeoff: restricting entry would better allow exploitation of scale economies, but would deny the benefits of competition and impose regulatory costs. Many economists, however, challenge the above portrait of the telephone industry (e.g., CRANDALL/WAVERMAN, 1995). They are wary of labeling too many costs as "fixed and common" and the industry as a ubiquitous natural monopoly. Moreover, to the extent there are fixed and common costs, current regulated prices do *not* recover such costs efficiently, nor promote universal service in the most efficient manner.[40]

5.2 Distortions in Current System and Tension with Competition

For many years regulators, supported by Congress, promoted universal service through a system of cross-subsidies among the markets served by a regulated monopolist and across different monopolists (BROCK, 1994). Through a complicated nationwide pooling of telephone costs and revenues, local exchange carriers (LECs), especially in high-cost rural areas, received substantial subsidies to keep their rates low. The subsidies were financed by setting artificially high prices for telephone equipment and long-distance calls. In addition, long-distance rates were set by geographic averaging: rates for routes of the same distance were set equal despite different traffic densities and therefore different costs. Subsidies probably also flowed from business to residential customers generally, by overpricing enhanced services (largely used by businesses) to support basic service.

This system was administered by AT&T, whose affiliate Bell System companies provided most local telephone service nationwide and virtually all long-distance service. The system came under strain once AT&T's virtual monopoly began to erode. Inflated prices for equipment and long-distance service attracted entry. The growth of competition in customer premises equipment (such as telephone sets) in the 1970s and later in long-distance service reduced the funds available for cross-subsidies. In response, after the breakup of AT&T the FCC introduced fixed monthly fees for all telephone subscribers, reducing the need for subsidies; the FCC and State regulators also instituted explicit access fees for all long-distance carriers originating and terminating calls on LECs' networks. These access fees today still help finance subsidies to rural LECs.

The inflated access fees, however, prompted large business customers to bypass the local exchange, and instead use private lines to connect their premises directly to interexchange carriers (IXCs). Such bypass again threatens the revenue used to cross-subsidize other services. Some local phone companies also allege that revenue from high-volume local business customers cross-subsidizes basic local service to low-volume customers, so that permitting entry into local service also will threaten cross-subsidies: entrants will siphon off the lucrative business customers and reduce the revenue available to keep rates low to other customers. Thus, there is tension between the current system of cross-subsidies and the continued growth of competition. Critics of the present system start by noting its distortions.

- *Distorted Entry Decisions.* As mentioned, access fees charged by LECs to IXCs have been substantially above costs and used partly to cross-subsidize service in rural areas and some basic service nationwide. Such pricing can distort entry in two ways: artificially high prices can encourage inefficient entry, and artificially low prices can discourage efficient entry. Correspondingly, the inflated access fees may have artificially encouraged the growth of competitive access providers that bypass local networks and link businesses directly to IXCs. The current system may also be discouraging efficient entry, and understating the potential for local competition. Artificially low prices for the subsidized incumbent's services can discourage even more efficient potential entrants from competing for such services (e.g., to serve rural areas), because under the current system only incumbents are eligible for certain subsidies.
- *Distorted Consumer Decisions.* The current rate structure also distorts consumer decisions. High long-distance rates subsidize telephone subscription but discourage calling; raising the fixed charge for telephone subscription and reducing the prices for calls would stimulate calling. The benefits from lower toll rates and expanded calling would make many consumers better off even after paying higher fixed charges.[41] Cross-subsidies from long distance to local service are sometimes defended on the grounds that low-income individuals use

local service relatively intensively; but the correlation between income and long-distance versus local calling may not be strong, and some studies have indicated that high toll bills often lead to low-income subscribers being disconnected for nonpayment. Better ways can be found to assist those with low incomes.

- *Lack of Transparency*. A necessary condition for good economic policy is to make costs and goals explicit and transparent. The true extent of cross-subsidies required to ensure goals such as universal service remains unclear because cross-subsidies are now implemented in opaque ways. Would many poor people really drop their telephone service if fixed charges rose but long-distance rates fell? Would entrants really shy away from rural areas, or would they serve them at only modestly higher prices? What portion of the inflated access fees charged for long-distance calls goes towards keeping basic local rates low as opposed to funding internal slack by local telephone monopolists?

5.3 Reforming Universal Service

One approach to preserving universal service and other social goals is to try and maintain a broad monopoly charged with meeting these social objectives, by legally prohibiting entry or by requiring all entrants to make substantial contributions to cover the incumbent's cost of providing below-cost services. The other is to permit widespread competition and develop alternative, market-based ways of funding legitimate social goals. Choosing between these paths reflects different world views about the current and likely future structure of the telephone industry. As technology changes, so will the extent of economies of scale and scope, and with them the efficient industry structure. If these economies become modest, there will be serious losses from restricting entry to protect inefficient cross-subsidies from cream-skimming entry.

The Act reflects a belief that technological change in telecom is so rapid and promising that competition may be possible and desirable in many segments now or in the near future. These rapid changes, and the Act's removal of legal entry barriers as described earlier, imply that attempts to promote universal service by maintaining cross-subsidies through regulated monopolies will become both increasingly inefficient and less tenable. Many economists favor giving competition freer rein and letting prices adjust to better reflect true costs (BROCK, 1994; CRANDALL/WAVERMAN, 1995)

Competition is likely to cut the cost to society of providing universal service, by fostering cost reductions and innovative ways of providing service, and by curbing market power and forcing down prices more effectively than does regulation. Competition may well reveal that most people would get affordable basic telecom services without requiring subsidies. To the extent that assuring universal service will require residual government intervention, such intervention is more efficiently done through mechanisms that are more transparent, more targeted to the particular goals, and that do not distort competition. It is difficult to believe, for instance, that cross-subsidizing local rates by long-distance rates through a system of regulated monopolies is really the most effective way to assist the poor. In reforming the present system we must tackle several issues.

- *What Should Be Included in Universal Service?* For many years there was only one basic service to be universalized or not: a telephone was a telephone. Today, however, telephone and other telecommunications networks are evolving to permit a much broader range of enhanced services. As conditions change, it will be important to review, perhaps on an evolving basis the range of services targeted for universal service, and be clear on what is meant by "sufficiently affordable" prices.
- *Who Should Be Eligible for Support?* Should support be given to all rural areas, including luxury ski resorts? Or should it targeted to the poor wherever

they reside? And how much should prices be allowed to vary to reflect differences in the costs of serving different regions?

- *How Should Universal Service Be Funded and Provided?* Once the goals have been clearly identified, funding mechanisms should be devised that do not distort competition. At present, subsidies to serve ostensibly unprofitable markets are not offered to all comers on an equal footing, but are reserved for incumbent monopolists and financed through surcharges on long-distance and other services. Alternative financing methods would be less distorting and more compatible with competition. An example might be a universal service fund, financed by charges levied on all telecommunications carriers, or more broadly. All eligible consumers, whether low-income individuals or designated regions, could draw on the fund, to help them pay for the provider of their choice. Alternatively, the right to provide subsidized service to a designated group could be allocated through competitive bidding by all qualified potential providers.

Revamping the funding of universal service is an integral part of increasing competition in telephone service. In the absence of explicit funding mechanisms, regulators might feel compelled to ensure universal service by protecting incumbents from "unfair competition".[42] For example, they might require entrants to offer a configuration of services, regional coverage, and rate structure very similar to that of the incumbent LEC monopolist. Such obligations could easily stifle competition. Forcing an entrant to be a clone of the incumbent could make entry unprofitable, because some--though not all--elements of local service may still be natural monopolies. Entry is more likely to occur and to be socially valuable if entrants are not forced to duplicate the incumbent's offerings but instead are given flexibility in choosing the technologies with which to serve customers and the mix of services so as to best exploit their comparative advantage.

5.4 The 1996 Act

The 1996 Act does not fully resolve these issues, but does take some useful steps. It requires the formation of a Federal-State Joint Board, representing State and Federal regulators and consumers, to review thoroughly the existing system of Federal support for universal service. The Board is to recommend reforms within 9 months of enactment. Within 15 months of enactment, the FCC is to complete a proceeding to incorporate the Board's recommendations and establish a specific timetable for implementation. The Act articulates some general principles to guide the reform.

Regarding *what* is included in universal service, the Act stipulates an evolving level of telecom services that the FCC shall define periodically in light of technological changes and market adoption patterns. A core principle in the Act is that *advanced* information services should be made widely available through public institutions such as schools, hospitals and libraries. To promote this goal, all telecom carriers are required to provide service to health care facilities in rural areas and to schools and libraries at discounted rates (more on this point below). Targeting the provision of advanced services through public institutions represents a sensible balance between the desire to make such services widely available, and the danger that mandating widespread provision at discounted rates could severely curtail the profitability of such services and stifle their growth.

Regarding *who* is eligible for support, a principle in the Act is that consumers "in all regions" should have access to telecom and information services whose quality, variety and rates are "reasonably comparable" to those available in urban areas.[43] But it goes further as regards interexchange and interstate telecom services--which include telephone service--by requiring the rates charged to residential subscribers in rural areas to be "no higher" than those charged in urban areas (this incorporates existing policies of geographic rate averaging). Many economists would balk at such a rigid requirement. After all, there are many offsetting benefits of living in rural areas; for example, we do not--and for good reason--try to equalize housing prices in urban and rural areas.

Regarding *funding* of universal service, the Act's principles call for making support mechanisms explicit, targeted and predictable, and for requiring all providers of telecom services to make nondiscriminatory support contributions. The FCC shall determine the appropriate contributions by interstate carriers (though it may exempt small carriers whose contributions would not justify the administrative costs involved); State regulators shall do so for intrastate carriers. Carriers that provide discounted service to public institutions as discussed above may elect to count these discounts as credits towards their required contributions to universal service, or to receive reimbursement from the support funds.

So far the Act's approach to universal service reform is consistent with moving from monopoly to competition, but there are potential tensions. While the Act requires States to remove barriers to competition, it allows them to "impose, on a competitively neutral basis...requirements necessary to preserve and advance universal service..." This language might permit States to require entrants to provide a very similar menu to that of the incumbent, such as similar regional coverage, in the name of preventing cream skimming that would threaten the incumbent's viability and thus universal service. As mentioned earlier, such seemingly "competitively neutral" requirements could nevertheless deter entry by depriving entrants of valuable flexibility. On the other hand, given the considerable uncertainty about the feasibility and imminence of competition in local phone service, a case can be made that States should be accorded latitude in determining their competition policies. Such experimentation may yield useful lessons. It remains to be seen how States will avail themselves of the latitude they have under the Act.

Another source of potential tension between universal service and competition lies in determining which carriers are eligible for subsidies to serve high-cost areas. At present, only incumbent telephone carriers qualify for many cross-subsidies. The Act changes tack in regions not served by rural telephone companies: if more than one carrier seeks to be designated as eligible for financial support to provide service in a particular (high-cost) area, the State PUC must designate all such carriers as eligible. In an area served by a rural telephone company, however, the PUC may designate an additional carrier as eligible only if it first determines this to be "in the public interest." This places a potentially high hurdle for the PUC, and tilts the playing field towards incumbent LECs. Placing the burden of proof on the advocates of competition is unfortunate, because it could delay competition even as technological advances, especially in wireless services, are increasing the scope for competition also in low-density rural areas.

On balance, however, the direction taken by the Act promises to better harmonize the goals of promoting competition and advancing universal service. Perhaps the safest verdict is cautious optimism.

6. Conclusion

Attempting a definitive "conclusion" in an area as volatile as telecom is clearly premature. Instead, I recap some of the main points. There is strong consensus in the U.S. that telecom regulation should be reformed to allow a greater role for competition. Many feel that such reform is long overdue. The 1996 Telecommunications Act, which passed with overwhelming bipartisan support, takes serious and wide-ranging steps in this direction.

It is important, however, not to become euphoric but to hold realistic expectations. Reform in telecommunications is not an easy business. The Act calls for implementation of many requirements. But calling for and actually implementing difficult, technical issues of interconnection and unbundling are two different things. "Nondiscrimination" is an elusive concept when the circumstances of different transactions differ in costs and quality (SCHWARTZ, 1986b). We can expect many hurdles ahead, especially in promoting competition in the huge telephone sector.

Indeed, cynics have called the Act "five pages deregulation, and two hundred pages regulation." They note that the Act does not eliminate much of the highly regulatory language in the 1934 Communications Act, while adding more requirements. A partial reply to this concern is that the Act (in section 401) also gives the FCC substantial discretion to refrain from enforcing certain provisions of the Act if it determines that enforcement is not necessary to protect consumers and would not be in the public interest. In making this determination, the FCC is to consider whether its regulatory forbearance would foster competition. This new regulatory flexibility is potentially quite important.

Cynics might reply, correctly, that this flexibility to make exceptions does not cover the Act's restrictions on the Bell companies' ability to offer unregulated services such as long-distance, nor requirements on local telephone companies to share their networks with competitors. In a sense the cynics are right: the Act does contain a lot of regulation of local telephone companies. But in a more important sense they are wrong: hands-off deregulation is not an effective way to introduce competition into local service given the prevailing monopolies over local networks. The key network externalities in telephone service mean that entrants will at a minimum need adequate, economical interconnection and there is no avoiding the need for a referee to enforce this.

Moreover, since long-distance service requires access to local networks, increasing competition in long-distance service is not a simple matter of letting the Bells enter with no competitive safeguards. As long as the Bells' access fees to long-distance carriers remain regulated--and there is a good reason for maintaining such regulation until strong competition develops in local networks--the Bells, if allowed to offer unregulated long-distance service, will have incentives to discriminate in access against long-distance carriers and shift profits towards their own affiliates. The 1984 AT&T breakup attempted to prevent such behavior, many would say quite successfully, through structural separation of the regulated and unregulated markets. To attain greater economies of scope and increase the number of potential players, the 1996 Act abandons structural separation. But without strucutral separation, preventing access discrimination may well necessitate increasing the regulation of Bell conduct.

One can look at the glass as half empty or half full. No doubt the road to competition will be bumpy. On the other hand--and all economists have at least two hands--these difficulties should subside once local competition takes root. Many of the problems are ones of transition, and stem from the initial asymmetries of the players. Denying interconnection is a potent strategic weapon for a dominant incumbent; it may be self defeating for a player in a symmetric industry structure. Accordingly, just as the presence of several competing long-distance networks has produced an active market in leasing bulk capacity, securing widespread access to local facilities will require far less regulation if local competition develops. A key question is how fast this will occur.

Overall, the Act holds much potential. It clarifies the rules for investors, thereby providing greater certainty. This alone should result in significantly increased investment. With the removal of artificial governmental barriers to competition but maintenance of proper competitive safeguards against private barriers, competition in telecom should flourish, bringing an explosion of innovation, new services, and lower prices. Of one thing we can be certain: the future in telecom will not be dull!

Endnotes:

* An earlier version of this paper was presented at the seminar on "Energy and Telecoms Regulation in Transition Countries," The Center for Economic Development, Bratislava, Slovakia, October 1995. The views in this paper do not purport to represent those of any United States government agency. I have learned a lot from discussions with Tim Brennan, Gerald Brock, Richard Clarke, Bill Corbett, Robert Crandall, Tom Hazlett, Evan Kwerel, Roger Noll, George Slover, Lisa Sockett, Joe Stiglitz, Jean Tirole, Scott Wallsten, Leonard Waverman, John Windhausen, Peyton Wynns and many others. I wish I could blame them for any errors, but unfortunately must take full responsibility.

[1] With government subsidies to a private or to a government firm, price can be set equal to marginal cost, allowing the first-best output solution; without subsidies, price can be driven to average cost, giving the second-best (Ramsey) solution.

[2] Several technologies may in the future offer economical alternatives to today's local telephone network. Cable companies are experimenting with upgrading their lines to deliver telephone service. Wireless technologies now used mainly for mobile communications might be used for ordinary telephone service as well if costs fall sufficiently. Fiberoptic lines, now used mainly by companies that specialize in providing access to long-distance carriers, could be extended to homes and businesses. Mobile telephone service from planned low-orbit satellites could eventually provide also basic local service. Similarly, large-scale competition to cable companies in delivering video services may come from various sources such as satellites, wireless land-based technologies, or telephone companies upgrading their networks.

[3] The 1996 Act states: "No State or local statute or regulation, or ... legal requirement, may prohibit or have the effect of prohibiting the ability of any entity to provide any interstate or intrastate telecommunications service." The Act defines "telecommunications" as "transmission between and among points specified by the user, of information of the user's choosing, without change in the form or content of the information as sent and received." At a minimum, this definition includes phone service.

[4] The original rationale was to protect cable companies from alleged unfair competition from phone companies (that controlled certain key facilities such as poles and conduits). Many observers, however, believe that protecting the cable companies was no longer a serious problem by the mid-1980s, once cable companies had wired most of the country. A possibly unintended effect was also to prevent cable companies from offering phone service. To offer phone service the cable company would have to become a common carrier, and the 1984 Cable Act prohibited provision of cable service by phone companies or any other common carrier.

[5] Operating as a common carrier, option (b), entails substantial regulatory obligations. A cable system, option (d), enjoys more flexibility, for example, as a private system it is not required to carry all others' programming on a nondiscriminatory basis. But it is subject to local franchise requirements (and rate regulation). An open video system, option (c), must offer nondiscriminatory access at "just and reasonable" terms to unaffiliated programmers and (when there is excess demand for channel capacity) may select the programming on no more than one third of channel capacity (substantially less than can a cable system), leaving the rest for independents. In exchange, it is exempt from regulation under Title II (governing common carriers) and from local franchising requirements, and qualifies for lighter regulation under Title VI.

[6] Section 310(b)(3) of the 1934 Communications Act limits foreign *direct* ownership interest in a radio license to 20%. Section 310(b)(4) limits *indirect* ownership to 25% but delegates discretion to the FCC to raise this limit up to 100%. As a practical matter, indirect ownership is a good

substitute, as it entails a simple transaction establishing a holding company. What the FCC could not waive were other limitations in Section 310(b) on the number of foreign officers and directors in companies that directly or indirectly hold radio licenses. An obscure provision in the 1996 Act (tucked in a section called "Elimination of unnecessary Commission (FCC) regulations and functions...") ends these limitations.

Meanwhile, in November 1995 the FCC issued rules on authorizing entry by foreign telecom carriers into U.S. markets, setting out conditions under which it would waive the 25% indirect ownership limit for radio licenses, and limits on investments in wireline facilities governed by section 214 of the same Communications Act. The rules apply to foreign carriers that possess market power (in practice, most foreign carriers); the FCC adopts a permissive policy towards investments by carriers that lack market power. In essence, the FCC will consider whether U.S. telecom carriers have effective opportunities to compete in the foreign applicant's home market. The new rules provide a clearer and more explicit entry standard to replace the FCC's previous case-by-case determination and they stress competitive factors as opposed to arbitrary criteria. Moreover, the rules also take into account foreign trade considerations, and the FCC has pledged that its policies will dovetail the Administration's position in trade negotiations. Thus, the U.S. will be able to implement any commitment to open up its telecom market to foreign trade and investment.

[7] Spectrum for government use is managed by the National Telecommunications and Information Administration (NTIA), a branch of the Commerce Department.

[8] Bidders are often interested not in a single license but in suitable blocks of licenses, which makes the values of licenses interdependent. For example, aggregating licenses over adjoining regions allows a PCS device to use the same spectrum frequency over a wider area and makes boundary coordination easier, features which can make licenses for certain spectrum bands complements. However, they also can be substitutes, because a bidder can reconfigure its proposed network to use a different set of frequencies. Designing auction rules to help bidders cope with this interdependence in the values of licenses can both promote economic efficiency and bring in greater auction revenue.

[9] The Senate bill would have: (a) extended the FCC's authority to conduct auctions through fiscal 1999; (b) with a few exceptions, *required* the FCC to use auctions for all spectrum bands sought for exclusive use by new applicants (not renewals), and not just for bands earmarked for certain uses as until now (the additional exceptions are for spectrum awarded free to public-safety radio services and to TV stations for digital broadcasting); (c) allowed private users to take over frequencies used by government users in exchange for compensating government users for their costs of moving to different frequencies.

[10] Some incentives could remain to blow the whistle since a broadcaster that provides important news may gain audience share and advertising revenue at the expense of rivals--in the same way that incentives exist to cheat on a cartel. But this incentive may not be strong enough when ownership concentration is moderate or high and when the stakes from the giveaway (as in spectrum) are high. Moreover, one would need a critical mass of media voices making the same point in order for it to be heard above all the normal "background noise" and resonate with the public.

[11] "Basic" service typically includes the signals (delivered over cable) of local TV broadcasting stations and public, education and government channels. "Expanded basic," also known as "cable programming" includes all other channels (such as ESPN), that are not offered on an individual channel basis. "Premium" programs include channels priced on an individual basis (such as HBO or Showtime), pay-per-view events, and in the future may include "interactive-on-demand" services, such as video-on-demand. "Premium" programs are not regulated. The problem of regulating firms that operate in markets with differing intensities of regulation and can manipulate

their mix of activities to evade regulation is paramount when discussing the issue of Bell entry into long-distance phone service.

[12] The Act includes other exceptions to the anti-buyout ban and also allows the FCC to waive the ban in so-called "failing firm" situations.

[13] Revenues from cellular and other phone service added about $16 billion, with some of the cellular revenues going to LECs.

[14] The so-called VIII(c) test in the 1982 consent decree allows a Bell to enter a new market (only) if there is no substantial possibility that the Bell or an affiliate could use monopoly power to impede competition in the market it seeks to enter. In principle, this could be demonstrated even without the emergence of local competition (e.g., if improved regulation could somehow prevent abuse of Bell market power in the entered market). But to argue against the "carrot" argument on the grounds that it is not included in the VIII(c) test is to miss the point that much of the pressure for the new legislation came from the Bells in order to vacate the 1982 decree. And if the decree is to be superseded by legislation, then we can start with a clean slate. In that case, invoking the carrot argument makes good economic sense, because it will be very difficult to obtain Bell cooperation with entrants in order to foster local competition if the Bells have no incentive to extend such cooperation.

[15] For a summary of these allegations and allegations of access discrimination against long-distance carriers, discussed shortly, see Judge Greene's opinion rejecting AT&T's request to dismiss the Justice Department's antitrust case (United States v. American Tel. & Tel. Co., 1981). See also Brock (1994).

[16] As explained in more detail shortly, cross-subsidies and discrimination inefficiently harm both consumers and competitors: (a) consumers pay higher prices and consume too little of a service; (b) consumers are denied the variety and innovation that competitors would offer; (c) potentially more efficient or innovative competitors are denied profit opportunities. Effect (a) can occur in regulated local service due to cost shifting and supply favoritism; or in unregulated long-distance service, because the preempted entrants may be more efficient, and because even equally efficient entrants could permit a regulatory yardstick hence tighter regulation of the monopolist in long distance.

[17] The monopolist then gains from access discrimination--using technological control of the bottleneck to disadvantage rivals in the unregulated markets, thereby allowing its unregulated affiliates to raise their prices. For example, under the 1996 Act the Bells' access fees to IXCs would remain regulated (until enough local competition develops), but the Bells could offer long-distance service to final customers at unregulated rates. The Bells then have strong incentives to discriminate against IXCs.

[18] Laffont and Tirole (1996) show that the Ramsey solution can be induced by choosing price weights that are proportional to the quantities that would be sold in the different markets in the Ramsey equilibrium, and choosing the cap to yield only zero profit. The focus of their paper is to justify including in the global price cap also the access prices charged by the local bottleneck monopolist (e.g., local phone company) to unregulated rivals (e.g., long-distance companies), as opposed to regulating access prices using other schemes such as the "efficient components pricing rule." In the U.S., the main issue has been that access pricing is regulated but the monopolist might operate in unregulated markets, creating the incentive for discrimination. The logic of their argument applies straightforwardly also to this case.

[19] Some go so far as to call the IXCs a cartel, and note that the "cartel's" job is made easier by FCC rules that required AT&T to file publicly with the FCC any changes in tariffs 45 days in advance, creating scope for price leadership. The obvious reply to this is to abandon such FCC

rules. Indeed, the FCC has recently reduced this 45 day requirement, albeit on the different grounds that AT&T is no longer a "dominant" carrier in long-distance service. Crandall and Waverman (1995) argue that long-distance prices are not perfectly competitive, but certainly not at cartel levels either.

[20] In 1985, approximately 50% more minutes were logged on calls from the U.S. than into the U.S., but this differential has increased steadily, and in 1993 exceeded 100%, suggesting an increasing gap between foreign and U.S. long-distance prices. Conceivably such an increasing gap might be due to lowering of access fees from LECs to IXCs mandated by the FCC, not to lower IXC margins brought about by increased competition. However, there is evidence that the decline in long-distance rates, at least within the U.S., has outpaced the decline in access fees.

[21] Allowing IXCs to offer (resold) local service while barring the Bells from long distance may seem unfair; but the fundamental asymmetry must be remembered--the Bells control the local network and therefore might threaten competition in long distance if allowed to enter. There is no similar danger of IXCs threatening local competition.

[22] It is possible to dilute the common interests of the different units, for example, by requiring that subsidiary managers be rewarded only based on the performance of their units, not of the company as a whole. However, to eliminate incentives for discrimination one would have to eliminate any commonality of interest (including through personnel rotation or central oversight), in which case it is difficult to see how any economies of scope between the various services would be preserved. That is, what would be the advantage of such regulatory quasi-separation within a firm over complete structural separation?

[23] Revenues of competitive access providers are included in the total access fees. Undoubtedly the absence of local competition is due in large part to past legal barriers, which the 1996 Act repeals. But entrants have also faced technological problems. Providing phone service over cable lines is still in experimental stages. Wireless may eventually offer alternatives to the wireline local loop to a customer's premises, but currently is higher cost, less secure, and lower quality than wireline connections.

[24] A good conceptual test for when sufficient local competition exists for competition alone (without regulatory intervention) to prevent Bell discrimination against IXCs would be to ask regulators if they feel confident enough in the strength of local competition that they would be willing to deregulate local rates. Under this test, a Bell entry into long distance would likely be a good few years away.

[25] A similar approach was proposed to judge Greene, who oversees the AT&T decree, by the Justice Department, AT&T and Ameritech (a Bell company). The so-called Ameritech proposal involved a trial in two of Ameritech's local service, Chicago and Grand Rapids. Ameritech would offer to rivals interconnection and access to unbundled pieces of its local network. In exchange, it could offer long-distance service in these markets, but only after proving there is some actual local competition, and only under DOJ's ongoing supervision to make sure it continues cooperating by opening up its local network to AT&T's entry. The Ameritech proposal is similar to the Act's, but likely contained stronger safeguards against Bell misbehavior.

[26] The meaning of "predominantly facilities-based" competitor is unclear. Precisely what facilities should the competitor be required to possess, as opposed to leasing from the local Bell? One criterion would be to focus on those facilities in which discrimination by the Bell against IXCs would be more difficult to prevent and insist that the competitor own such facilities, while counting on regulation to prevent Bell discrimination in access to the other facilities. For example, if discrimination in switching functions were more difficult to prevent than discrimination in access

to loops, the test for a "predominantly facilities-based" entrant would require an entrant to possess its own switching facilities--so as to provide competitive alternatives to the Bell in that function.

27 In an earlier version of the House bill, before being allowed to offer long-distance service in its region a Bell company had to face a local competitor that was "*comparable* (to the Bell) in price, features, and *scope*." Such a requirement of strong local competition would have reduced the need for regulatory and antitrust discretion in approving Bell entry. However, this "comparable scope" language was dropped after furious lobbying by the Bells, presumably recognizing that it will take a long time for such a competitor to emerge.

28 Similar sentiments have recently been expressed by the main telephone regulator in the U.K., Oftel director-general Don Cruickshank (Lewis, 1995).

29 Both resale and partial-facilities entry can be viewed as unbundling the LEC's integrated functions: resale involves the unbundling of only the complete package of LEC network functions from the marketing and other customer-related functions. Partial facilities entry involves unbundling certain of the network facilities or functions. But "unbundling" is sometimes used to describe only partial facilities entry, as distinct from resale.

30 Consumers may pay lower prices for two reasons. (1) Even entrants that are no more efficient than the LEC in providing the entered services (e.g., marketing functions in the case of resellers, and trunk lines in the case of some partial-facilities entrants) can help to cut prices, since regulation alone is unlikely to constrain the LEC's pricing down to cost. (2) Entrants may be more efficient than the LEC in their entered segments, resulting in cost cutting and price cuts to consumers. In (1), social gains arise due to the expansion in output. In (2) they arise for this reason and because of lower costs.

31 Concrete advantages of one-stop-shopping can include: (a) assuring customers that there will be "no finger-pointing" between different providers when something goes wrong (recall that resellers assume complete responsibility for satisfying their customers); and (b) offering simplified pricing structures e.g., providing quantity discounts and guaranteed prices on all of a customer's calls without having to distinguish whether calls are intraLATA or interLATA. Simplified pricing is said to be important to customers, a claim consistent with the surge in advertising of such plans by wireless telephone providers ("one price anytime, anywhere"). Entrants also might bring other innovative services, such as call forwarding, voice mail, call waiting, etc. It is difficult to anticipate in advance what such new services might be, but that is the beauty of competition: it spurs firms to be creative and innovative in responding to perceived customer demands long before these are recognized by economists or other outsiders.

32 The requirements imposed on all incumbent LECs are very similar to those imposed on Bell companies as a precondition for being allowed to offer long-distance service originating in their service regions. The implementation process also is similar: parties are encouraged to reach interconnection agreements through voluntary negotiations, but either party may petition the State PUC to intervene at any time, and may seek binding arbitration during the period 135 days to 160 days after the incumbent LEC received the interconnection request. The State PUC must resolve the issue in a manner consistent with the FCC rules within 9 months of the original interconnection request to the incumbent LEC. If necessary, the PUC must impose a solution.

33 The Act opted for the former: it requires incumbents to offer their services to resellers at wholesale rates, defined as retail rates less the costs avoided by incumbents.

34 Recall that in 1994 LECs' revenues from wireline phone service, including about $28 billion in access fees paid by IXCs, were about $100 billion; IXC revenues net of access fees were about $35 billion.

[35] "Cross-subsidization" was used earlier in the paper to describe attempts by a regulated monopolist to evade cost-based regulation by misattributing costs of its unregulated business to the regulated business, where the higher "costs" can be used to argue for a higher price. Here the term refers to the underpricing of some services financed by overpricing others, but at the initiative of regulators rather than the firm, seeking to finance certain social goals or favor certain groups.

[36] Fixed and common costs create economies of *scope* between the different markets, but need not create a natural monopoly if the fixed costs are small relative to market demand and thus create only small economies of *scale*. In such a case competition is feasible, and the allocation of common costs does not require regulatory intervention. For example, large economies of scope exist in producing wool and mutton, as both share the common cost of sheep. But if there are no economies of scale in raising sheep, there will be many sheep producers, the wool and mutton markets will be competitive, and the resulting prices will allow producers to just cover the common cost of raising sheep.

[37] Selective assistance to some groups may be necessary if there are large asymmetries in incomes. However, one can question the magnitudes of positive externalities from phone ownership, and the implicit assumption that many consumers would choose not to subscribe to phone service if forced to pay competitive market rates. Finally, as discussed later, even if many consumers would drop out, it is probably more effective to assist such consumers directly than by cross-subsidizing entire service classes (e.g., basic phone service to all).

[38] Efficiency could be in a narrow Ramsey sense, or might involve additional considerations such as income redistribution; Ramsey pricing places equal weights on a dollar to each consumer.

[39] The upshot could be higher prices in the remaining markets in order to cover the fixed costs; or worse, that these fixed costs simply can no longer be covered, the monopoly becomes non-viable and some markets go unserved. In both cases, consumption in the previously cross-subsidized markets is inefficiently reduced relative to the regulated situation.

[40] According to critics the current pricing structure may not reflect any broad public policy consensus, but regulatory capture: some groups might be more effective than others at influencing the regulatory process, e.g., all residential customers are sometimes subsidized by business customers. Or there may once have been a legitimate reason for a particular pricing pattern but conditions have changed, yet the pricing pattern survives due to political inertia.

[41] The introduction of fixed monthly fees after the breakup of AT&T caused some to worry that marginal consumers would drop their telephone subscription. But the fall in long-distance rates worked in the opposite direction, to increase the value of phone ownership. From 1984 to 1993 the number of long-distance calling minutes rose by 130% (local minutes rose only 17%), and telephone ownership in the U.S. increased from 91% of all households to 94% (14.6 million subscribers were added).

[42] Texas recently gave rural phone companies three years of added protection against entry and Wyoming ten (Pearl, 1995).

[43] In designating consumers "in all regions" the Act adds low-income consumers regardless of their location to the list of consumers to whom access should be provided.

References:

BRENNAN, T.J. (1987), "Why Regulated Firms Should Be Kept Out of Unregulated Markets: Understanding the Divestiture in *U.S. v. AT&T*," *Antitrust Bulletin*, vol. 32.

BRENNAN, T.J. (1995), "Is the Theory Behind *U.S. v. AT&T* Applicable Today?" forthcoming, *Antitrust Bulletin*, vol. 40, no. 3.

BROCK, G.W. (1994), *Telecommunication Policy for the Information Age: From Monopoly to Competition*, Harvard University Press.

BROWN, S.J. and D.S. SIBLEY (1994), *The Theory of Public Utility Pricing*, Cambridge University Press.

CAULEY, L. (1995), "Calls Waiting: Rivals Are Hung Up On Baby Bells' Control Over Local Markets," *The Wall Street Journal*, Tuesday, October 24, 1995, A1.

CONGRESSIONAL QUARTERLY (1995), "The Fine Print: A Side-by-Side Comparison of the House and Senate Telecommunications Bills," Special Report, September 23, 1995, vol. 53, Supplement to No. 37.

CONGRESSIONAL RECORD (1996), "Conference Report on S. 652, Telecommunications Act of 1996," and "Joint Explanatory Statement of the Committee of Conference," vol. 142, January 1996, H1078-H1135.

CRAMTON, P.C. (1995), "The PCS Spectrum Auctions: An Early Assessment," mimeo, Department of Economics, University of Maryland.

CRANDALL, R. and L. WAVERMAN (1995), *Talk Is Cheap: The Promise of Regulatory_ Reform in North American Telecommunications*. The Brookings Institution, Washington DC.

ECONOMIC REPORT OF THE PRESIDENT, February 1996, U.S. Government Printing Office, Washington DC.

GORE, A. (1995), "Remarks Delivered by Vice President Gore to the Networked Economy Conference," Washington DC, September 12, 1995.

KWEREL, E.R., and J.R. WILLIAMS (1993), "Moving toward a Market for Spectrum," *Regulation: Cato Review of Business & Government*, 1993 No. 2, 53-62.

LAFFONT, J.-J., and J. TIROLE (1993), *A Theory of Incentives in Procurement and Regulation*. MIT Press, 1993.

LAFFONT, J.-J. and J. TIROLE (1996), "Creating Competition Through Interconnection: Theory and Practice," manuscript, Institut d'Economie Industrielle, Toulouse.

LEWIS, W. (1995), "Telecoms Regulator to Face Public," *Financial Times*, Weekend, November 18/19, 1995, page 4.

PEARL, D. (1995), "Good Connections: Despite Deregulation, Rural Phone Subsidies Are Likely to Survive," *The Wall Street Journal*, Thursday, November 30, 1995, A1.

SCHWARTZ, M. (1986a), "The Nature and Scope of Contestability Theory," *Oxford Economic Papers*, vol. 38 Supplement, November 1986, 37-57.
SCHWARTZ, M. (1986b), "The Perverse Effects of the Robinson-Patman Act," *Antitrust Bulletin*, vol. 31, no. 3, Fall 1986.
United States v. American Telephone and Telegraph Company, 524 Federal Supplement 1336 (1981).

G. Regulation and Tariff Policies in the Energy and Telecommunications Sectors in the Transition Countries: The Case of the Czech Republic

*Zdenêk Hrubý**

1. Introduction

The transition in the Czech Republic is characterised by wide-spread liberalisation and privatisation. These processes bring a necessity for new regulation structures. This paper describes briefly the situation in the energy and telecommunications sectors, and the state of regulation with an emphasis on tariffs. Each sector has its own particular problems and requires careful regulatory reforms. The paper explores these problems and comments on recent regulatory changes. The paper commences with section 2 describing the features of the energy sector. Section 3 concentrates on price regulation and the design of tariffs in the energy sector. Section 4 analyses the characteristics of the telecommunications sector. Section 5 describes the regulatory environment and the development of tariffs in telecommunications. Section 6 provides a brief comparison of the regulatory positions in the selected CEC and EU countries. Section 7 presents conclusions.

2. Features of the Energy Sector

The state and development of the Czech economy and of the energy sector have their roots deep in the past. Due to its tight political and economic links with the Soviet Union during the whole period after the Second World War, the Czech economy was oriented to an extensive development of high energy and materials intensive sectors. This called for providing an ever growing supply of energy and raw materials. This resulted in, among other things, low energy and raw material input prices and therefore a reduced ability to generate revenue to cover the import of high-quality energy sources and to carry out modernisation of the energy economy.

The unbalanced orientation of Czech foreign trade towards the former Soviet Union excluded the possibility of diversifying across imported energy sources. The result of the Czech economy's separation from the world markets was, among others, the impossibility to utilise in the Sixties the effects of world crude oil prices for starting up a dynamic development, as was the practice in industrially developed countries. The economy is still predominantly dependent on pipelines from Russia, but the new government has decided to build an alternative pipeline from Germany. The agreement with the German government is complete and the pipeline is expected to be operational by the end of 1995.

The energy economy was for a long time oriented on the utilisation of indigenous coal. The extremely high share of coal, particularly of the low-quality lignite, was negatively affecting (and is still negatively affecting) the efficiency of the combustion processes and, in particular, is placing an unbearable load on the environment. In 1993 conventional power plants generated 76% of power, nuclear plants 21% and hydro plants 3%. Power and heat generation, based mainly on the combustion of lignite, was generating crisis situations due to the high rate of noxious emissions and of solid combustion products, mainly in the Northern parts of the country. The large extent of the use of lignite and briquettes for local space heating in built-up areas with low houses was resulting in a bad emissions situation in densely populated districts.

Adverse economic conditions, together with distorted prices and state subsidies to fuel and energy prices, did not stimulate energy conservation either by producers or consumers. Lack of finance, particularly in convertible currency, which was partly due to the low efficiency of Czech industry, had a negative feedback effect in the low standard of manufacturing technologies and high energy consumption, low technical level of energy appliances, distribution systems, thermal insulation, measuring and control systems, etc. All this again led to higher demand for energy.

Until 1989, the Czech gross domestic product was increasing at an average yearly rate of 2.2% and the consumption of power was increasing at the same rate. The consumption of primary energy sources was growing until 1987, when it reached its peak, at a rate of 0.8%. Since 1987, yearly consumption of primary energy sources has decreased, but at a lower rate than GDP. The same applies to the consumption of electric power since 1989. In the years 1990 to 1992, GDP has diminished by 23% at constant prices, but at the same time final consumption of fuels and energy in industry, civil construction, transport and agriculture decreased by only 13.1%. A decrease of 14.8% was noticed at the same time in commerce, services and households. Due to this fact, the energy intensity of GDP rose by 9% in the year 1992 against 1990. Solid fuels maintain their dominant position in the consumption of primary energy resources.

Imports of fuel and energy were almost exclusively from the former Soviet Union (hard coal for power generation was also imported from Poland). The lion's share of imports was liquid fuels, especially natural gas. The foreign trade balance as concerns fuel and energy has been considerably passive during the whole time. The share of imports of primary energy sources in 1992 was 28%.

After 1990 the state enterprises in the energy sector were transformed into joint stock companies. These companies were partly privatised in the first wave of voucher privatisation, and are being privatised at present in the second wave. Privatisation of these enterprises is performed on the basis of privatisation plans. The situation on the stock market reflects the positions of each enterprise. While share prices of hard coal mines are very low, because of high costs, the shares of CEZ, state protected virtual monopolist producer of electric power (that owns public power plants which participated in the total generation by almost 89%), are traded at one of the highest prices on the domestic stock market. The position of

CEZ is significant economically because of its very ambitious investment plan for 1990s, probably the biggest investment plan in the Czech Republic for this period.

3. Price Regulation and Tariffs in the Energy Sector

In the Czech Republic there does not exist any regulation of the kind familiar in developed Western countries. The overall characteristic is direct state control. There are no established regulatory bodies or institutions, and no exact regulatory rules have been implemented. The regulatory environment is rather weak and ineffective. The only law that applies to the energy sector is the general "Price Law" of 1990. The law stipulates that the seller of a product cannot obtain excess profit. The Law specifically states that:

- The seller cannot use his position to gain undue benefits by selling for a negotiated price which would include unjustifiable costs or unfair profit.
- Economically justifiable are costs which correspond to the valid wage regulation, technologically necessary direct and indirect costs, and distribution costs; the determination of economically justifiable costs is found in similar activities, allowing for differences associated with particular product.
- Fair profit is, over the long term, the usual profit associated with a particular business activity on the domestic market, and which is justifiable with a view towards economically justifiable costs, with allowances made for the quality of the product, the usual risks associated with production or distribution, and development of the product.

On the other hand, there does not exist any genuine competition within the sector, except in the coal industry. As we will describe below, this situation has specific reasons (see section 5).

The actual prices are proposed by the Ministry of Industry and Trade, and approved by the Ministry of Finance. The tariffs are set by the government, which also authorises new investments for the entire energy sector. It means that all the pricing policy and important financing decisions are totally in the hands of the central state administration. Therefore, prices are distorted. However, the energy lobby is one of the strongest in the Czech Republic and only companies like CEZ (Ceske energetické závody, one of the largest companies in the Czech Republic with 15.5 thousand employees, public power plants operator) has the necessary background and professional knowledge to provide information for government's energy policy.

Development of fuel and energy prices in the Czech Republic until 1989 reflected the approach of the centrally planned economy, striving for general accessibility of energy. It reflected a policy of so called cheap energy. Prices of energy carriers were defined and regulated almost independently of prices on world markets and, particularly with respect to prices of energy for households, did not take into consideration the objective value of energy sources.

Although prices of fuels and energy (for production sphere) were gradually adapted toward their real economic value, this process was never completed. On the other hand, the prices development for households was characterised by an artificial stagnation and single augmentations, most recently having been realised in 1979, which level practically persisted until June 1, 1991. During the whole period, however, the economic value of fuel and energy was growing, prices in the production sphere were gradually increased, thus creating an ever growing difference between prices for households and in the production sphere. The latter were higher and the difference was covered from the state budget in a form of a negative turnover tax, or by a price subsidy. In the case of natural gas and electric power there were also cross-subsidies within the framework of tariffs for organisations and for households.

In the framework of the transformation of centrally planned to market economies, after 1989, interventions into fuels and energy prices were made in the framework of general price liberalisation to make them more real and consecutively to liberalise them fully. These prices modifications are subject to government prices regulation policy, being for a transition period set either as maximum prices or objectively regulated (calculations including only economically justified costs and reasonable profit).

On December 1, 1990 a single prices increase was realised for fuels and energy for production sectors (coking hard coal by 44%, coke by 47%, hard steam coal by 50%, natural gas by 65%, electric power by 53%, centrally supplied heat by 74%), this, however, could not cure fully the distortions in the level and pricing of energy resources. Household prices, where the situation was worst, were subject to a single augmentation as from May 1, 1991, which, however, due to the necessity to provide social acceptability, was not able to provide an economic level of fuels and energy prices. Therefore, prices of all kinds of fuels and energy are continuing to be changed in 1992 and 1993 in the framework of the general prices liberalisation started on January 1, 1991.

Due to the more rapid liberalisation of prices for the production sphere and to the general rise of materials and other input prices for energy processes, the cross subsidies in tariff systems for gas and electric power persist, losses on prices for households being compensated by higher profits on prices for production consumption, as well as by direct subsidies from the state budget in case of centrally supplied heat for households. The price of the latter is defined as the maximum of the difference between the price invoiced by the heat supplier and that charged to households, and is paid directly to the owner of the heated facility.

In addition to price liberalisation measures, the growth of prices is also subject to other influences. It is for instance price increases of imported energy resources, influenced by the transition to world prices after the collapse of COMECON, and by devaluation of the Czech crown against western currencies, and the growth of profit rate in fuels and energy prices, which should compensate for the lack of capital caused by the absence of foreign capital, and by the restrictive credit policy of government and other influences.

The influence of world prices has had its greatest impact on the development of liquid fuels and natural gas, and partly also on the prices of hard coal and coke, which are influenced by prices of Polish and Australian hard coal. Therefore, the most important changes happened in prices of liquid fuels, where the increase for instance of motor fuels reached, in 1993, 130-180% as against 1989.

Although energy prices for households are subsidised, there has been a large rise in their prices, particularly due the fact that they had been originally very low, as for example in the case of centrally supplied heat, its price rising by 471% over the period 1989-1993. Prices of solid fuels for households grew in 1993 as against 1989 in average by 170% (hard coal) to 280% (lignite). As from January 1, 1993, value added tax was introduced, as a part of the measures taken to unify and simplify pricing methods and in view of standard approaches applied in advanced countries with market economies. VAT for energy sources is 5%, while the normal level is 23%. Another tax item, applied mainly in liquid fuels, is the consumer tax.

Even though direct state control over energy prices is implemented, both the price level and tariff structure show a kind of paradox. Let us look at the electric power subsector. Concerning the price level, it is not adequate simply to compare the prices in the Czech Republic with developed western countries, because of the large difference between the exchange rate (ER) and purchasing power parity (PPP) exchange rate. The relationship PPP/ER is Exchange Rate Deviation Index (ERDI). Comparison of electricity prices for industry in Germany, France and the Czech Republic is shown in Table G1.

Table G1: Comparison of Electricity Prices for Industry in Germany, France and the Czech Republic (1993, DEM per MWh, rounded)

Country	Price (ER)	Price (PPP)	Price (ERDI)
Germany	150	150	1.0
France	90	100	1.1
Czech Republic	85	230	2.7

Note: Czech ERDI is a gross estimation.

These prices show the paradox that the price in the Czech Republic (computed over ER) is just a little under that of France and (computed over PPP) high above that of Germany. Current prices are targeted to cover the extensive investment program to meet the financial requirements of the sector and are at the upper limit of acceptability. High average price levels mean that the tariffs reflect marginal (LRMC) rather than average costs (as usual) with expensive consequences for the Czech economy. The difference between average costs and LRMC is considerable. Payments for useful consumption of electricity (consumption * price) in the country are (relative to GDP) more than twice the usual level in OECD countries.

The paradox of the tariff structure is that wholesale price of electricity is 1500 CZK/MWh (rounded 50USD), however, residential retail price is only 800-900 CZK/MWh (rounded 26-30USD). This is a result of the demand for financial resources in the power sector and the lack of resources in households. Cross subsidisation of consumer groups is a common practise in most countries, but not on this scale. The effort spent by the government in correcting this paradox has been negligible, since such steps would be highly unpopular and the utility company is mostly interested in increasing power sales.

4. Telecommunications Sector Features

Czech telecommunications, particularly the PSTN, are a typical example of the state of telecommunications in CE countries. The current situation, can be characterised above all by systematic long term under-investment. The reason for the under-estimation of the value of developing telecommunications in the past was systematic under-estimation of the role of the whole services sector in centrally planned economies. The reason for it was the comparatively low weight of services in output structure of CE countries. This fact results from the preferences of state administration for pushing the primary and secondary sectors and from a system of measuring national income with little weight to consumption services.

Despite this systematic pattern of under-investment, an under-developed technological base and under-provision of telecommunication services, these services served in the past as a stable source of revenues in both national and foreign currencies (from international services) for their owner, namely the state. These revenues were swallowed up by redistribution within the central planning system.

In the period between 1980 and 1990 the number of telephone mainlines in the CSFR increased from 1,616,000 to 2,334,000 (according to OECD statistics), i.e. from 10.6 mainlines per 100 inhabitants to 14.9 mainlines per 100 inhabitants. Within this period the CSFR slipped from the first position among Central and Eastern European countries to the third position, being below Bulgaria and former Yugoslavia. The waiting list grew during the same period from 196,000 at the beginning of the 1980s to 334,000 in the late the 1980s, and 606,000 in 1993.

The situation in the Czech Republic, as a part of CSFR, was in stark contrast to the advanced OECD countries. Among the 24 OECD countries, the penetration rate of mainlines was 43.3 per 100 inhabitants in 1990, which was almost three times the rate in the CSFR. At the same time, compared with some other small industrial countries, one finds even sharper distinctions (e.g. Sweden 67, Switzerland 59). Among the other CE countries the CSFR was, as was noted above, in third place according to this measurement. Moreover, the country has traditionally had advantageous conditions for the development of telecommunications relative to other CE countries (i.e. industrialised country, density of population, large industrial agglomerations, relatively little damage to the network during the Second World War, and geographic position in the centre of Europe). Qualitative characteristics of network performance (number of faults, fault maintenance, etc.) depend heavily on the technical state of the network and installed equipment; and in comparison with average OECD data these characteristics were substantially worse in the CSFR.

Up to autumn of 1990 the following structure of the field of posts and telecommunication services could be found:

- A central organ on the federal level, i.e. the Federal Ministry of Posts and Telecommunications, covering the whole territory of CSFR. It elaborated PTT policies and provided for their implementation. It co-ordinated unification of rules for activities and operations of posts, telecommunications and the postal distribution of periodical press. It regulated and executed state administration and state inspection in the PTT domain.
- Two state enterprises called "Administration of the Posts and Telecommunications" with spheres of activities in the Czech Republic and the Slovak Republic and two state enterprises called "Administration of Radiocommunications" for wireless communications with the same spheres of activities.

With effect from January 1, 1993, i.e. after the splitting of Czechoslovakia and the establishment of the Czech Republic, the Ministry of Economy assumed all responsibilities with the sphere of telecommunications activities in the Czech Republic.

The state administration section was supplemented with a regulatory section and both together represent the Czech Telecommunication Office, CTO. The CTO has been arranged into the Ministry of Economy.

In order to achieve the necessary improvements in telecommunication services and to enable the privatisation of telecommunications, a completely new operating environment had to be established. On January 1, 1993, the former state enterprise "Administration of Post and Telecommunications Praha" was split into two new state enterprises - the Czech Post and the SPT Telecom. Together with the increasing of postal tariffs in October 1992 it ended the cross-subsidy by telecommunications of the operating loss on the postal service. As part of the on-going privatisation programme, certain former commercial subsidiaries, construction and assembly activities and some production activities were hived off and have been privatised already. To privatise the main activities, the Telecommunications Privatisation Project is realised in the second wave of voucher privatisation and has been done step by step. As the first step the state joint-stock companies were created with effect from January 1, 1994. In the second wave of the voucher privatisation, 70% of stocks will remain in state ownership, 26% will be for individual shareholders, 3% for restitution, 1% for the National Foundation. The process will continue through finding foreign participation. The shares of SPT Telecom, the major carrier of the infrastructural telecommunication network and the monopoly telephone service operator with the major role in some other telecommunication services, are highly interesting in the second wave of the voucher privatisation. They are expected to be one of the highest prices on the domestic stock market, which position is similar to the position of CEZ, the major and practically monopoly power plants operator.

To develop Czech telecommunications as quickly as possible, SPT must realise in the 1990s an economically and technically very ambitious Telecommunication Development Project (framework is Digital Overlay

Network). The main aim is to modernise the network and increase penetration rate twofold.

There are several other minor operators in the Czech Republic's telecommunications. Czech Radiocommunications carries on broadcasting networks and radiocommunication transmission networks. EuroTel and Radiokontakt Operator are joint ventures concerned with mobile services. As far as the public telecommunication services are concerned the following licences have been issued already:

- Audiotex information services (9 licences)
- Voice Mail services (1 licence)
- coin and card boxes (11 licences)
- data services (5 licences).

Permits for private telecommunication services were issued as follows:

- mobile radiotelephone networks (7)
- data networks (5)
- VSAT networks (17).

5. Regulation and Tariffs in the Telecommunications Sector

The regulatory environment and processes in the Czech Republic is more advanced in telecommunications than in the energy sector, but is still underdeveloped compared with western standards. However, competition in main areas of telecommunication services -public fixed voice telephone- does not exist. The former Federal Ministry of PTT had prepared, and in April 1992 the Federal Parliament approved, the Amended Telecommunications Act. In accord with the EC Green Paper the Act enables:

- demonopolisation and liberalisation of telecommunications;
- the starting of privatisation of telecommunications and;
- the separation of the state administration from economical activities.

The Act separated regulatory and operator functions and established the possibility to open telecommunication services for competition. Herewith the Act opened possibilities for private sector access to telecommunication services provided for the public in the market environment.

By decision of the Czech Parliament, the validity of the Act was accepted for the Czech Republic, too. The regulation of telecommunications has been divided between two Ministries - the Ministry of Economy regulates the admission of contractors to telecommunications. Tariffs for inland services are prepared by the Ministry of Economy in negotiations with the operator, but the final regulation decision comes under the responsibility of the Ministry of Finance. For the time being the range of regulation is determined only generally. A new proposal of an amendment to the telecommunication law is in an approval process specifying the system of the state regulation more clearly. According to the proposal we can say already that the Department of telecommunications of the Ministry of Economy:

- Determines and manages the state policy of telecommunications development;
- determines basic conditions for the system of the state regulation in telecommunications and rules of the licence policy;
- represents the state in international organisations;
- determines the principles for frequency spectrum management;
- determines national numbering plans for telecommunication services;
- issues licences for public networks and services;
- approves tariffs for international telecommunication services;
- reviews decisions issued by the CTO.

Czech Telecommunication Office, CTO:

- Issues licences for private networks and services;
- proposes rules of tariff policy;
- proposes tariffs for international telecommunication services;

- approves technical characteristics of telecommunication equipment and issues certificates of ability;
- manages the frequency spectrum;
- executes the state administration and the state inspection of telecommunications.

Concerning tariff regulation, one of the basic reforming steps in the economic transformation from the central planning to market oriented mechanism was the price liberalisation proclaimed at the beginning of 1992. But it did not concern certain price groups on which the state regulation was put, e.g. public transport, energy and fuel prices, rents and prices for postal and basic public telecommunication services, e.g. telephone, telex, telegraphy and public facsimile services and leased lines.

New tariffs for the basic public telecommunication services have been prepared by the operator. After discussion in the ministry responsible for telecommunications (formerly it was the Federal Ministry of PTT, at present it is the Ministry of Economy) the proposal of tariffs has been moved to the Ministry of Finance. Having reached agreement among ministers, the Ministry of Finance has issued the final decision.

Public mobile service and public data service via the public packet switched data network as well as the other services, typically all Value Added Services, are not price regulated.

Until October 1992 the Czech Republic had telecommunication tariffs valid from 1979. With effect from January 1, 1993, VAT of 5% has been introduced into telecommunication tariffs. And finally in September 1993 the new tariffs were approved. But only partial changes were made. Tariff structure remained the same. The price of one impulse was increased by 50%, reduced rates of 50% were introduced for direct dialling long distance telephone services and timing for local calls was agreed, too. This new tariff structure is still far from being a cost-based structure and prices are still well below OECD average. See Table G2.

Table G2: Telephone Traffic Domestic PSTN Tariffs in the CR Compared with OECD 1992 Average, USD

	Tariff CR	OECD average
Charge		
Installation charge:		
Business	166.6	197.79
Residential	66.6	186.63
Rental charges:		
Main line	1.66	16.18 Business
Dual line	1.00	
Group line	0.50	
Handset monthly	0.20	
Local calls per minute	0.008	0.03
Public pay phone per 6 minutes	0.011	
Long distance calls:		
I zone peak	0.050	0.08
I zone off peak	0.025	
II zone peak	0.100	0.20
II zone off peak	0.050	
III zone peak	0.150	0.32
III zone off peak	0.075	

Note: Tariffs excluding tax.

6. Situation and Approaches in CEC and EU

Table G3: Main Regulation Features in PSTN in the Czech Republic, Poland and Hungary, 1993

Feature	CR	Poland	Hungary
Separation of regulation and operation	yes	yes	yes
Regulator	Ministries, Agency as a body of the ministry	Ministry	Ministry
Organisational separation of post, broadcasting and Telecom services	yes	yes	yes
Privatisation of PTT envisaged	yes	no	yes
Amount of foreign direct investment	0	0	30% of equity
Number of players in the national network	1	1	1
Length of license	unlimited	unlimited	20 years
Number of players in the international network	1	1	1
Length of license	unlimited	unlimited	unlimited
Rules against cross-subsidisation from monopoly telcom service level	no	no	minister's ad hoc decision
Regulation of end user prices	yes	yes, not used	yes
Liberalised CPE markets	yes	yes	yes

Table G4: Main Regulation Features in PSTN in Selected EU Countries, 1993

Feature	France	Germany	Greece	Italy	Portugal	U.K.
Regulator	direct govt.	direct govt.	govt. assisted by NTC	direct govt.	govt. and ICP	OFTEL
Regulation of end users prices	direct govt.	direct govt.	committee of 5 ministries	ministry	govt. approval of ICP and operators proposals	price-cap (RPI-x) set by OFTEL
Private operators	no	no	no	no	no	yes
Effective comp. on CPE market	yes	yes	no	nearly no	yes	yes
Foreign investment	no	no	no	no	no	possible
Compliance with EU directives	reasonable	reasonable	bad	bad	reasonable	always ahead of Brussels

As tables G3 and G4 show, the regulatory positions in the Czech Republic and Hungary are not very different in telecommunications. The only important difference between the CR and Hungary is entry of a foreign partner into Telecoms. However, in this case is the Czech Republic only delayed compared with Hungary. The basic approach is the same and entry of foreign partner to the amount of approximately 30% of equity is expected in the near future. The same is valid for the entry of local players, which is expected starting November 1994.

For Poland, the situation is slightly different in privatisation, but this process is under discussion.

In the electricity sector, the development is retarded. Looking at the EU countries, the U.K. is a rare exception among EU countries with its radical approach to liberalisation and connected new regulatory framework. The other countries, although formally complying with EU directives, are moving along this path more slowly. In practically all countries the reform attempts are accompanied by strong opposition from interest groups - utilities, investors, employees, see e.g. the case of Germany.

Not surprisingly, the original ideas from the beginning of transition in the CEC are being replaced by more conservative and slower approach to the liberalisation. The examples from EU countries encourage the opposition against reforms and even strengthen political (return of postcommunist governments in CEC) and interest (foreign investors) pressures.

7. Conclusions

Both sectors, energy and telecommunications, are typical examples of very important public utilities vital for the development of the Czech economy. The present state indicates the necessity of restructuring, modernisation and improvement in both cases. Moreover, both sectors show practically monopolistic structures in the main and most sensible areas. The main tool for the negotiations of state administration with these sectors should be regulation. Nevertheless, the present state of regulation and approaches to it are different in both cases. The question is why?

The strategy of the Czech Republic's transition places macroeconomic stabilisation as a top priority. To achieve this aim, the government prefers consensual development without a radical restructuring of industrial production in the first years of transition, leaving major changes to new private owners. Radical restructuring connected with re-orientation after the destruction of the former COMECON market would add turbulence to the economy in the first, economically and politically most difficult years of reforms.

One of the most important features of this strategy is a very slow and conservative approach to the bankruptcy of inefficient enterprises to avoid a general collaps due to the mutual insolvency of many of them. Delayed bankruptcies and restructuring have kept unemployment in the CR at a low level (3.5%) and this is the lowest among all the transition countries, and even among developed countries. Nevertheless, this situation brings a negative influence on labour productivity and the overall efficiency of Czech producers. Most of them are unable to compete in markets in terms of quality and technological development, and therefore compete through labour costs. This fact lead the government to the implementation of wage regulation. To make the living standards of the majority of the population acceptable after decrease of real incomes, the government has decided to continue its direct administrative control over some key, and for households, sensitive items, i.e. rents, energy prices and public transport. These items are administratively controlled rather than regulated in framework of market conditions.

The existence of many households on low incomes means that it is not possible to raise the above mentioned prices without the introduction of an adequate safety net.

There is also another side to the energy sector problems, which results, among other things, (e.g. environmental damages) in a lack of regulation in this sector. Looking at the unchanged heavy structure of Czech industry (in principle, the country is over-industrialised) with high energy input demand, it is clear, that energy sector is very important. Moreover, the heavy industry exporters (e.g. metallurgy) stand behind the Czech export success. The fact lacks of appropriate domestic primary energy sources means that the energy sector is a key problem for the government, also with consequences for most problematic, highly populated industrial regions with concentrated restructuring and environmental problems (e.g. Northern Bohemia).

The government made the energy sector a top priority by defining and approving policy made in 1992. Top priority for state guarantees for energy programme loans have also been extended by international financial institutions (e.g. IBRD). This contrasts with telecommunications where the policy is not yet defined and approved. The international credits were delayed because state guarantees for energy credits have been preferred.

In the above described situation, the government is not interested to implement any regular and independent, market oriented regulatory mechanism in an effort fully to control the sector. Inherent in the current situation is the interest of CEZ to establish clear regulatory policy in the country. The lack thereof decreases its credit worthiness. The recent report by Standard & Poor on CEZ gave it a BBB Minus rating. The report said that the minus rating is necessary primarily because of the lack of clear regulatory policy in the country. On the other hand, it is in the company's interest to be able to influence this potentially new regulatory body. Judging by the possible staffing of this body, CEZ may be able to exert great influence upon this new body.

The above mentioned situation in the energy sector does not mean that the development of telecommunications is not a priority in the Czech Republic. Underdeveloped telecommunications create barriers for dynamic development in the economy, particularly for the fast growing service sector and newly established small- and medium-sized enterprises. The first democratic government defined very early its interest in the fast development of telecommunications so as to encourage economic growth.

However, telecommunications have been a long-term profitable sector, even in the period of under-investment under central planning. Tariffs were very low and unchanged for long time, thus there was a room for improvement. Moreover, telecommunication tariffs are not so socially weighted and sensible by the population compared with rents and energy prices. Telecommunications do not have a serious impact on the environment or on employment in problematic regions.

While the energy sector brings quite unique problems into the economy of the Czech Republic, telecommunications development questions are more standard, and responsible bodies look to precedence in the past and present development of Western countries (e.g. in Great Britain).

This results in a more sophisticated approach by the government to telecommunications, by granting more autonomy to the sector in the establishment of regulatory rules and institutions (establishing the Czech Telecommunication Office, and amending the Telecommunication Law).

However, some of the original ideas have not been fulfilled. In an effort to enable fast development of basic services in the whole territory of the CR through concentration of profits and capacities into one operator, which also increases the interest of foreign partners and the credibility by financial institutions, no competition is allowed. Originally, some level of competition was discussed (duopoly). As proclaimed currently by the Minister of Economy, this approach should be used for a longer period of dynamic development of telecommunications, i.e. approximately for the 1990s.

The Czech Telecommunication Office status is far from the idea of the OFTEL model, as an independent body with wide responsibilities. Now, the CTO is a body of the Ministry of Economy and not all regulatory responsibilities are placed there. Those examples are close to the government's effort to build an effective centralised state administration rather than system of independent bodies with distributed responsibilities and authorities. (Such effort is not proclaimed, but realised).

The regulation of telecommunications in the near future will make use of some standard formulae to control price adjustments. The rules will accommodate assessment of productivity gains, rebalancing constraints and revenue requirements for funding new investment. The telecom operator will nevertheless have considerable autonomy, particularly regarding investment. The new approach by the government will use for the first time in the CR an explicit formula to govern price adjustments. It is expected that the approach will lead to both rate of return and price-capping regulation.

* I should like to thank Michaela Erbenová, and Frantiõek Hesoun.

References:

DOYLE, C., HRUBÝ, Z., MÜLLER, J. (1993), Czech Telecommunications and Transition: How fast and what form should deregulation take?, Cambridge: Department of Applied Economics, University of Cambridge.

DOYLE, C., HRUBÝ, Z. (1992), Czech Telecommunications: Bureaucratic Bungling Amid Constitutional Change, Stuttgart: Paper presented at ITS European Regional Meeting.

HESOUN, F. (1994), Liberalisation, Regulation and Telecommunication Market in the Czech Republic, Praha: Mimeo.

HRUBÝ, Z. (1994), Regulation and Tariff Policies in the Energy and Telecommunications Sectors: The Case of the Czech Republic, London: Paper presented to the conference Regulation of Utilities in Economies in Transition, London Business School.

HRUBÝ, Z., Mobile Telecommunications: The Czech Republic, in: Schenk, K.-E., ed. (1994), Mobile Telecommunications in Western and Central-Eastern Europe: Integrative and Institutional Dimensions, Norwood, Massachusetts: Artech House. Forthcoming.

HRUBÝ, Z. (1992), Telecommunications Investment and Tariff Policy in the CSFR, Paris: OECD.

HRUBÝ, Z. (1992), Telekomunikacní infrastruktura v CSFR, Praha: Prognostický ústav CSAV.

LEITE, A. P. N., LUCENA, D., SEABRA, C. (1994), The Design of Regulatory Institutions in the European Union, Lisbon: Paper presented to the Utilities Regulation Network Conference.

MITCHELL, B. M., VOGELSANG, I. (1991), Telecommunications Pricing: Theory and Practice, New York: Cambridge University Press.

SUK, J. (1994), Price Regulation in the Energy Sector: The Czech Republic, Lisbon: Paper presented to Utilities Regulation Network Meeting.

Telecommunications of the Czech Republic: Information (1994), Praha: Ministry of Economy of the Czech Republic.

Telecommunications: Pressures and Policies for Change (1993), Paris: OECD.

H. The Process of Systemic Transformation and Reforms in the Energy Sector: The Regulatory Issues of the Reforms

Andrzej Szablewski

1. Introduction

Two issues are the subject of this paper. The first one regards the relation between the process of implementation of the Economic Transformation Program and reforms in the entire energy sector. The second one is related to this issue of the reform in the energy sector which comes to have a key meaning for further development of those reforms, especially in reference to the power sector. That is a creation of the new legal and regulatory framework for the energy sector supplementing a new model of operation of the energy sector.

2. Systemic Transformation and Reforms in the Energy Industry

There is a need resulting both from the theoretical and practical points of view, to review and analyze the characteristics of the relationship between the process of transformation of the whole economy and directions of reforms in the energy subsector, with a particular look at grid subsectors. In these subsectors the market oriented direction of reforms causes the biggest number of controversies, doubts and at last objections, which are raised more and more strongly as the process of reforms is proceeding. There is nothing strange in the fact, that a discussion on the forms of reform of the power sector is very lively and heated, despite the fact that a similar type of reform was recommended for the Polish gas sector.

The majority of the analysis presents an opinion, that reforms of the Polish energy sector constitute an integral part of the transformation process. Moreover, in the first period of that process, when a concept of reform was originated, the dominant opinion was that deep changes in principles of operation and in the way of organization of all energy subsectors are not only an important element of the process of passage from a centrally planned economy into a market economy, but even a necessary condition for the success of those processes due to the important position of the energy sector for the Polish economy. There is no doubt that it had to influence not only the process of studies on the concept of reforms but also the fact that, finally the most radical program of changes was adopted, if the point of reference for its evaluation is taken as a standard model of operation of the energy sector in the majority of countries of western Europe.

In the beginning of the nineties in Poland arose a dilemma which is faced by all countries with economies in transition. The choice, if the final goal of reforms in such a sector as energy is a traditional model of operation, is between vertically integrated monopoly and deep public intervention or reforms which should somehow be related to proceeding or suggested changes aimed at activation of market mechanisms and limitation and reduction of the scope and rationalization of methods of public regulation, was made in favor of the second solution.

It is worth paying attention to another very important and very rarely raised problem, that of the relationship between the process of transformation and reform in the energy sector and those in the other network industries. There is a common opinion that reforming these areas is a very difficult task due to the necessity of breaking through various barriers which are:

- After the long period of communist rule, strong monopolistic organizational structures and related interests had developed. There were strong feelings of identification with the traditional model and a lack of understanding that the departure from that model is a necessary reaction to changing external conditions as well as changing conditions in the sector itself. It is necessary to point out, that in the past the development of vertically integrated monopolistic structures created very convenient conditions for the fast development of the power sector in Poland, it increased the stability of the system and made possible a reduction in costs followed by a reduction in prices of electric energy.

- A strategic meaning of those subsectors for the state, economy and society, which in a natural way indicates the need for care in undertaking reformatory actions and in that way makes easier a defence of the *status quo* since radical reforms may have serious, often very difficult to repair, consequences in macroeconomic and microeconomic dimensions as well as in social and economic areas.

Considering the above, it is stressed that the chances of the radical reforms weakening the monopolistic character of the power sector to the benefit of the market mechanisms are small and they are usually connected with substantial changes in economic policy towards the liberalization and privatization of the economy. That is what happened in such states as Great Britain, Chile and Argentina. In that context, the process of transformation of the economy and especially its first stage can give an advantage in implementing such reforms. Decisions made in that period will probably have long-term consequences, because as time passes by the implementation of transformation will be more and more difficult.

An example of this rule can be very well seen in Poland in the process of reforming the gas and power sectors. The breakthrough in the restructuring of the energy sector achieved in the beginning of the process of transformation, activated the process of reforms which, despite several delays in the implementation of particular stages, lasts until today and there is a serious chance that the Polish power system will be reset into a new system of operation, based on competitive and regulated market impulses. The lack of such a radical change in the case of the gas sector, reduced the scope of reforms in that sector, first of all slowing them down and in that way putting a huge question mark over the possibilities of their implementation.

3. The Model of an Independent Regulation

I share the opinion of those who, at the present moment, attach substantial meaning with respect to future implementation of reform in the energy sector to the introduction of new law, especially in the energy sector where reforms are most advanced and where new laws will make it possible to establish a new system of regulation. The lack of appropriate legal solutions and of an institutional basis for the operation of the regulatory system may have serious consequences for the power sector. Two factors linked with each other play a key role here.

The first one is the lack of an appropriate legal and institutional framework, making impossible the introduction of new mechanisms of operation and financing of the energy sector. The use of such mechanisms would allow the full utilization of the possibilities created by the reforms performed so far. Main issues are controversially judged to be the breakthrough restructuring of the power sector and a concept of the market in that sector.

The second factor is an extension of the discussion stage and the resulting lack of decision-making, at least regarding the framework of the regulatory system. This creates the threat of a return to the traditional economic and organizational solutions, which still have a lot of followers, and hence the wasting of all the reform results so far achieved. The problem is that the existing highly irrational mechanism of operation of the energy sector is treated by many people not as a result of the delays in the process of transformation but as the permanent consequence of the already implemented reforms, thus making it easier to attack the entire concept of reform.

3.1 The Concept of Reforms in the Energy Sector

The processes of the regulatory system associated with these controversies and resistance to proposed solutions have to be judged in the broader context of goals adopted by the government at the beginning of the nineties program of deep reforms of the entire energy sector. The purpose of the reformatory actions is to achieve the strategic goals of the state in providing for energy security of the state, improvement of efficiency in production and utilization of energy, and reduction of the harmful impact of the energy processes on the environment through the:

- introduction into the energy sector of economic mechanisms based on competitive market elements and also the regulated market; a mixed character of the market solution, and preservation of state intervention constitutes evidence of the acknowledgement of the specific status of the energy sector (a strategic role of the sector and existence of elements of natural monopoly);
- attraction of private capital, including foreign capital due to the size of the demand for such capital in the sector.

Achievement of the goals mentioned above, requires the establishment of conditions allowing for activation of efficiency pressures and entry of private capital. The concept of reform assumes the necessity of undertaking several actions concerning the ownership and organizational structure as well as the legal and economic basis for the operation of the energy sector. The program of necessary actions covers, besides those mentioned above, restructuring linked with the commercialization of energy enterprises which is also very difficult due to social and political reasons planned for several years operation of granting to prices of energy an economical character. Finally it is planned to create a new legal framework and institutional solutions to maintain state supervision over the energy sector in the new way. The key element of the proposed changes is ultimately the establishment of a new, to some extent independent from the influence of current policy, system of economic regulation of energy enterprises. For that purpose it is proposed to establish a separate Energy Regulatory Authority equipped with appropriate powers.

3.2 The Justification for Independent Regulation

The proposal to establish an independent system of regulation results from the new approach to the position and role of the state in the energy sector. A new formula anticipates that the state, maintaining its necessary powers with respect to the energy security of the state, environment protection and protection from the consequences of the existence of monopolies, shall act also to the benefit of the development of good functioning of market mechanisms in the energy sector. To that purpose it was decided to guarantee in the energy law the principle of the institutional separation of three functions which the state is to perform in the energy sector. Those functions are: creation of energy policy, ownership of the energy enterprises, and regulation. Three factors support the necessity of the separation of the regulatory function, which also covers the proposal to establish an independent regulatory institution:

- Merging in one state body the functions of the regulator and conductor of energy policy containing social, political or macro economic elements, must lead to conflict between different ratios and in consequence difficulties in compromising with respect to prices, which has to mean interference in the operation of the energy market. In that case it is necessary to point out that the idea of an independent regulation does not (necessarily) mean withdrawal of the state from the control of prices (the regulatory institution is also an organ of the state) but also does not exclude a possibility of conducting an active social policy for example in the form of openly subsidizing the poorer households through for example free bonds to pay for power bills. The goal is to release prices of energy from the function of a social policy instrument. We have to add that prices are a very harmful instrument of social aid, because they are not a targeted instrument. The beneficiaries of this aid are both poor and rich. An important benefit of independent regulation is an increase of transparency in the actions of the state, disclosure of the costs of moving away from economics based solutions, and due to this, increase of the responsibility for taken decisions.

- The institution of the independent regulator allows equal treatment of state and private enterprises, which is hard to ensure when the same organ fulfills ownership and regulatory functions.
- The establishment of an independent regulator introduces an element of stability in the process of regulation, which is very important in the case of the energy sector.

In the assumptions of the reforms the independence of the regulatory organ is understood as providing some decision autonomy, albeit within the framework given by government policy and energy law. This organ is to enable the smooth operation of the energy market. In practical terms, this means first of all promotion of competition (within the scope defined by energy policy), mainly by facilitating access to the market and withdrawal from the regulation of prices in those areas where competition exists, and secondly supervision of the proper operation of the regulated market within areas where monopoly exists. In that second case the activity of the regulatory institution shall be based on two principles.

The regulator does not take in any case the position of an institution managing energy enterprises through intervention in the decision-making process or decisions themselves. The task of the regulator is assessment of whether energy enterprises are obeying the determined rules, in particular with respect to the formulation of the level and structure of prices.

Regulatory actions shall aim to maintain the proper balance between the interests of energy enterprises (with no distinction as to their ownership) and the interests of customers.

4. Controversies Surrounding the Model of Independent Regulation

From the beginning of work on the energy law, the principle of an institutional separation of the regulatory function constituted the key element of the new law. But transmission of this principle into legal language faced a lot of difficulties. Those difficulties were not only connected with the necessity of finding an appropriate legal formula for relations between a state institution, energy enterprises and customers, but first of all were of a political nature. Due to the above mentioned strategic importance of the energy sector, the state is to maintain a strong authority. So any change in that respect, especially when connected with a reduction in the position of the state in some areas of its activity, must create controversies. Proposals for the establishment of a regulatory body separated from the structures of the state and enjoying a independent status were charged with the danger of a loss of the real control of the state over the energy sector and transfer of those powers to a not very precisely defined organ.

Another factor causing doubts as to the concept of an independent regulatory system is the fact that such a system varies from solutions applied in the majority of European countries and it is based on the experiences of such countries as United States where this system has a very long history, and Great Britain, where the establishment of independent regulatory organs was an integral element of the reforms implemented in network industries. Furthermore it is often said that the development of regulatory structures threatens a basic goal of the reforms - liberalization of the energy sector. In that context it is often said that existence of a regulatory authority doubles the position of the Anti-monopoly Office.

There is no doubt, that the most important problem connected with the above mentioned list of doubts regarding the system of independent regulation is a problem of the status of the regulatory authority. The goal is to find a solution which on the one hand provides for the necessary independence of the authority, meaning isolation of the authority from the pressure and influence of the current political interests, and on the other hand guarantees that this institution acts in a reasonable manner. In the first stage of work on the energy law it was stated that the independence of the regulator shall be provided for through the:

- joint management of the regulator by a board of five;
- the principle that board members are appointed by the Prime Minister for a defined term of the office;
- a special procedure for dismissal of a board member (only in the cases determined by the act - serious sickness, gross misconduct, crime committed or resignation);
- the rotation of members, making it impossible to replace the entire board at once.

On the other hand, reasonable activity of the regulatory organ was to be guaranteed through the:

- principle of transparency given the duty to publish decisions in a special bulletin;
- requirement to prepare an annual report;
- possibility to appeal decisions.

As to the argument that the soundness of the proposed regulatory solution should be doubted because it is novel, and little known in Europe, it is worth making two comments.

In market economies, especially in those where a traditional model of the energy sector is dominant, there exist various systems of regulation. So there is no single standard of regulation, which can be used as a point of reference for the establishment of a regulatory system in Poland;

The necessity to create a system of independent, economic regulation is closely related to market oriented type of reforms in the network based industries; it seems necessary to add that there is a foreseen possibility to introduce a kind of economic regulation even at the international level as a consequence of developments within the European Union.

As to the opinion that the establishment of a regulatory authority threatens to increase the scope and degree of regulation in a way that is contrary to the principle of promotion of competition, it necessary to say that in grid-based industries regulation and competition do not have to exist as alternative solutions. Moreover, it is pointed out that in those industries development of competition can take place only upon the existence of regulation and development of the competitive market may increase the need for regulation. This of course does not mean that there is no danger of overregulation and the resulting negative consequences for economic efficiency of energy enterprises. That is shown in discussions on the weaknesses of the American model of regulation and a discussion on the British style regulation.

5. Regulatory Authority and Antimonopoly Office

It seems necessary to discuss the wider charge that due to the existence of the Anti-monopoly Office the establishment of a new energy regulatory authority is evidence of an unjustified increase in economic bureaucracy. Moreover, the supporters of this kind of argument may call forth examples of some countries, where anti-monopoly institutions play the role of a regulator of network industries.

There are three arguments which can be called upon in order to justify the establishment of a regulatory institution for such areas where, due to the fact of existence of a joint network, elements of a natural monopoly are very much visible and where there is no doubt of the need of independent regulation. Those arguments are closely related to the particular concept of regulation recommended as a part of the market oriented reforms. This concept says that the regulator is to take actions enabling development of competition, and in those places where there is no economic justification for competition, introduce a regulatory regime. This regime is to substitute for the regime of the market competition, which does not exist in some areas of network industries.

This kind of regulation is a permanent process, which means that it requires regulatory institutions to engage directly in the process of awarding licenses, day to day supervision over the enterprises operating in a monopolistic position, and making regulatory decisions.

Regulation is a complicated process of searching for a balance on the one hand between the possible and demanded scope of the competitive market and necessary scope of monopoly, and on the other hand between the interests of suppliers and customers of energy where this balance cannot be achieved as a result of a pure market mechanism. Regulation of enterprises in a monopolistic position is there because the abuse of monopoly power is not only harmful to consumers of energy but also to other energy enterprises, whose entry into the market is conditional on, for example, access to the transmission or distribution grid. This indicates that regulation requires a deep and very specialized knowledge, covering technical and organizational aspects of energy processes as well as a very broad spectrum of the economic and legal basis for operation of enterprises.

In reference to the above, it can be proved that effective regulation is a function of the appropriate number of specialized bodies, and their potential for permanent regulation. This kind of potential rarely exists in the anti-monopoly institutions, because they are the institutions within whose jurisdiction remains the entire economy. Moreover, the purpose of the anti-monopoly institutions is to enforce the anti-monopoly law, which means that their activity is concentrated on intervention in situations where this law was violated. It means that elements of permanent regulation are not typical for such institutions and these institutions are not prepared to monitor the operation of enterprises on a regular basis, which is so much required in the case of economic regulation.

However, this is not to conclude that the required solution is to grant to the regulator the status of the Antimonopoly Office in the area of energy and hence to deprive the Antimonopoly Office of the right to intervene on the basis of the Act on Counteracting Monopolistic Practices. There should be some kind of "partition of work" indicated first of all by the scope of the competitive market in particular subsectors. In those areas where such a market exists, the Antimonopoly Office should fulfill its statutory functions - protecting competition and counteracting monopolistic practices, counting in that respect on the cooperation of the regulator.

I. Restructuring and Regulatory Reform in the Polish Energy Sector: an Assessment

George Yarrow

1. Introduction

Interest in liberalization in the electricity and gas segments of the energy sector is a relatively recent phenomenon in Europe, and, although such interest developed somewhat earlier in the United States, the legal and political constraints that inhibit major structural change in North America have meant that developments there have proceeded at a relatively slow pace.[1] In Europe, Britain is the country that has implemented the most radical policy reforms and that has taken the greatest strides towards establishing competitive markets. Even in Britain, however, the reform process dates only from the early 1980s and the first pieces of liberalizing legislation -- the Oil and Gas (Enterprise) Act 1982 and the Energy Act 1983, which provided for third-party access (TPA) in gas and electricity respectively -- were relatively modest steps that did not have major impacts on the ways in which energy markets functioned.[2] Major regulatory reforms were only finally implemented during and after the privatizations of the gas industry (1986) and the electricity industry (1990 and 1991).[3]

Elsewhere in the European Union progress has been much more limited. The European Commission has put forward proposals for limited third-party access to gas and electricity grids for very large purchasers of electricity and has embarked upon a course of trying to use existing Articles of the Community Treaty -- such as those concerned with the free circulation of goods and with the protection of competition -- to open up the networks of incumbent utilities.[4] As occurred in Britain in the early 1980s, however, even relatively modest TPA proposals are being resisted by major utilities, although that resistance varies from country to country (for example, German electric utilities appear now to be much less opposed to liberalization than, say, is EdF in France).

As a consequence of this general situation it is not possible to evaluate the reform process in the energy sector in Poland against the background of a history of long-established, liberalized markets either in Europe or in other OECD countries. For the most part, either traditional policy approaches based upon monopolistic utilities continue to prevail or else reforms have been in place for only a relatively short time.[5] And, in respect of reforms, it not currently possible to point to a specific set of policies which represents an international consensus as to how energy sectors should best be regulated or deregulated.

Notwithstanding these points, what is very clear from current developments is that there is a worldwide trend toward more liberalized energy markets, with different countries proceeding at different paces and adopting slightly different mixtures of privatization, industrial restructuring, and regulatory reform.[6] Polish reforms should, therefore, be evaluated against this trend, not against the particular positions that have currently been reached by other countries. This is a particularly important point in the light of the time it can take for major reforms to be implemented: "models" of energy policy in other countries that are already considered untenable will be even more outdated by the time that re-orientation of the Polish energy sector is complete.

The general situation regarding the state of energy sector policies elsewhere in the world raises both difficulties and opportunities for policy-makers in Poland. The difficulties are the obvious ones associated with the lack of an established, universal "model" of energy markets and energy regulation that can straightforwardly be applied to Polish conditions. The opportunities arise from the fact that, since arrangements elsewhere are in a state of flux, Polish policy makers are not faced with the task of simply catching up with economies in, say, Western Europe. Rather, there exist real possibilities for Poland to "overtake" many, more established market economies in terms of the effectiveness of national energy policies. This could be accomplished by first analyzing where the trend of European Community policy-making is taking the energy sectors of the Member States and then proceeding to move Polish policy quickly in the same direction.

The possibility of overtaking arises from the relatively fluid political situation that is associated with the process of systemic transformation in Poland and in other countries of Central and Eastern Europe. This means that radical changes that could take years or even decades to achieve elsewhere (because of entrenched interest groups and the protections afforded by established systems of property rights and established divisions of political responsibility) might feasibly be implemented much more quickly in transforming economies. To give just one example, policies calling for major industrial restructuring are much easier to implement within a context of centralized state ownership than under decentralized private ownership, since private ownership properly calls for respect to be given to existing and established rights. A good illustration of the relevance of ownership in this context is the comparison between the progress made in providing for third-party access to electricity transmission grids in Britain and the USA. In Britain a centralized public organization was quickly converted into a privately owned grid with a single set of rules for access throughout the country. In the USA the fragmented ownership of transmission assets means that it is much more difficult to ensure that an electricity generator wishing to supply a customer in another part of the country will be able to gain access to the relevant transmission facilities (which may be split among several different owners).

Of course systemic transformation gives only the *opportunity* for rapid progress: there is no guarantee that such progress will occur. Whether opportunities are realised or not will depend upon the particular circumstances prevailing within a given country. Relevant factors here include the skills and energy of those responsible for the conduct of public policy and the political will of the government to drive reform forward.

In at least some respects the current position in Poland is not dissimilar to that in the UK before the major reform programme of the 1980s and early 1990s. At the beginning of the 1980s Britain was well behind the United States in the development of competition in electricity and gas markets (energy was supplied by state-owned monopolies whose managers were responsible to the Minister of Energy). Once it had been decided that a reform path was to be taken, however, circumstances in Britain were much more conducive to speedy progress than they were in the US. First, the structure of parliamentary government was such that it was possible to develop, introduce and implement new legislation relatively rapidly. Second, as already noted, the fact that the industries were state-owned meant that existing and established property rights did not hold back the process. Third, there was a strong political will for reform at the highest level of government, backed up by the necessary skills to develop and implement path-breaking legislation.

The first and second of these factors meant that opportunities for rapid progress existed in Britain at the relevant time. However, that the existence of such opportunities was not in itself sufficient to induce major changes in policy is demonstrated by the fact that major regulatory reforms were *not* implemented in the 1960s and 1970s, both decades in which there was considerable criticism both of the performance of the state-owned enterprise sector and of the way in which it was regulated. Although options for reform were evaluated from time to time during this earlier period, the missing factor was commitment to the reform process at the higher political levels. The importance of such commitment is therefore abundantly clear from the historical record.

The result of the conjunction of conditions that existed in Britain in the 1980s is that the UK has now "overtaken" the United States, and also the other major countries of Western Europe, in the move toward liberalized energy markets.[7] Whereas in the 1980s British economists and administrators tended to look to the United States for "models" of how liberalization might proceed, nowadays American colleagues are showing a growing interest in the British "model", seeking to learn from it in the development of their own policies, particularly in electricity. For example, the influence of the British experience can be seen in a number of the proposals for reform that have been put forward and developed by the California State Commission. If, therefore, the opportunities are seized, it may not be too many years before Poland is receiving western experts not to advise on the conduct of energy policy but rather to learn from, and even perhaps to copy, parts of the Polish experience.

2. The Energy Sector in Poland

The Polish energy sector at the beginning of the 1990s was characterised by gross inefficiencies in energy production and energy use, with an energy-to-GDP ratio almost twice the average of OECD countries. Efficiency was negatively affected by factors such as artificially low energy prices, soft budget constraints and weak incentives, outdated technologies and poor management. As a consequence, during the period of transition to a market economy there is considerable scope for improving performance, leading to benefits in terms not only of increased GDP but also of reductions in environmental damage.

In 1990 the energy sector accounted for about 10% of GDP, 12% of total investment and about 5% of the working population. Coal, which accounted for around 10% of export earnings in the same year, and imports of oil and natural gas were together responsible for about 15% of total import expenditures. All gas imports and nearly all oil imports have traditionally come from the Soviet Union. Although around a third of gas consumed in Poland is produced there, the heavy reliance on a single source of imports means that Poland faces a gas supplier with strong market power;[8] a characteristic of the market position which, at least for the moment, constrains future policy developments.

Table I1: Shares of Fuel Consumption

	1990	1993
Hard coal	63.5 %	62.2 %
Brown coal	13.3 %	13.9 %
Petroleum	12.5 %	13.4 %
Natural gas	9.5 %	9.0 %
Others	1.3 %	1.5 %

Coal is the predominant fuel in the Polish economy, accounting in 1990 for over: 95% of primary domestic supply, 75% of total fuel consumption (see Table I1) and 40% of final energy consumption. Over 90% of electricity generation is from solid fuels and there is no nuclear power. Since, in general, it is easier to promote competition in electric systems that rely largely on fossil-fuel technologies rather than nuclear technologies, the medium-term prospects for market liberalization appear to be more favourable in electricity than they are in gas.

In the recent period the organization of the Polish electricity and gas industries has been based upon public ownership, central planning and high degrees of both horizontal and vertical integration. In these respects the organizational pattern resembles not only other industrial sectors of the domestic economy but also the electricity and gas systems of some other countries (e.g. France and pre-privatization Britain).

Before 1982, however, in the gas industry there existed various local and regional (state-owned) enterprises. In that year the industry was re-organized by establishing the Polish Oil and Gas Company (PGNG) as a single legal entity responsible to the Ministry of Industry. PGNG encompassed all activities in the natural gas industry from wellhead to burner tip, including activities such as exploration, equipment manufacturing, installation, repairs and pipeline construction. The company functioned with twenty two main operating entities, none of which was a separate legal entity or produced separate accounts. The degree of internal integration was therefore very high.

The arrangements were somewhat different in electricity. The Power and Brown Coal Board (WEWB) -- which was liquidated in 1990 and replaced by the Polish Power Grid Company (PPGC) -- did not function according to the usual centralized model. Rather, there were a variety of distinct enterprises operating in the industry, including:

- 28 enterprises responsible for the various forms of electricity generation (see Table I2 for an outline of the production structure);
- 33 electricity distribution enterprises (also responsible for some smaller generating units embedded in their networks);
- 6 enterprises responsible for manufacturing equipment for power stations;
- 6 enterprises responsible for power station maintenance; and
- 12 enterprises responsible for the maintenance of transmission and distribution systems.

Each enterprise operated as a separate financial centre, preparing its own accounts and paying its own taxes and dividends to the Ministry of Finance (an arrangement that continued after the creation of PPGC). WEWB played an overall co-ordination role and, for example, set the transfer prices at which electricity was purchased by the distribution companies (prices which played a major role in determining how overall industry revenues were divided among the principal enterprises).

Table I2: Electricity Generation Capacity and Production, 1990

Fuel type	Number of stations	Capacity (MW)	Output (MW)	Share of
Lignite	5	9103	52.6	38.6
Hard coal				
• Power	16	13842		
• CHP	34	3881		
Pumped storage	5	1546	2.0	1.5
Hydro	121	459	1.3	1.0
Total (grid supply)	181	28831	128.2	
Autogeneration	221	3168	8.1	5.9

3. Policy during the Initial Phase of Economic Transition

Early political decisions made it clear that Poland wished to follow the worldwide trend toward less monopolised energy markets. Thus, the Government's Letter of Intent addressed to the World Bank in the first phase of the economic transition expressed a desire to move toward a market oriented system in the energy sector based upon:

i. competition among suppliers where such an arrangement was feasible; and
ii. regulation of the core, naturally monopolistic businesses of gas and electricity transport via pipes and wires.

At the same time it was recognised by both the Government and the relevant international agencies that, given problems with government finances, the energy industries would, over time, need to attract private investors. This position implied at least partial privatization of energy supplies, although it did not necessarily imply that existing enterprises would need to be transferred to the private sector. Indeed many energy systems comprise a mixture of privately owned and publicly owned enterprises. In the Norwegian electricity industry, for example, very major changes in competitive conditions and in regulatory structures were made without any accompanying privatizations of enterprises: the bulk of the Norwegian power sector therefore continues to be state owned.

General aims such as the promotion of competition, the effective regulation of monopoly, and the introduction of private capital are, however, much easier to state than to achieve. Implementation requires that very specific local conditions be taken into account, and here the inheritance from the socialist period has been unhelpful in a number of respects. One immediate set of problems concerned the structures and levels of energy prices, which bore little resemblance to underlying costs.

The "theoretical" response to such a price-structure problem is simple: if prices are out of line with costs there should be an early reform to make prices better reflect costs. In practice, of course, this solution to the problem may not be easy to implement. First, it may not be at all clear what the relevant costs are: one of the failures of central planning is its inability to reveal underlying costs. Britain again provides an example of the potential consequences of such a failure. Prior to privatization, the UK government substantially underestimated the costs of generating electricity from nuclear power stations and simultaneously overestimated the costs of distributing electricity. These errors in cost estimations had at least two embarrassing political consequences:

- nuclear power had to be withdrawn from the privatization programme at a late stage when the magnitudes of the costs gradually began to be revealed, and
- since estimates of likely future distribution costs were used to set the (regulated) level of charges for electricity distribution, the overestimation of costs led to final electricity prices that were significantly higher than was necessary for economic efficiency, particularly for domestic customers.[9]

In this context, it should be noted that regulators of natural monopolies face similar information problems -- albeit generally on a lesser scale -- than central planners. It would, therefore, be naive to believe that this type of issue disappears once a system has been reformed and once its naturally monopolistic parts have been subject to regulation by a new agency. The available evidence on the effectiveness of regulation is not altogether encouraging on this point, and it is worth remembering that "effective regulation" is something that is very, very difficult to achieve. On the other hand, where it is feasible, the introduction of competition is likely to give much better results in respect of:

- Revealing information about costs; and
- making prices reflect costs.

A second problem with price reform -- and probably the most important problem in practice -- arises from the political consequences of rapid shifts in price levels and price structures. Such shifts lead to changes in the distribution of income which, when transparent, will tend to generate resistance from those groups who are made worse off by the change.

Long study of the conduct of public policy in general, and of regulatory policies in particular, has consistently shown that these redistributive effects of reforms are of crucial importance in explaining the development of policy.[10] The introduction of competition can only occur if there is a political willingness to accept the associated changes in price levels and structures. Where there is major resistance to the redistributive consequences of new policies, any reforms implemented, even when they appear to involve major changes in existing arrangements, are likely to be only tangential to the central economic issues.

The socialist inheritance meant that, at the outset of the reform process, Poland potentially faced a much larger price adjustment problem in the electricity sector than, say, the countries of Western Europe. While the latters' electricity industries have generally been used to favour one or other group of consumers, on average the levels of implicit subsidies appear to have been more modest than in Central and Eastern Europe. Presumably in response to the sensitivity of distributional issues in the energy sector, the Polish Government decided that a gradual process of energy price adjustment was to be preferred to a more sudden shock.

While the political judgments underlying this decision are fully understandable, it does stand in contrast with the much more rapid lifting of price controls that took place in other sectors of the economy. And, more significantly here, the decision to adjust prices gradually has served to slow down progress toward the aim of a market oriented system in the energy sector based, where possible, on the forces of competition.

4. Industrial Restructuring

The initial Polish approach to industrial restructuring in the energy sector has been based upon an idealised set of principles established by economists for policies toward the electricity and gas industries. In particular, it has been based upon the recognition that the electricity and gas industries can be viewed as comprising a number of different businesses, some of which are naturally monopolistic and others of which are not. Traditionally these different businesses or activities have been bundled together and in the past the two industries have been characterised by high degrees of both horizontal and vertical integration. The general trend in policy worldwide, referred to in the introduction, is now towards lesser degrees of horizontal and vertical integration. However, the extent of the planned reduction in integration can and does vary from country to country. For example, in the vertical dimension the aim (in descending order of the degree of integration) may be:

i. accounting separation, whereby separate accounts have to be produced for each business or activity but otherwise the regulated company can operate in a fully integrated way;

ii. business or structural separation, whereby the businesses or activities have to be managed as distinct concerns but common ownership can be retained (for example, the separate businesses could be distinct subsidiaries of the same company); and

iii. full ownership separation.

In general, of course, (ii) implies (i), and (iii) implies (ii) and (i).

It appears, however, that not all of the different businesses than can exist in the electricity and gas industries were fully recognised in the early stages of the reform process. In particular, policy recognised <u>three</u> vertical stages in the supply chain:

- Production and wholesale supply (extraction of gas and generation of electricity);
- transmission (of gas at high pressure and electricity at high voltage); and
- regional and local distribution.

In fact, there is a fourth, very important stage, <u>retail supply</u> (which is also sometimes referred to as electricity/gas <u>trading</u> or <u>marketing</u>). This fourth stage involves the sale of energy to final customers, including activities such as invoicing and payment collection.

Retail supply is a trading activity: there is no need for the supplier to be either a producer of electricity/gas or to be a transporter of electricity/gas (in the sense of operating a transmission or distribution network). Thus, where there is open access to the relevant distribution and transmission systems, it is possible for suppliers to (a) purchase electricity or gas from generators or offshore gas producers, (b) pay transmission and distribution companies for the use of their transport system, and (c) sell their electricity or gas on to final customers with whom they can negotiate terms and conditions of sale, including prices.

The confusion between distribution and retail supply is probably associated with the fact that, even in highly liberalized markets such as Britain's and Norway's, local and regional distribution companies also act as suppliers of electricity. That is, there has not been an enforced (vertical) separation of ownership of the two activities, only accounting and regulatory separation.[11]

Nevertheless, the distinction between distribution and retail supply is a crucial one. Unlike distribution, retail supply is not naturally monopolistic and it is through the choice of supplier that the final customer is able to exert a direct influence on the market. Without alternative retail suppliers, the industry presents a monopolistic face to its final customers, an outcome that will inevitably affect public perceptions of the nature of the reform process.

4.1 Restructuring in Gas

The initial proposals to restructure the Polish gas industry are a good example of the attempt to turn economic principles into practical policies. The proposals called for the monolithic state-owned industry to be divided into three parts:

- one or two upstream companies;
- one gas transmission company; and
- six regional gas transmission companies.

The scheme of vertical separation here is clear enough, and commercialization of the industry through the creation of new joint-stock companies could be expected to have beneficial effects on the ways in which the industry is managed. However, the importance of such restructuring for competition and regulation is not immediately obvious. First, both transmission and distribution are naturally monopolistic and so the restructuring of these parts of the industry could not in itself be expected to have a major direct impact on competition. Second, given the dependence of Poland on imported supplies of gas, it is far from clear that separation and division of upstream activities would have itself materially affected competitive forces. Much the most important determinant of market conditions upstream is the dominant position held by Russia, and major changes in competitive conditions will depend chiefly upon international developments (eg. de-monopolization in Russia) rather than domestic actions.

The principal policy gains from such a scheme of restructuring would lie less in promotion of competition and more in the facilitation of more effective regulation. Business separation simplifies regulatory monitoring, and the existence of several companies tends to increase information available to regulators and makes yardstick regulation possible. For these benefits to be realised, however, there need to exist regulatory agencies which can make good use of the available information. In the Polish case the institutional framework of regulation has not yet been established. Thus, even though the initial proposals for gas restructuring have not been implemented, it is doubtful if this particular outcome has had much of an impact on economic performance so far.

It is also useful to examine the gas proposals in the light of the actual experience of gas and electricity privatization in Britain where:

- British Gas was privatized as a fully integrated transmission and distribution company with some interests in upstream production (the company owns some gas fields but buys most of its inputs from the major oil companies operating in the North Sea);
- the National Grid Company, which owns the high voltage electricity transmission network, is itself owned by the Regional Electricity Companies which are responsible for local distribution (although the RECs are now contemplating selling the grid).

In neither of these cases was there full separation of transmission and distribution.

It is true that the British Gas privatization was widely criticised because of the lack of associated steps to promote competition, including the lack of prior restructuring, and that it can be argued that Poland should not simply duplicate the mistakes made in Britain. On the other hand, the subsequent record of British Gas shows that full vertical separation is not a pre-condition for developing effective competition in the supply of gas. Separation may help, but competition can be developed without it, and what appears to be of most importance is the existence of a regulatory framework capable of ensuring third-party access to existing pipeline networks.

4.2 Restructuring in Electricity

The restructuring issue is of greater importance in electricity than in gas, largely because of the much greater scope for competition and efficiency gains upstream in electricity generation. In this case the initial Polish proposals called for:

- a separate grid company;
- the consolidation of the thirty three distribution enterprises into roughly half that number of independent companies;
- the consolidation of the twenty eight state-owned generating enterprises into:
  - backwards with lignite mines;
  - nine combined heat and power (CHP) companies; and
  - four or five groupings of hard coal generating stations.

Opposition from distribution companies has prevented the implementation of the proposals for their consolidation, but, as in the gas industry case, little has been lost by this reversal. If it later turns out that there are major efficiency gains to be had from merger then, in a more commercial environment, the companies themselves will be capable of achieving the necessary consolidation. These are decisions that can safely be left for a later day.

Opposition to restructuring in generation is a more serious issue. The initial proposals have been blocked, and there have been pressures from the industry and the Ministry of Industry for a scheme that would have only one or two hard coal companies. In the latter case each of the two companies would be regionally based, compared with the original proposal that groupings should be regionally dispersed in order to prevent the creation of regional monopolies.

The move away from the initial restructuring proposals in the hard coal generating segment of the industry clearly represents a negative development so far as competition is concerned, but, once more, it is not obvious that the loss is a major one. In a liberalized market major sources of competitive pressure would remain even if the hard coal stations were consolidated into two, regionally based companies. First, the costs of high voltage transmission are low enough that the two companies would be able to compete for customers in each other's areas (although the Antimonopoly Office would need to be vigilant against geographic market sharing and other forms of collusive arrangement). Second, hard coal stations account for only a little over half of capacity and there would therefore be scope for competition with lignite and CHP generation. Third, an important (and possibly *the* most important) source of competitive pressure is the threat of new entry, and this would be largely unaffected by the degree of consolidation in the hard coal segment.

It can also be noted that failure to agree on a scheme of re-organization in generation is potentially more favourable to the development of competition than either of the alternatives that have been advocated. As noted in section 2, the *status quo* is characterised by the existence of many electricity generation enterprises, each of which can operate with a significant degree of autonomy. In this respect, therefore, the situation is very different from that in many other countries, where the *status quo* in the absence of reforms is a highly concentrated generation structure. In Poland, for example, it is possible to envisage consolidations that are desirable on efficiency grounds being achieved by a series of voluntary mergers between enterprises, subject only to the normal oversight of the process of concentration by the competition authorities. That is, major mergers would need to be approved by the Antimonopoly Office, which could withhold its consent in cases where the amalgamation could be expected to have damaging effects on competition.

Part of the rationalization for the original decision to form four or more groups of hard coal generating stations was the desire to develop short-term price competition among generators via a pooling mechanism in which each company would bid to get its sets dispatched. Since the operating characteristics of lignite and CHP generation render these technologies relatively inflexible in the short term, competition for dispatch was viewed as being restricted largely to the hard coal stations.

In practice, however, it is probably unwise to rely heavily on hour-by-hour price competition for dispatch. Historically, power pools have been used by utilities to reduce their supply costs rather than to bid for custom. That is, they have been inter-utility markets, not markets in which suppliers and final customers interact. This, of course, is changing -- the encouragement of short-term price competition is one aspect of the global policy trend -- but it remains to be seen how the new forms of electricity pooling actually perform in practice. Game theory suggests that the repeated nature of the bidding competition will facilitate implicit collusion.[12] If prices are bid on a half-hourly basis, say, then 1440 different "competitions" for dispatch will take place in a normal month, and in a year the total will rise to 525,600. Given that the population of firms playing such pricing games will change only slowly, there will be ample opportunities for inter-enterprise signalling through the bidding process. If one firm seeks to acquire extra volume today by bidding lower prices it must take account of the fact that this is likely to lead to an immediate reaction (tomorrow) on the part of competitors who want to regain lost business. The short-period for which the pricing advantage lasts before rivals can retaliate is therefore a severe disincentive to price-cutting. Nor is the disincentive likely to be radically affected by whether each firm faces only one, two or three rivals. It is, therefore, very much open to question whether increasing the number of bidders from two major generating groups to, say, four major groups would in practice actually have very much of an impact on prices.

In contrast, economic analysis suggests that competition via contracts is much less vulnerable to collusion. A firm which attracts a customer on the basis of, say, a three-year contract to supply will know that the customer cannot quickly be regained by its competitors. Thus, the ability of competitors to retaliate is reduced by longer-term contracts (the "retaliation lag" is lengthened), and hence the incentive to set lower prices to gain business is greater. What this suggests is that it is competition via contracts that is the most important form of price competition in the market; and this form of competition will be much less affected by the number of hard coal generators than will day-to-day competition for dispatch.

4.3 Conclusions on Structural Decisions

That there has been some retreat from the initial, radical plans for restructuring of the electricity and gas industries is a matter for concern less because of the likely economic impacts on future energy markets and more because it tends to signal a weakening in political resolve. In particular, it could signal a downgrading of the political priority given to the objective of moving to a market oriented system in the energy sector, based, where feasible, on the forces of competition.

A considerable amount of academic work on the determinants of economic performance in different industries has tended to suggest that the degree of concentration is not the most important of those determinants. For example, large numbers of cross-sectional studies of the relationships between profit margins and various measures of industrial structure indicate that links between concentration and margins are generally rather weak.[13] Nor is there an authoritative body of evidence suggesting that lesser degrees of vertical integration have substantially beneficial effects on performance. Rather, what the evidence seems consistently to indicate is the importance of entry conditions in determining industrial behaviour.

The British evidence from the energy sector does not to date contradict these conclusions from the more general economic literature. The UK government was criticised by many economists for only creating two fossil-fuel (coal and oil) generating companies at the time of privatization, rather than the five to ten companies advocated by some analysts. Nevertheless, the subsequent labour productivity growth of the two major generators has surpassed that of any of the major privatized companies in other industries, not least because the generators quickly came under competitive pressures from new entrants into the market.[14] And although the development of competition in generation can be held to be an example of the beneficial effects of vertical separation (of generation from transmission), it is also true that the last four years have seen the development of strong retail competition even though retail supply and distribution remain vertically integrated.[15] That is, contrary to the views expressed by some analysts, control over distribution wires has not enabled distribution companies to prevent the emergence of competition at the retail level.

In conclusion, then, reform of the horizontal and vertical structures of the electricity supply industry is not in itself the key to promoting competitive pressures and economic efficiency. Other factors, and in particular entry conditions, are generally of greater significance. That there has been a retreat from early, radical proposals for restructuring the industries is not, therefore, necessarily a great loss.

The more important consequences of industrial structure in electricity and gas are connected with their impact on the conduct of regulation. Here there is no doubt that horizontal and vertical arrangements in the industries have a material effect on the information that is available to the regulator. Generally speaking, the greater the degree of horizontal and vertical separation the greater will be the transparency of conduct in the industry and the greater will be the information available to the regulator. Whether or not regulators are capable of making effective use of the extra information is, however, another matter.

5. Regulation

The most serious setbacks in the reform of the Polish energy industries have been the delays and changes to the proposals for establishing a set of regulatory arrangements for the sector. Regulation is a crucial determinant of performance not only because it has to serve as a substitute for competition in the naturally monopolistic parts of the industries (transmission and distribution) but also, and perhaps more importantly, because it is the key to promoting competition in the non-naturally monopolistic parts of the industries (production and supply).

In section 4.3 above it was noted that, empirically, entry barriers appear to be a more important determinant of competition and industrial performance than, say, the degree of market concentration; but in electricity and gas entry conditions are very heavily influenced by regulatory policy. For example:

- Regulators are generally entrusted with the task of licensing operators in the relevant industry. A liberal licensing policy will greatly facilitate new entry whereas a restrictive licensing policy will create barriers to entry.
- Entry into the electricity and gas markets to supply final customers depends upon potential competitors being able to gain access to existing wire and pipeline networks on terms that do not discriminate against them. Since regulators are responsible for setting or policing the terms for such third-party access, regulatory decisions effectively determine the significance of this major source of potential entry barriers.

5.1 The Problem of Delegation

Polish developments confirm the general experience that regulatory governance is one of the most difficult exercises for governments to get right. Although a system based upon an independent regulatory agency or agencies is one that has obvious merits and is widely advocated, it is easy to forget how difficult it is for politicians to set up such bodies.

In Britain the new regulatory system that was first established at the time of telecommunications privatization in 1984 followed almost forty years of controversy about, and failed attempts at reform of, the regulation of state-owned utilities (which had the legal form of commercial corporations). To illustrate with just one event, in the mid-1950s the then Conservative government considered proposals to establish an independent body that would be responsible for many aspects of regulation of the electricity supply industry. The proposal was rejected on the basis that it would mean that the government would lose control of vital decisions in the industry![16]

A similar political resistance to the delegation of powers is something that appears to have held up progress here in Poland also. In particular, politicians have been reluctant to give up control over the level and structure of energy prices. The problems of (even partial) government withdrawal from the energy sector are compounded by the particular economic characteristics of the sector. Whereas in some industries it is possible to argue for simple deregulation, in the energy sector there is a delicate balance to be struck between market forces and regulation. Indeed, as argued above, regulation is itself a key to the development of competition. If the state simply privatized the industry and withdrew completely the likely outcome would be monopolization, not competition, as a result of the effects of unrestricted exercise of market power by the new companies which would own and control the naturally monopolistic parts of the transmission and distribution networks.

5.2 The Trade-offs

One of the principal concerns about the Polish situation is that, while the pressures for delegated regulation of the energy sector may be accommodated via the establishment of an energy agency, the agency will be operated in a way that retains substantial day-to-day influence of government ministers over particular decisions. That is, the form but not the substance of independent regulation will be created. This would be a typical political compromise.

It should be stressed, however, that the political side-effects of creating independent regulatory bodies with substantial delegated powers are by no means uniformly negative. If ministers choose to control prices then, in a democratic system, they are likely to be held responsible for those prices. That may be fine so long as prices are low, but it is not something a minister will necessarily welcome in periods when energy prices are set to rise. And, of course, there is a limit to the extent to which politicians can hold prices down: Subsidies absorb central government financial resources, and therefore detract from the ability of politicians to make themselves popular via alternative uses of those resources (i.e. expenditures on other activities) or via lower taxation.

Whereas the first phase of developing a reform programme for the energy sector in Poland seems to have occurred in something of a political vacuum -- with public officials and their advisers making the running -- subsequent developments have been characterised by more familiar political considerations. In particular, analysis of regulatory processes suggests that decisive moves toward regulatory reform will tend to depend upon the emergence of circumstances in which the relevant politicians perceive that they themselves will benefit directly from change. International experience suggests that one of the most important factors affecting the balance of political advantage is the general state of public finances. When the financial position of the state is problematic -- for example, when there is a large fiscal deficit -- there is a much greater incentive for politicians to cede part of their control over industries such as gas and electricity so as to distance themselves from the price increases that may be required to reduce the burden of the industries on the national budget.

5.3 The Consequences of Delay

The delay in creating regulatory institutions with substantial delegated powers is the single biggest weakness of the Polish energy reform programme to date. The precise form of such institutions is of lesser importance, provided only that they have a degree of independence from ministers. Such independence would be difficult to achieve if regulation were conducted from within the Ministry for Industry. On the other hand, issues of whether to have:

- a single energy agency or separate agencies for electricity and gas (and possibly other energy industries);
- one regulator or a regulatory commission; and
- the regulatory agency as a "stand-alone" body or as a branch of the Antimonopoly Office,

are, although significant, only of second-order importance.

Until the question of regulatory governance is satisfactorily resolved the Polish energy sector will be liable to all of those weaknesses associated with state-ownership of industry. Thus, on past experience, the central government will likely have difficulty in establishing a clear and consistent longer-term policy for the sector and decisions can be expected to be vulnerable to short-term political influences. The performance record of industries operating in such conditions is all too apparent, and it is not a good one.

As well as the difficulties posed for Polish managers by the uncertain political environment, delays in establishing a credible regulatory framework also have negative consequences from the point of view of encouraging inward foreign investment into the sector, particularly for larger-scale projects. Private investors will simply not be attracted by the prospect of having the returns on their investments be vulnerable to shifting political priorities. As a consequence, there will either be little inward investment or, if such investment is nevertheless actively encouraged by the government, Poland will find itself having to offer foreign investors extremely favourable terms in order to attract funds. Put another way, establishment of credible, independent regulation is a good way to lower the cost of capital; failure to achieve this objective is a good way of increasing the cost of capital.

6. Conclusions

In the early stages of the reform process the Polish government set out a series of proposed reforms based upon best practice international experience. As in other countries, however, political pressures have seriously impeded the implementation of best practice techniques. The result is that policy formulation and implementation now have to proceed in a second-best way, taking account of all sorts of constraints that limit what can be done in a particular period.

Such an outcome is not a cause for despair. In fact, second-best policy development in an environment of political and social constraints is the norm in the energy sector: It was the Polish situation in the immediate aftermath of the collapse of the socialist regime that was exceptional. The situation does, however, call for careful setting of priorities, and economic theory textbooks are unlikely to be much of a guide in this task. They are full of illuminating analyses of particular problems but rarely deal with the precise interactions of issues that are to be found in specific, policy-making situations.

In my view, available evidence indicates that the highest priority for the Polish energy sector should be to establish a credible, independent regulatory framework. Industrial structure (in the form of the degree of horizontal and vertical integration) also matters for economic performance, but it matters much less than regulation. One reason for this is that the most important single influence on the degree of competition in industries such as electricity and gas, the extent of barriers to entry, is itself dependent upon the effectiveness of regulation -- for example via the impact of regulation on the terms and conditions for third-party access to networks and on the general attractiveness of the markets to investors.

Once credible, independent regulation is established, other pieces of the jigsaw will start to fall into place. Private capital will flow more vigorously into the energy sector, and there will be better prospects for privatization of existing enterprises. The entry of new capital will be a source of competitive pressure that will stimulate incumbent firms to improve performance, and eventual privatization may lead to restructuring of parts of the industry (such as distribution) on a voluntary, agreed basis.

Until such a regulatory framework is developed, however, the energy sector is liable to stagnate. Objectives of domestic enterprises will be unclear, the sector will be unattractive to new capital, and competitive spurs to better performance will remain weak.

Endnotes:

1 See, for example, LeBeouf, Lamb, Greene and MacRae, Competition, Structural Change and Regulatory Reform in the US Electric Utility Industry, London: Centre for Regulated Industries, 1994, for a relatively recent summary of the situation in the power sector in the USA.

2 See J. Vickers and G. Yarrow, Privatization: An Economic Analysis, Cambridge: MIT Press, 1988, for an assessment of the legislation of the early 1980s.

3 For an analysis of gas privatization, see Vickers and Yarrow, op. cit. The later electricity privatization is explained and assessed in J. Vickers and G. Yarrow, "The British Electricity Experiment", Economic Policy, 12, 188-232, 1991.

4 See L. Hancher, EC Electricity Law, London: Chancery Lane Publishing, 1992.

5 See International Energy Agency, Electricity Supply Industry: Structure, Ownership and Regulation in OECD Countries, Paris: OECD, 1994.

6 Examples of the international experience are set out in The World Bank's Role in the Power Sector: Policies for Effective Institutional, Regulatory, and Financial Reform, Washington DC: World Bank Policy Paper, 1993.

7 Norway is the other European country that has implemented a major programme of liberalization; see E. Hope, L. Rud, and B. Singh, "Markets for Electricity: Economic Reform of the Norwegian Electricity Industry", SNF Working Paper No. 12, 1993.

8 It is also relevant that the market share of domestically produced gas is declining and is forecast to continue declining.

9 On the excessive levels of electricity prices after privatization, see G. Yarrow, "British Electricity Prices since Privatization", Essays in Regulation, Oxford: Regulatory Policy Institute, 1992; and G. Yarrow, "Power Sector Reform in Europe", paper presented at the ASSA meetings, Washington DC, January 1995.

10 See P.L. Joskow and N.L. Rose, "The Effects of Economic Regulation", in R. Schmalensee and R. Willig, Handbook of Industrial Organization, Amsterdam: North Holland, 1988.

11 Regulatory separation means that the two businesses (distribution and supply) are treated as distinct for regulatory purposes (e.g. when setting price controls).

12 See J. Tirole, The Theory of Industrial Organization, Cambridge, Mass.: The MIT Press, 1988, especially chapter 6.

13 A comprehensive survey of this empirical work is to be found in D. Hay and D. Morris, Industrial Economics and Organization, Oxford: Oxford University Press, 1991.

14 Labour productivity in electricity generation increased by over 80% during the four-year period following the reorganization of the industry,

15 Following reorganization of the industry, the average market share of each Regional Electricity Company in the supply of electricity to large industrial users in its own home area fell from 100% to less than 50% within two years.

16 See Vickers and Yarrow, Privatization, op. cit.

References:

HANCHER, L. (1992), EC Electricity Law, London: Chancery Lane Publishing.

HAY, D. and D. MORRIS (1991), Industrial Economics and Organization, Oxford: Oxford University Press.

HOPE, E., RUD, L. and B. SINGH (1993), "Markets for Electricity: Economic Reform of the Norwegian Electricity Industry", SNF Working Paper No. 12.

INTERNATIONAL ENERGY AGENCY (1994), Electricity Supply Industry: Structure, Ownership and Regulation in OECD Countries, Paris: OECD.

JOSKOW, P.L. and N.L. ROSE (1988), "The Effects of Economic Regulation", in: SCHMALENSEE, R. and R. WILLIG (Eds.), Handbook of Industrial Organization, Amsterdam: North Holland.

LEBEOUF, LAMB, GREENE and MACRAE (1994), Competition, Structural Change and Regulatory Reform in the US Electric Utility Industry, London: Centre for Regulated Industries.

TIROLE, J. (1988), The Theory of Industrial Organization, Cambridge, Mass.: The MIT Press.

VICKERS, J. and G. YARROW (1988), Privatization: An Economic Analysis, Cambridge: MIT Press.

VICKERS, J. and G. YARROW (1991), "The British Electricity Experiment", in: Economic Policy, 12, 188-232.

WORLD BANK (1993), The World Bank's Role in the Power Sector: Policies for Effective Institutional, Regulatory, and Financial Reform, Washington DC: World Bank Policy Paper.

YARROW, G. (1992), "British Electricity Prices since Privatization", in: Essays in Regulation, Oxford: Regulatory Policy Institute.

YARROW, G. (1995), "Power Sector Reform in Europe", paper presented at the ASSA meetings, Washington DC, January 1995.

J. Energy Law Developments in the European Union and Poland

Anita Ronne

1. Introduction

This paper is based on my presentation at the conference although revised considerably in order to reflect the last developments and amendments to the Polish draft Energy Law. The paper is divided into 3 sections: First a description is given of the energy policy developments in the European Union and Poland as the political framework indicates the intentions for the establishment of the legal regulatory system. Chapter 2 will discuss the Community law of specific relevance to the energy sector and the Polish proposal for a new Energy Law. The paper is completed by a conclusion under chapter 3.

2. Energy Policy Developments

2.1 European Union

Today no one questions the central importance of the energy sector and its far-reaching economic effects on free trade. It is therefore more astonishing that little in the way of concrete actions implementing a common energy policy has been achieved in the European Union until just recently - despite the fact that powers to regulate various aspect of the energy market has been vested in the different institutions within the Community since the fifties.

The first European Community, that for Coal and Steel, was founded by the Treaty of Paris in 1952, and the European Economic and Atomic Energy Communities followed with the signature of the Treaty of Rome in 1957. In contrast to the Treaty establishing the European Coal and Steel Community and the Euratom Treaty which apply to specific sectors of the Community, the EEC Treaty (now EC Treaty) covers practically the whole economy of the Member States. Especially following the signing and ratification of the Single European Act in 1987 and of the new Treaty on European Union in 1993 (usually known as the Maastricht Treaty) the process of integrating the different national policies into a common framework has accelerated. The last legal framework development, which should be mentioned is that of the Energy Charter. It is obvious that the historic changes in Central and Eastern Europe have important energy dimensions. Through the Energy Charter the European Community is seeking to ensure that energy provides a source of stability during a difficult transition period.

The general reluctance of the national governments to transfer any sovereignty to the Community institutions contributed to the continued failure on the part of the Commission to make any significant progress on energy questions. The falling oil prices of 1986 and their effect on prices of other fuels created, however, pressure for competition at the national level. Large energy consumers began to question the large price differentials for similar fuels in the different Member States. They looked to the American experience of deregulation of natural gas markets as evidence of the benefits of greater competition.

But first of all, the decision of establishing a single European market for goods, services and capital has resulted in a radical change of the interests of the Community in regulating the energy sector.

The Energy Council's Resolution from 1986 (86/C241/01; cf. O.J. C 241/1) on the Community's energy objectives for 1995, which was followed by the publication of the Commission's working document on The Internal Energy Market (COM.no. 238 (1988)) marked the concrete change of attitude. The energy objectives for the period 1987-1995, expressly recognized the need for "greater integration, free from barriers to trade, and creation of an internal energy market with a view to improving security of supply, reducing costs, and improving economic competiveness" and 4 sets of actions have been identified. These include first the opening up of public procurement, harmonization of technical standards and approximation of indirect taxation.

Second and most importantly, the Commission commits itself to a determined application of the provisions of Community law, which refers to the provisions on free movement of goods, on state monopolies, on competition and on state aids.

Thirdly there is expressed a commitment to seek a proper balance between energy and environmental objectives. Finally, two specific priority areas are outlined - energy pricing and infrastructural development for the poorer regions of the Community.

2.2 Poland

Under the framework of the national economy transformation program, the Polish Government is attempting to reconstruct and modernize the energy sector towards a market oriented system. The aim is to reduce costs and the negative impact on the environment and furthermore to improve energy security, and an efficient use of energy and fuels.

The structure of the energy sector has been characterized by vertically integrated State-owned monopolies, State subsidies and artificially fixed energy prices. The new tasks are therefore comprehensive and include:

- transformation of State enterprises into commercial companies and eventually privatization;
- increased competition by de-monopolizing the most important sub-sectors;
- liquidation of subsidies and basing prices on justified costs;
- creation of a new legal framework which can match the new developments.

In other words the expressed policy is in line with that of the European Union. But one thing is policy, - another - the implementation of the same.

3. New Legislation

3.1 European Union

Community law, like any other legal system, is a system in permanent development, and the case law of the Court of Justice is a clear example hereof. Community Law (in the energy sector) consists essentially of:

- primary EU law (general provisions of EC, ECSC and Euratom treaties),
- secondary law (regulations, directives and decisions),
- case law of the Court of Justice,
- international law (e.g. GATT rules) and
- so-called soft law (acts without legally binding character but still having importance).

3.1.1 Primary EU Law

Primary EU law is composed of the three treaties - the EC Treaty and the ECSC and Euratom Treaties. In this respect it should be noted that the ECSC Treaty expires in 2002 but essential parts will be incorporated in the EC Treaty (e.g. state subsidies for coal). The EC Treaty rules apply fully to all economic sectors including the energy sector and aims at the foundation of a common market with free movement of goods, persons, services and capital. As already mentioned the Commission commits itself to a determined application of the provisions of the Community law and the following articles should be highlighted in this respect (taken essentially from the EC Treaty, but refer also to corresponding articles of the ECSC and Euratom Treaties):

- Art. 30-36: (elimination of quantitative restrictions): Quantitative restrictions on imports and exports of any product, quotas and all measures having equivalent effect are prohibited between Member States. In the energy sector the concept of security of supply may play a role in the application of these articles. Any restrictions, however, may not constitute any arbitrary discrimination on trade between Member States.

- Art. 37: (State Monopolies): The exclusive rights over imports, exports and wholesale distribution have always been considered incompatible with rules on free movement of goods and more particularly with Art. 37, which requires the adjustment of national monopolies of a commercial nature. This obligation applies to all goods, eg. petroleum products. Exclusive transport and/or distribution necessarily constitutes an obstacle to intra-Community trade in these products and is only acceptable if and so far as, it is justified by imperative requirements of public interests. All these rules imply in principle a right for any operator to construct its own lines, at least whenever the operator of the network does not want to or cannot offer its service.
- Art. 52-66: (Freedom of establishment and to provide services): The right of establishment of nationals of a Member State in the territory of another Member State and the free movement of services rest on the prohibition against discrimination and is also relevant for energy companies.
- Art. 85-89: (Common rules on Competition): In brief terms, the competition rules prohibit agreements between undertakings which prevent, restrict or distort competition and any abuse of a dominant position. The competition rules apply equally to the energy sector.
- Art. 90: (Public enterprises and enterprises with exclusive or special rights): If Member States confer certain special rights to undertakings - public or private - which is the case in many Member States with respect to production, transportation and distribution of gas and electricity, these undertakings should still comply with the rules set out in the Treaty and especially those on non-discrimination and free competition.

- Art. 92-93: (State Aid): Aids granted by States and which affect trade between Member States is incompatible with the common market. e.g. lower (subsidized) tariffs, rules for the coal sector (ECSC Treaty);
- Art. 129b: general principles of Trans European (energy) Networks;

The Court has defined goods as "objects transported over a national border to be subject to a commercial transaction" and electricity and natural gas is considered to be covered by this definition. In 1991 the Commission decided on this legal basis to open infringement procedures against several Member States which maintain in their legislation exclusive rights over imports and exports of electricity and gas. In 1994 six actions against Member States have been referred to the Court of Justice.

3.1.2 Secondary Legislation

Secondary energy law is a long list of regulations, directives etc, such as those on the Single Energy Market, transparency, public procurement, State aid regime for coal, Euratom safeguards measures etc. The following will concentrate on the new initiatives concerning the Single Energy Market.

The Community has implemented a package of new legislation in order to establish the internal energy market. A gradual approach has been applied and to secure a so-called first stage two directives on transit of gas and electricity, and a directive on price transparency have been adopted. Under the second stage, the Commission has proposed directives according to which the energy market should be further liberalized by way of the introduction of a system of third party access, unbundling of management and accounting, and more freedom in network construction. The development of the third stage depends on the experience gained under the second one.

As is well known, gas and electricity trade is characterized by the existence of a great number of national monopolies controlling the infrastructure through which these goods are transported. Although there is an integrated network of gas and electricity lines connecting most West-European States, the border-crossings are few and limited. The directives on transit and price transparency are the only substantial results of the coming into being of an internal energy market as it is not likely that the proposed directives on the internal electricity and gas market will be adopted as they stand today.

In the conclusions made by the E.U. Council on 29 November, 1994 the Council "confirms the necessity of completing the internal energy market ..."; that it is "essential to work towards secure, open, transparent, efficient, competitive and environmentally acceptable electricity and gas markets..."; and that the Council is "convinced that the completion of the internal electricity and gas markets require flexible solutions, which, however, must .... lead to equivalent and non-discriminatory opening of markets". The Council furthermore applies a procedure which will first concentrate on the electricity sector (in fact the same procedure which was implemented when preparing the transit directives). An agreement has now to a large extent been reached towards the power production sector whereas further examinations are needed for the rest of the market.

The *transit directives* gas (91/296/EEC; cf. O.J. L. 147/37) and electricity (90/547/EEC; cf. O.J. L. 313/30) provide for the opening up of cross-border rights of transit. The directive applies only to major transmission grids, if the grid of origin or final destination is situated in the Community and this transport involves the crossing of at least one Member State's border. Transit conditions shall be non-discriminatory and fair for all parties concerned, but may not endanger security of supply nor quality of service.

Furthermore a special system of procedure is introduced. Member states are required to ensure that the Commission is notified of any requests for transit.

It must also be assured that negotiations on the transit conditions are opened and that the Commission and the national authorities are informed of the conclusion of a transit contract as well as failure to conclude such contracts. In case of failure the directives provide for a special conciliation procedure. If the reasons for not entering an agreement appear insufficient, the Commission shall implement the procedures provided for by Community law. It can therefore be concluded that the transit directives do not provide individuals, including the industry and energy distribution companies with any right of transit, nor do they give the Commission any new substantive powers.

The starting point for the proposal for the *price transparency directive* (90/377/EEC; cf. O.J. L. 185/16) was the observation that transparency of gas and electricity prices was not satisfactory on the industrial market of the Community and thus endangered the completion of the internal energy market. The directive requires the utilities to publish details of rates charged to customers.

Finally, the Commission has presented *two proposals for liberalization in the areas of gas and electricity* (COM.no.643(1993)) which are characterized by three main features. First the creation of a *transparent and non-discriminatory system for granting licences* for the construction of electricity lines and gas pipelines, where the aim is to open up investment in production and transport to independent operators.

Secondly, the concept of *unbundling* is introduced. This principle implies, that management and accounting of operations connected to production, transmission and distribution have to be separated, so as to ensure transparency of the different operations in vertically-integrated undertakings. It does not, however, affect ownership structures.

Thirdly, the introduction on a limited basis of *third party access* whereby the transmission and distribution companies are obliged to offer access to their network at reasonable rates, within the limits of available transmission and distribution capacities. It is proposed that the qualification criteria for this second stage shall be such as to include:

- large industrial users whose consumption exceeds 100 GWh (Giga Watt per hour) of electricity or 25 million cubic meters of gas on a site.
- distributors of a certain size in each of the Member States, being those who supply respectively at least 3% of the electricity, or at least 1% of the gas, consumed in their Member State. Distributors who do not reach the threshold can join together in order to qualify.

As already indicated it is not likely that the proposed directives will be adopted as they stand today.

Poland is striving for complete integration with the European Union, and it is therefore essential to investigate whether the new legislation follows these briefly described trends. Due to limited space, I will concentrate on the proposal for a new Energy Law.

3.2 Poland

Following a very long legislative process and 14 different official versions a draft Energy Law was finally approved by the economic committee (KERM) of the Council of Ministers (the Government) on November 10, 1994. The next step will be the approval by the Government and then the draft will be submitted to the Sejm (Parliament). The discussion in Parliament is expected to take place during the spring of 1995.

The proposal for a new Energy Law has as its prime objective to create an economic, stable and transparent market for energy and fuels to attract investors into the system. The Law intends to assure energy security of the State and efficient use of energy and fuels, to increase competition in order to reduce costs and improve the efficiency of operation by the energy enterprises. The Law is specifically designed for customers that are connected to a grid supply of energy or fuel and the goal is to ensure a proper balance between the interests of the consumers and the energy enterprises in order to provide the people of Poland with a reliable and satisfactory quality of service.

The Energy Law covers all energy sectors except the exploitation of natural resources (The Geological and Mining Law) and the use of nuclear energy (The Atomic Law). It defines the responsible administrative bodies, the tasks to be undertaken by the different players within the energy sector and the objectives to be taken into consideration in this respect. The Energy Law neither prefers nor discriminates against various forms of ownership.

Regulatory bodies

The reform assumes liberalization of the energy market. Although it is envisaged that the large state energy enterprises are to be replaced by a number of competing companies, some natural monopolies will remain (the grid based systems) because of lack for an economic basis for constructing parallel systems. In those areas where a monopoly will exist public regulation is needed in order to protect the customers, to provide a substitute for market conditions and to provide a pricing policy that balances the customers' interests with the financial requirements of the suppliers. The Law provides these forces by establishing a regulatory body which is independent from the traditional ministerial system.

A distinction between the regulation of heat and other kind of supplies has to be made in many respects. Thus the restructuring of the heat supply systems will be launched on the background of a planning system to be organized by the *gminas* (communes) and supervised by the *voivoda* (counties). The heat supply plans have to be consistent with the national energy policy. Whereas supplies of power, gas, oil, coal and renewable fuels are to be regulated at a national level, heat will be handled locally by the *voivoda*. The regulatory body for all other energy sub-sectors will be a new Energy Regulatory Authority (ERA).

The draft Law has comprised different solutions as to the management of ERA. Pursuant to the latest version, ERA is to be established as a central State administrative organ, managed by a President to be appointed and removed by the President of the Council of Ministers. There is no statutory limitation as to a possible removal. ERA reports directly to the Council of Ministers and is required to submit to the Ministry of Industry and Trade all information regarding its operation. Further details about the organization and mode of activities will be put forward in a statute issued by the Council of Ministers.

The independence of the regulator has gradually become more and more weak. First it was decided to exempt the heating sector from the Energy Regulatory Authority and vest the task in the *voivoda*. The *voivoda* is, however the local representative of the central government and consequently not independent from the political level. Thus the heating sector does not follow the main concept of separation of political and regulatory functions.

The second change was to delete the financial independence of ERA. Then the board management of the regulator and the qualified grounds for dismissal were deleted, and it was decided that the President is not to be appointed for a fixed time, but can be removed at any time. Furthermore it was introduced that the tasks of ERA can be suspended or entrusted to another organ if the energy security or economic balance of the state is endangered.

All these changes have as a consequence that the draft no longer assures continuity and stability within the regulatory organ.

The Licence System

A central element of the Energy Law is the establishment of a licensing system permitting private and state-owned companies to produce, supply, buy and sell energy and fuels. The rights and obligations of the energy enterprises are found in several chapters of the act, but whereas the act itself lays down the framework within which activities are to be carried out, the more specific terms are issued in the form of a licence. The basic concept of the licence is that each licensed company will enjoy, for a certain period a right to perform specific activities. In return, the licensees must pay fees and commit themselves to certain obligations in carrying out their activities. Decisions on the grant of a licence and disputes have to be published.

Main rights and obligations

Pursuant to the Energy Law the overall requirement of the energy enterprises is the responsibility of maintaining a continuous and reliable supply and to connect customers if it is technically and economically feasible. It puts a planning obligation upon the electricity and gas enterprises in order for them to meet supply demand. To improve transparency energy enterprises are required to keep separate accounts for different activities. In order to ensure an efficient and safe use of energy and fuels, the Law indicates special demands to the equipment and installations to be used.

Pricing

A crucial parameter for the transformation process is energy pricing. The main idea behind the introduction of a new pricing concept is that within energy markets where competition exists, the market should establish the price. Within areas where monopolies prevail the prices will need to be regulated in order to assure a balance between the interests of the consumers and the companies. Pursuant to the Law the regulated company will be free to establish its own pricing and tariff structure, based on the principles determined in the Law and the secondary legislation, but subject to the direction and approval of the Energy Regulatory Authority (ERA) and control by *voivoda*.

The current draft of the Energy Law stipulates, however, that all licensed activities will be subject to price control by the regulator. The main problem with this is that there may be activities which require technical or quality regulation but where price control is not appropriate. One of the basic intents of the new regulation system as also expressed in the Law is to develop competition and only regulate prices where monopolies or monopolistic practices exist. Thus the draft is not consistent with this philosophy.

The Law assures a regulated company with sufficient revenues to reflect justified costs of operations and raw materials, allocations for maintenance and new projects and a return on investment. In return, however, the company must prove its operation is prudent and costs are reasonable.

Subsidies to the customer can be necessary but should be provided for through social policy which does not affect the market pricing or distort the energy market. The abandonment of artificially fixed energy prices was, however, weakened by determining that the Minister of Finance retains in a period of transition the right to fix the prices to the final consumer.

Third Party Access

With a view to improving security of supply, reducing costs and improving economic competitiveness, the Energy Law seeks to open the market for different supply companies by introducing a system of third party access. Energy enterprises operating a grid are thus obliged to provide equal access to the grid if technical and economic conditions so allow and the involved parties can agree upon the terms. If the parties cannot reach an agreement the regulatory organ has the power to set the conditions.

4. Conclusion

Energy is an integral part of our economic, environmental and political life, and it has a major part to play in fulfilling peoples' social and economic aspirations for a better life. The European Union - like Poland - is faced with a number of challenges in this respect. Progressing towards a political union, an internal market or a transformation from a centrally planned economy towards a market oriented system will require a contribution from the energy sector because of its fundamental importance in achieving economic and environmental goals.

However the concept of creating an internal market in the European Union and within this an internal energy market, has been from the outset very ambiguous. It is fair to say, that a more realistic level has been found today.

Poland has expressed its aspiration quickly to join the European Union. Implementation of the strategy to restructure the energy sector will properly bring Poland closer to that objective. The success of the reform program is dependent upon the development of a suitable legal and regulatory framework which is compatible with the major changes the energy sector of Poland will undergo over the coming years.

The legislative process has been a general problem. The submittance of the draft Energy Law to Parliament (Sejm) has been postponed continuously and the draft has been subject to new amendments. First and foremost these difficulties have been related to the lack of an agreed policy. Although a policy letter was issued in 1991 it is the impression when reading the comments from the different ministers to the draft Energy Law that the necessary agreements to or common knowledge of the basic concept of the new regulatory system does not exist. Opposition to the independence of the regulator and to the principle that energy prices should reflect real costs and not include social aspects are important examples in this respect.

The new regulatory system is based on a concept of introducing market mechanisms in order to obtain an economic, stable and transparent market which can attract domestic and foreign investors. Thus competition should be promoted as far as possible. Whenever such a competitive market exists the regulatory involvement should be restricted to creating conditions to allow the most efficient, cost effective operations to prosper, and wider issues of safeguarding concerns such as energy security and the environmental impact of energy production and use. The new concept therefore should also result in a redefinition of the roles and positions of the government, the energy enterprises and the customers. The draft Energy Law, however, assigns to the Ministry some important regulatory and operational functions and all the changes as to the concept of the regulator have as a consequence that the draft no longer assures continuity and stability within this organ. The link between price control and licences has the result that the regulator will be faced with a stark choice between imposing no regulation on the activity, which may for example jeopardize the safety of the grid or compromise quality, or exercising price control which is incompatible with the concept of free competition.

The complete regulatory regime is not developed but exists only in certain fragments. The Energy Law is only one piece of transition means and contains only the main principles of regulation. Details will be executed partly in the secondary legislation, partly in the licences and partly by practice to follow. Consequently, there is a risk that consistency in the full regulation system is not properly ensured. The present draft Energy Law raises important questions regarding the regulator's role in approving and controlling tariffs, the method of economic price regulation to be applied and whether there will be an open licence policy, based on objective, non-discriminatory and transparent criteria.

In this respect it should be stressed that establishing a proper legal framework is a continuous process. Experience will eventually highlight weaknesses of the system, lack of rules and needs for adjustments. After a period of time this can activate amendments to the Law or a change of executive regulation. The time has come, however, where it is absolutely necessary to fill the gap of regulation and as soon as possible.

K. Institutional Choice or Muddling Through - Problems of Transformation in the Electricity Supply Industry in Russia

Petra Opitz and Wolfgang Pfaffenberger

1. Institutional Choice in the Phase of Transformation

In this paper we will analyse the options of institutional choice in the phase of transformation for a sector with special characteristics. Network-oriented sectors like electricity supply, natural gas supply, telecommunication, railways, etc., even in a market economy have some peculiarities which have led to special arrangements. Frequently, instead of the general rules prevailing in the market, special rules, excluding almost all elements of competition, have been introduced. In extreme cases the systems were state-controlled: The structure of the state-owned electricity supply industry (ESI) in Great Britain was very similar to the one of the former GDR.

On the other hand, in addition to the economic characteristics of this sector there are other reasons for state interference. Electricity supply is partly considered an element of a general infrastructure which requires state interference. Frequently such government interference may become unnecessary in the long run, however it tends to survive due to the many special interests connected to this status.

Currently the question of deregulation or re-regulation of network-oriented industries is discussed in many countries. As a result of these discussions and the many changes that have occurred in the various countries the scope for institutional choice has widened. It seems interesting to apply the Western discussion to the process of transformation in a former centrally-planned economy. There are significant parallels between the transformation of the ESI in

Russia and the reform in Great Britain. In Britain the privatisation of the formerly state-owned ESI led to the introduction of a new set of institutions in which elements of competition in the fields of production and distribution were mixed in a new way with the monopoly elements of transport of electricity.

A logical analogy may not offer good advice for real world transformation problems: Is it realistic to assume that transformation processes in a country like Russia can be considered as the beginning of a rational choice of institutions? On the way to a market economy new sets of rules are necessary and the behaviour of all participants in this process has to correspond to these rules. The interaction between the formation of institutions and the formation of changed behaviour is complex and requires a process of development. Ideally, the formation of rules and economic behaviour should develop in an interactive process over time. Transformation thus needs time and has to be built up out of the changes within microstructures of the economy (*bottom up*) rather than through new rules developed by central rule setters (*top down*) alone.

Looking at Russia we can see over-activity in the creation of new state rules. Almost all areas of life require new regulations. In view of the general economic and social crisis currently prevailing in this country the pressure to introduce new regulations is immense. On the other hand the effectiveness of these rules is rather poor. Thus we will assume in the following that currently there is a wide gap in Russia between the small autonomous sphere of markets on the one hand and state regulation on the other hand. This assumption will provide the basis for an analysis of the development in the field of ESI. First the ongoing discussion in the West will be treated. Then a short overview will be given of the state of the energy sector and the electricity supply sector of Russia. Finally we will sketch out some elements of an analysis of the ongoing process. The authors have just started an empirical analysis of this process. [1]Thus the results presented in this paper are preliminary. Transformation requires a change in the structure of property rights, the market conditions and, last but not least, a change in the inner organisation of the enterprises themselves. The last topic will not be treated in this paper. We hope to be able to present some results regarding this topic at a later stage of this project.

2. Models for Network-oriented Energy Sectors in the West

2.1 Special Characteristics of the ESI

The institutional set-up in the ESI is based on the following characteristics:

- Electricity is network oriented. The availability of power requires a special transport and distribution network.
- Transport and distribution networks are to a large extent subject to the conditions of natural monopoly: a single integrated network is cheaper to run than several competing networks if the owner of the network acts as if he were in a competitive environment. As this is highly unlikely, there is a need for political regulation.
- Production and consumption of power occur simultaneously; it is practically impossible to decouple production and consumption by setting up stocks.
- The availability of electricity is a precondition for the use of a number of basic goods and services (e.g. electric light). In the central part of the electricity market the possibilities of substitution are very limited.

These characteristics have considerably influenced the development of the ESI.

Frequently network orientation has led to vertical integration of the system ranging from production, over distribution to the ultimate buyer.

The fact that production and consumption occur simultaneously is a strong incentive to mix various demands within one network because this helps to economise on the costs of production and the network. This might also be an incentive for different producers to cooperate in a joint network because in this way changes in demand can be pooled; thus reserve capacities can be minimised.

The limited possibilities of substitution of electricity to some extent explain the readiness of political system to intervene in the field of power supply. This can be observed in many countries.

If the development of the electricity market is left to private enterprises it is likely that regions of low demand will not be developed. A high investment into the distribution network would be required which private enterprises would consider too risky. This would be acceptable, neither for the population of these regions nor for the suppliers of electricity-oriented products. In the history of the ESI we often find state-guaranteed supply regions. In these regions the investment risks are minimised. In most cases the contracts with private suppliers stipulated that the supplier had to supply to all potential consumers. Frequently local governments themselves supplied electricity because private enterprises were not ready to do so or did not do the job in a satisfactory manner.

In a way at this stage of development electricity was considered a merit good. To make it generally available special government intervention was necessary.

The fact that it is rather difficult for the consumer to substitute electricity opened the door for monopolistic abuse. Therefore it was to be expected that special measures of price regulation were introduced in this sector.

2.2 Elements of Competition

For power suppliers the regional monopolies were a great advantage because the investment risks were very low. This low-risk situation provided the basis for the expansion of this sector during the phase of development. Due to strong scale-economies during this phase of development the actual electricity prices continuously decreased for a long time so that questions of market structure and regulation were not of great political interest. The scale economies of the generation side, however, reached their limits some time ago and the electricity markets in highly developed economies only grow at an average or a below-average rate. Now it is believed that the introduction of elements of competition will lead to a greater efficiency in this sector. Positive effects might result from a better co-ordination in this sector or from different behaviour of those running the enterprises or of the consumers.

Figure K1: The Electricity Supply Industry

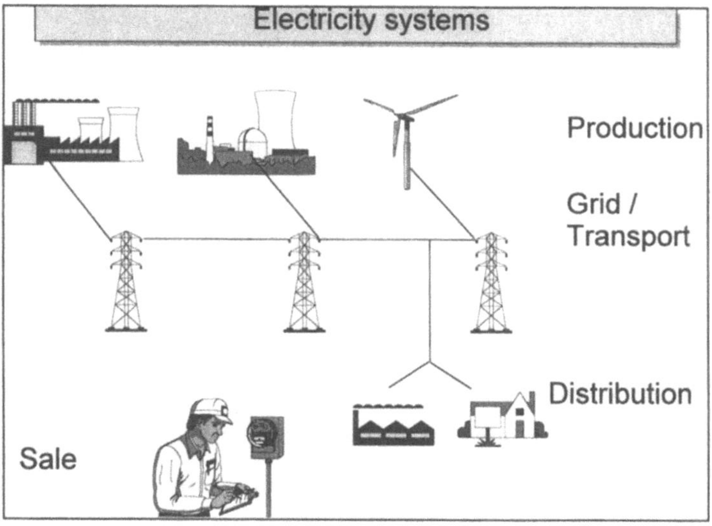

The electricity system consists of production, transport, distribution (in the physical sense) and sale. Market orientation may even begin on the production level. This requires that competing producers have access to different distribution networks to transport their product. In order to achieve this while maintaining private property rights in the transport network, it is necessary to define rights of access to these networks. Such models of "Third Party Access" can lead to different set-ups depending on whether the access rights defined apply only to utilities or also to the consumers of electricity.

In any case a strong degree of interference into private property rights is required. The scope of possible interference depends largely on the legal framework and the prevailing political system. At a later stage we will give some examples of this.

In a way "Third Party Access" is based on a fictive assumption regarding the physical nature of the processes. A number of different and separate generators feed power into the system. Different distributors extract power to transport it to their customers. Since production and consumption occur simultaneously, it is impossible to guarantee that power will physically flow according to the stipulations in the access contract between the different parties involved. Instead, a kind of mixed product develops in the transport system which is composed of the production segments of different producers After the mixing process this product is transported to the consumers. Thus the definition of general rights of access is not a good basis for a general model of competition in the ESI. It is hardly possible to make sure that all quantities and conditions that have been agreed on in bilateral agreements between producers and consumers can actually be guaranteed. This is presented in Figure K1. If there is a contract between the coal-fired power station and the distributor and if due to some technical defects the coal-fired power station can only supply power at a reduced rate, the distributor will continue to consume. An adjustment will occur automatically, resulting in a compensation by other producers. However, there is no contract between those other producers and the consumer who has a contract with the coal-fired power-station. If a model of general access is implemented we must anticipate strong external economies between the parties involved. This would lead to a distortion of market results. The model of access is easy to implement on the wholesale level where there are only a small number of actors and where it is relatively easy to internalise the external economies through additional contract clauses. In a way the definition of access rights can be considered a first step towards a more market-oriented set-up of the ESI.

In a generalised model it must be considered that supply and demand occur simultaneously and that the consumers rather buy the product "network electricity" than the product of an individual producer. In order to achieve this an institutional distinction must be made between producer and transport system. The services of the transport and distribution network are subject to the conditions of a natural monopoly. To achieve an economically efficient solution it is necessary that all actors may gain access to the system on similar conditions. It is also necessary to define rules stating the duties and rights of all parties involved. However the network system does not have to be state-controlled. But it seems plausible that a separation between production and the transport network would be helpful in order to avoid discrimination against certain producers or consumers.

It seems logical to define the network as a pool for buying electricity and turn it into the market place of the system. For this market place general rules can be defined so that suppliers with the lowest marginal cost will have the possibility to feed power into the system. On this basis an optimisation of the system as whole is possible. This is an important point because the argument that has always been put forward in favour of a centralised system was that in a centralised system a general optimisation is possible. If such a pool is defined the advantage of having independent suppliers can be combined with the advantage of centralised optimisation. In this case the processes can be organised similarly to a stock exchange.

Today distribution and sale of electricity is usually integrated. In an open model it seems plausible to separate these two activities and to allow independent trade. If access rights are being defined this already implies a separation of trade and distribution (as a physical service) because producers delivering directly to consumers also serve the trading function by doing so. This function can be taken over by independent traders. This seems logical in a system where the actors decide individually on the basis of cost and utility. The question is whether the distributor should also be allowed to engage in trade. It is necessary to separate both functions clearly in order to avoid cross subsidies at the distribution level. [2]

Pricing

The relative scarcity of power depends on the extent of demand at a given point in time in the network. Therefore scarcity-oriented prices have to be time variable. Time-oriented pricing can also be used in electricity systems without competition. The practical possibilities to apply time-oriented pricing depend strongly on the information and transaction costs connected to such pricing models. For consumers who only acquire small quantities transaction costs may be too high. In these cases it is possible to use the basic idea of time variation in simplified models in order to express day- or season-dependent changes in the load factor.

Time-oriented pricing is generally used in this sense. Frequently we find tariffs composed of two parts. Partly they are based on load (kW) and partly on energy (kWh). The incentive effects of such tariffs are similar to those of genuine time-oriented prices so that they may be considered a practical solution to the problem of transaction cost.

In a more competitive environment we would expect a greater variation in pricing because the models used for pricing could themselves be an important parameter of competition between different traders.

2.3 Transformation in Different Political Systems

2.3.1 Germany

In the present institutional set-up in Germany competition is excluded from all important parts of the system. The system is mainly based on regional monopolies. A number of vertically integrated wholesale companies supply about 85 % of all electricity and are responsible for running the transport network, each for its own region. Local and regional distributors produce only a small fraction of the total production. The transport network is run by the wholesale companies under a private agreement which sets up rules for the reserve and safety function of the network. Within this network the wholesale companies also operate a kind of spot market. This part of the system is not subject to government regulation. This is an interesting aspect of the institutional set-up: it shows that the quasi-public good "network service" can be established without state interference.

Figure K2: Structure of German ESI

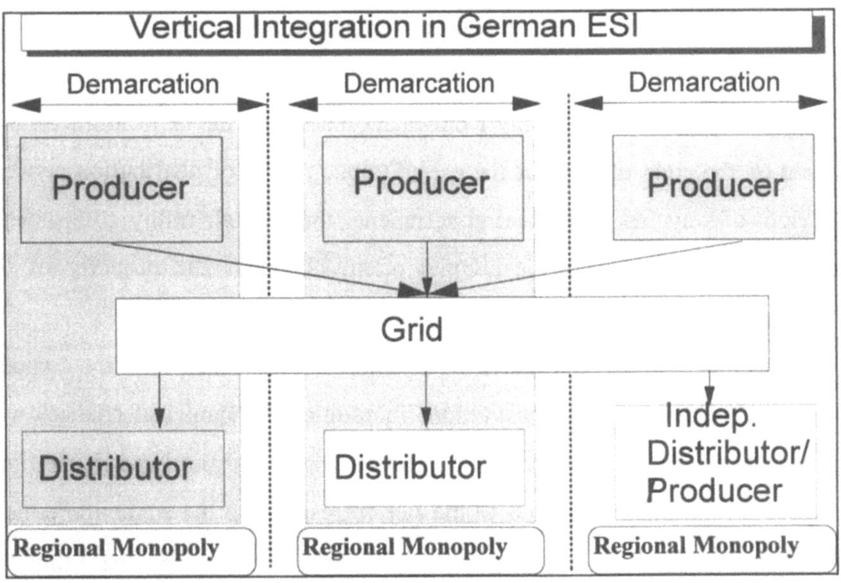

The distribution of power occurs in regional monopoly zones, either by wholesale companies themselves or by regional or local distributors. The structure is presented in Figure K2. Regional separation between wholesale companies and the regional separation of distributors correspond to each other. Although there is an exchange of load between wholesale companies, generally each wholesale company is responsible for keeping up enough capacity for its own area of reach.

There is no general optimisation of production throughout Germany; instead each wholesale enterprise basically optimises for its own region.

Regional monopolies are based on the obligation to supply; this is accompanied a number of government regulations: investment and prices require government permission. However, the power suppliers have the right to serve large consumers on the basis of an individual price agreements. These special agreements are not subject to price control.

The general order of markets in Germany is based on open competitive structures (contracts that exclude competition are generally not allowed[3]). Therefore special exemptions from this rule have been granted for the ESI. Usually they occur in two forms of contracts: franchising between local governments and utilities to supply electricity in a region and contracts between utilities to delimit areas of supply from each other. The role of local governments is based on the right of way for the use of public roads for distribution networks. This right of way lies with local governments. In general, utility companies are organised as private companies. Quite often, however, the property of these companies is held by public bodies.

Opening up the German ESI to competition is restricted by competing powers within the Federal Constitution. In addition, fundamental changes would render it necessary to interfere with property rights (institutional separation of transport and production) which would not correspond to the economic policy of competition in Germany. The concept of reform developed by the Ministry of Economics in 1994 tries to define a reform within these restrictions in the following way:

- Franchise contracts between local governments and utilities will be maintained but local governments will have the possibility to grant the right to supply to more than one company.
- Access rights to the transport network will be given to certain customer groups (Third Party Access).

- Investment regulation will be removed in order to allow free setting-up of power stations and networks.[4]

2.3.2 USA

In the USA the basic conditions are similar to those in Germany. Basically the distribution of power is performed through regional monopolies. Because of the size of the country the conditions within the transport network are more complex. In the United States there are independent producers apart from local government distributors which supply electricity within a certain region or in several regions.

Political interference used to be stronger than in Germany. A first step towards change was the Public Utilities Regulatory Policies Act of 1978 which was expected to help independent producers to feed electricity into the system.[5]

In this context it was often required to advertise new capacities so that external suppliers could gain access to the supply region of the advertising utility. "If more independent companies take over production within one region, a separation of production and transport will be the result."[6]

This first step towards reform was followed by another one in 1992 with the Energy Policy Act. This Act created a systematic basis for access to the network. Producers and distributors can generally negotiate bilateral contracts. Those utilities whose networks are necessary to fulfil those contracts are required to supply the necessary network services. Regulation of all those activities is done by the Federal Energy Regulation Commission and the State Regulatory Commissions.

Political restrictions in the United States are similar to those in Germany (federal state). On the other hand the reform is much more advanced in the USA. Just like in Germany a complete change of the system was not possible in the USA. However, in the United States a gradual opening-up of the market was achieved by implementing small-step reforms. Currently access rights are restricted to utilities. However, it can be anticipated that access rights will also be defined for final consumers.[7]

2.3.3 European Union

The range of institutional set-ups in the countries of the European Union is quite extensive. The centralised, state-owned system in France is one extreme; another one is the liberalised model in Britain. In most countries we have mixed models, although almost everywhere regional monopolies and, frequently, vertically integrated enterprises provide the basis for the organisation of the power market. The European Commission had the task of developing proposals that could help to achieve a more homogenous structure, however, without interfering with national rights and national political institutions. The delimitation of areas of competence within the Union is in itself a conflicting issue.

The Commission's proposals were based on the general idea of competition in the European Treaty and focused on the application of this idea to the network-oriented industries. Lately national autonomy in defining economic sectors that could be exempted from competition has gained in importance.[8] A homogenous market-oriented structure with similar interior structures in all countries seems impossible at the moment. The Commission's proposals are based on the idea of a unified system similar to the American reform proposals and as such can be formulated within the political restrictions for the Commission. The foundations of this reform are:

1. *Liberalisation of the setting up of power stations and networks.*

   This is necessary to achieve competition among producers. A free setting-up of power stations enables newcomers to gain access to the market and a free setting-up of networks gives suppliers and consumers of electricity the possibility to gain direct access to the network.

2. *Unbundling*

   Vertically integrated enterprises are supposed to separate their activities in production, transport and distribution. The Commission has no possibility to prescribe a complete separation of property rights. Therefore only separate accounts are required. This unbundling is supposed to make different cost accounts available in order to achieve a greater transparency for the calculation of pricing for network and distribution services.

3. *Third Party Access*

   Distributors and certain consumers will have the right of access to the network. This will lead to the creation of a wholesale market like in the USA. Direct access of consumers is essential for the creation of a consumer market. The different proposals of the Commission stated different rules or definitions of those rights. Certainly they are not based on the interior logic of the system but rather on the attempt of the Commission to achieve a majority on a political level. This, however will not be discussed in further detail in this paper.

4. *Central dispatch*

   As regards transport and distribution network dispatchers in charge have to be named for each region of distribution. It is the task of those dispatchers to co-ordinate activities with other networks and to co-ordinate the production of the various generators by applying rules of cost minimisation.

5. *Single buyer*

   The necessary compromise between countries which are centrally organised (like France) and those that want to achieve an open market model is to be solved by introducing the so-called single-buyer concept. In centralised countries specific large consumers will have the right to import electricity. Technically this will have to be carried out by the utility company (single buyer) that is supposed to pass the advantage of such deals on to the consumer. It is quire clear that a centralised monopoly has a number of parameters to avoid the effects of genuine competition because of the low transparency of prices and costs. Therefore the single buyer concept can only be considered part of a political process of establishing a European market rather than being a logical step in itself.

2.3.4 Great Britain

After World War II the British system was nationalised. It consisted of a centralised agency for production, and transport and regional distributors. When the system was privatised production was divided into two separate companies (with the exception of nuclear power), the regional distributors became separate companies; the transport sector was also divided into two separate companies which were privatised. Most of the shares are held by regional distributors. Thus privatisation was used to create an open market structure. Transport and distribution remained monopolistic systems. Both areas are subject to strict regulation regarding the conditions of supply and pricing. Distributors have to fulfil an obligation to supply all consumers with net services according to unified prices.

Consumers have the right to buy electricity from any of the producers. Complete access will be introduced step by step with large consumers being the first to gain it. In this way general consumer competition will gradually be achieved and trade with electricity can develop independently of the holders of regional monopolies.

In the area of production, competition will arise from the fact that several suppliers offer electricity. The system is not very transparent due to the many private contracts between distributors and consumers. Apart from these contracts a general pool was set up which provides the basis for the network company to decide on the use of generating resources. Suppliers offer their services for half-hour intervals. On this basis the network company calculates the marginal cost price for each time interval and allocates generators correspondingly. This newly-created market place is likely to affect all decisions regarding the system.[9]

3. The ESI in Russia

3.1 Basic Energy Situation

Russia is very rich in energy resources. These are concentrated in the eastern part of the country, in Siberia, which in 1990 produced more than 77 % of the energy.[10] The Siberian share in natural gas, the most important resource of Russia, amounted to about 93 %. The main consumers of energy, i.e. the industrial enterprises which are located in the western territories use about 80 % of the entire energy produced.

Consumption of energy per capita is amongst the highest in the world and also was the highest within the former USSR and the COMECON (see Figure K3).

Figure K3: Primary Energy Consumption

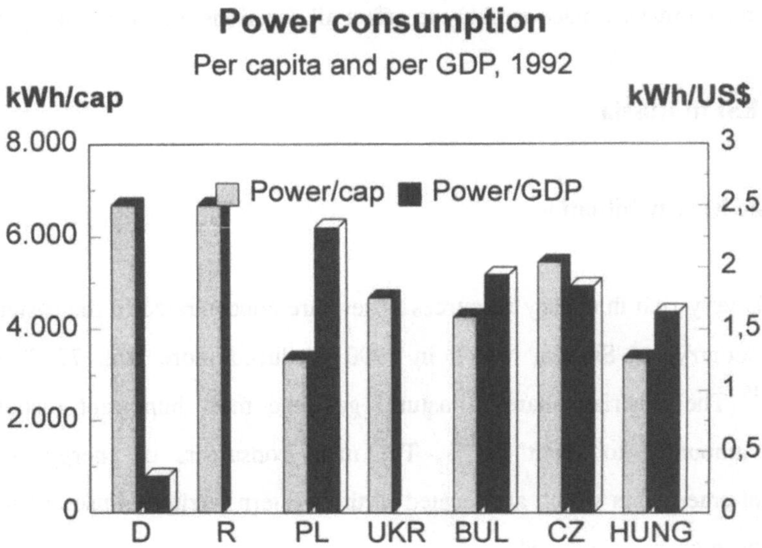

Source: OECD. No GDP für R and UKR

The supply of energy for the western territories is primarily based on the transport of primary energy resources from the eastern regions. Only 29 % of all electricity used in Russia is produced in Siberia.

Table K1: Industrial Production, Energy Consumption and Power Production in Russia (1990 =100)

	1991	1992	1993
Industrial Production	92	75	62,6
Energy Consumption	96	88	
Power Production	99(100,3)	93(96,3)	(91,2)

Sources: Nekrasov, u.a.(1993): Al'ternativy razvitia rossijskoj energetiki, in: Problemy prognozirovania, Nr.6, 1993, S. 15; DIW-Wochenbericht 47-48/1994, S.808. () = Figures by DIW.

The production of oil, natural gas and coal has decreased dramatically since 1990. This is due to different reasons. One reason is the over-use of the capacities in the 1970s and 80s without proper replacements. Also growing export of energy resources was meant to stop stagnation and decay of the former Soviet economy. The dramatic decrease in industrial production has also led to a reduction in the consumption of electricity. The security of supply is still a problem. Energy conservation is not yet an issue. On the contrary, the energy intensity which in the past was twice as high as in the OECD countries has increased further.[11]

Figure K4: Power Consumption per Capita / GDP

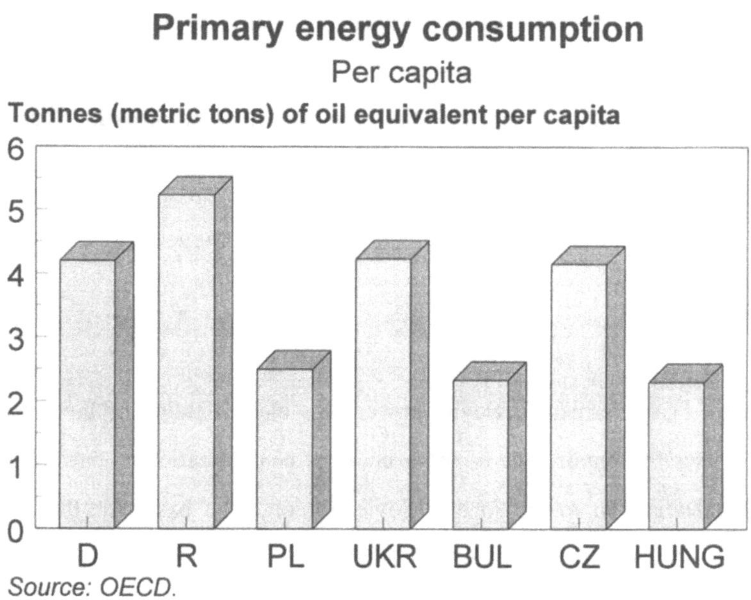

Although the production of fuel and electricity has been decreasing Russia is still the major producer of natural gas and the second largest producer of electric power.[12] Export of energy resources is an important source of foreign exchange. Export quantities were increased so that the falling level of international prices could be compensated for. Export of energy resources in former republics of the Soviet Union also serve a political function. The regional imbalance has led to industrial disproportions (in Siberia there is an excess of energy-intensive sectors) and also has led to high environmental burdens in Siberia (79 % of the total emission of all industrial sectors). The industrial monoculture had negative impacts on the population: social services are by a factor 1.5 - 2 below the level of the European part of Russia.[13]

In the past, energy pricing rendered energy conservation uneconomic so that a large part of the current energy use can be considered to be wasteful. Merenkov assumes that this amounts to about one third.[14] Other experts estimate the potential of energy conservation at 40 - 45 % of the present consumption, one third in the energy and transformation sector itself, another third in industry, 20% in the residential sector and 10 % in transport.[15]

In comparison to other East European countries in transformation Russia has by far the highest consumption of energy as can be seen from the indicators in Figure K3 and Figure K4.

3.2 State of the ESI

Lenin's statement "Communism is Soviet power plus electrification of the whole country" symbolises the importance of the process of electrification as part of the process of industrialisation in the former Soviet Union. The basis for this was established with the GOELRO plan of 1920. The high priority given to electrification led to enormous capacities in the power sector and a highly developed electricity network covering almost all of the country. This electricity supply was available in almost all regions. The total consumption of electricity per capita is on the same level as the one in Germany and thus is of the highest rank internationally. In view of the backwardness of Russia in many other areas this seems strange. The reason for this development probably is the rich endowment with energy resources and the extensive nature of the growth process of the past, two factors that appeared to set no limit to the expansion of the electricity industry. A market-oriented economy puts a higher weight on the relative scarcity of different factors. For the energy industry this means a qualitative change from an extensive development to an intensive one, which puts more weight on the conservation of energy and technical and organisational improvement. The decrease in the demand for electricity in the current economic crisis can help to facilitate this process. A further expansion of the sector is not necessary. Because of the high share of large generators, the advantages of falling average costs have been made available already.

The capital stock of electricity production is outdated. About half of the power stations are more than 23 years old [16] Since 1991 no additional capacities have been made available.

Figure K5: Power Production by Source

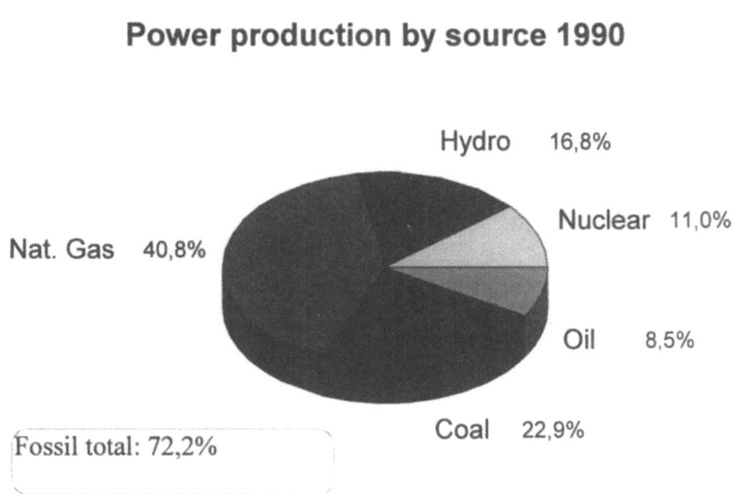

Source: Nekrasov, a.a.O., p. 13

The structure of production is illustrated in Figure K5. Fossil resources (natural gas and coal) are most important. The share of combined heat and power production is great because the residential sectors in the cities are usually connected with a central heat supply. Although the level of consumption is comparable to Western countries the structure of consumption is quite different. The share of industry in consumption is much higher than in Western Europe (seeFigure K6). Thus the energy intensity of industrial production is much higher in Russia. It reaches a level of about 1.25 kWh/US$ against 0.66 kWh/US$ in the OECD countries.[17] Due to the time characteristics of industrial consumption the load factor of Russian producers is very high. Frequently, problems of supplying electricity to residential customers had to be solved by switching off part of the industrial load.

Figure K6: Power Consumption by Sector

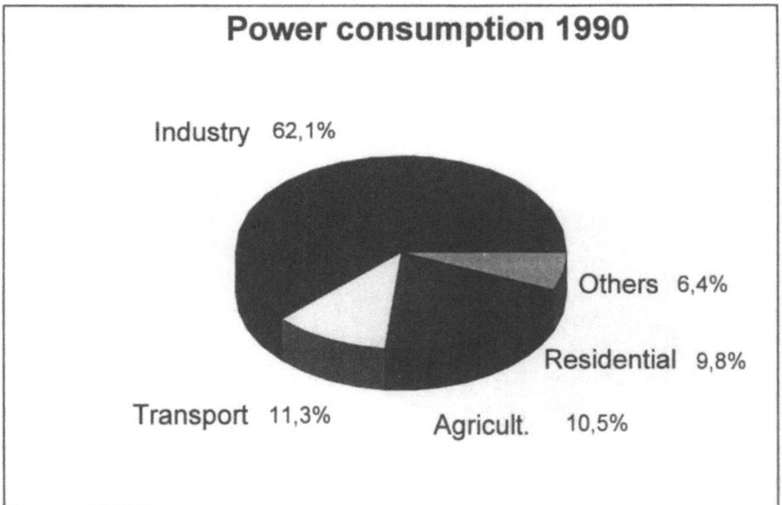

Source: OECD

3.3 State of the Institutional Reform of the Sector

In the Soviet economic system the power sector was run by two ministries: the Ministry for Energy (Minenergo, responsible for the whole non-atomic supply) and the Ministry for Nuclear Energy (Minatomenergo, in charge of nuclear energy). Both ministries were co-ordinated by the central planning agency GOSPLAN.

The ESI was organised as a state monopoly. It was called the unified energy system of the USSR. It comprised the whole sector from production to transport and distribution including all additional services for construction and research and development. In the past the supply of heat and electricity was part of the basic social service that the government had to guarantee for the population and the economy. Therefore tariffs were part of the social policy. This system that persisted over many decades has influenced present-day thinking and expectations. A change in prices to conform with cost structures and as part of a competitive market economy requires a longer process.

The unified energy system consisted of eleven regional electricity administrations. Six of them were located in the area of the former Russian Republic, i.e. today's Russia. They are identical with the statistical division into economic zones, (Northwest, Central, Volga, North-Caucasus, Ural, Siberia). The far east was not connected to the system. The five remaining electricity administrations were responsible for one or several of the former republics of the USSR (see Figure K7).

Like all other sectors of industry the ESI was controlled on the basis of some central quantitative indicators and by the principle of the central allocation of resources and distribution of final products. Investment was centrally planned and financial means were taken from the state budget. Thus economic agents were highly dependent.

Figure K7: Power Network

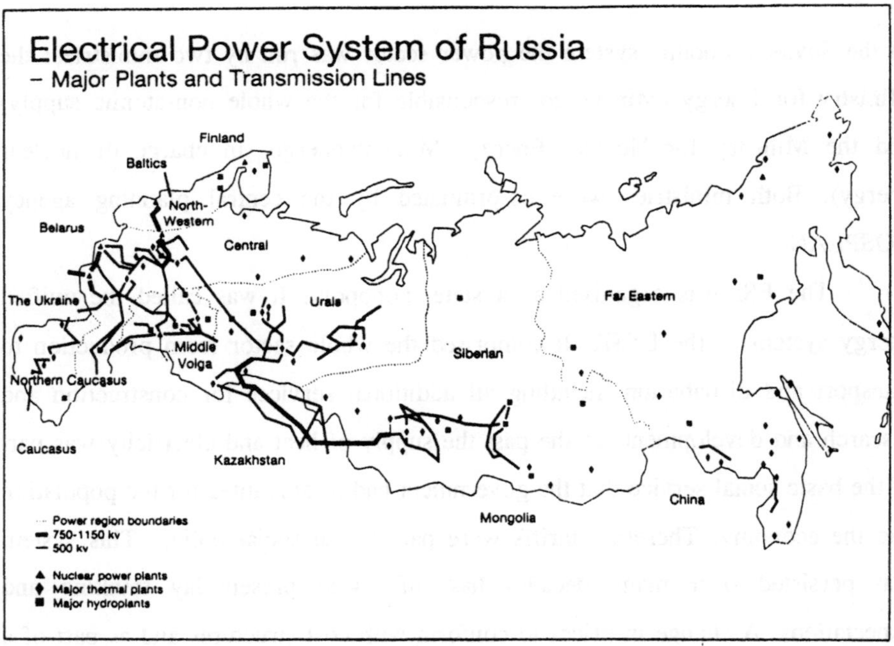

In the present-day reform in Russia the ESI is still a sector which is given high political priority. The Ministry for Energy and the former Ministry for Oil, Natural Gas and Coal, were combined into the Ministry for fuel and power (Mintopenergo). The concentration of these sectors - in Russia they are often called Fuel and Power Complex (TEK) has strengthened its economic and political influence.

Privatisation and liberalisation of prices mark the beginning of the reform in this sector. In both aspects there are certain restrictions compared to other sectors.

Privatisation

Former state property can be allocated to the federation, to regions or to local governments. In this context the assets of the ESI were principally allocated to the federation. According to the program of privatisation in sensitive sectors the state has to retain a share of 25.5 % or 51 % of the total assets for at least three years. After three years it is possible to renew this state ownership or to sell those shares. The ESI was considered such a sensitive sector.

On the basis of the presidential decree no. 923 (Aug. 15, 1992) "organisation of ESI of the Russian Federation under the conditions of privatisation" and no 1334 (Nov. 5, 1992) the Russian electricity sector was privatised and united to form the joint stock company called Russian Company for Electricity (RAO EES Rossii).

The tendency to maintain highly centralised structures in the energy sector with Rosugol for coal, Gasprom for natural gas, was thus continued. RAO EES consists of the following parts:

1. The 72 regional energy systems (RES); these have some production capacity of their own as well as the regional distribution networks and are responsible for the supply of consumers in their region).
2. Central and regional dispatch units;
3. The transport system of 330 kV and more;
4. Large generators (among them all thermal power stations producing more than 1000 MW, all hydro power stations producing more than 300 MW). This is about one fourth of the whole capacity and
5. Construction units and research capacities.

All these parts were supposed to become independent joint stock companies of which RAO EES was supposed to retain a majority of 50 % or more. Thus it was planned to turn RAO EES into a kind of holding company. The main owner of RAO EES retaining 51 % of the shares is the government of the Russian Federation. As with many other Russian enterprises most of the remaining shares are retained by the employees. [18] The main criterion seems to have been the ability to maintain the unified structure. The decision of the setting up of RAO EES was strongly influenced by the central administration; the regional energy boards almost had no influence on this decision.

When the regional companies were formed they were expected to give 50 % of their shares to RAO EES. In some cases all of the shares were given to RAO EES. Frequently, however, the companies refused to do so. In some cases no shares were transferred at all, in other cases considerably less than 50 %. Altogether the state company is in control of almost the entire ESI. Only a small number of large generators have established independent companies.

The production of nuclear energy remains with the Ministry of Nuclear Energy and is not a part of RAO. The electricity produced in nuclear power stations is sold to RAO according to the terms of a contract between Mintopenergo and Minatom.

Prices

Prices are subject to state regulation. The principles for price regulation are laid down annually in a special decree issued by the Committee for prices of the Russian Federation after consultations with Mintopenergo and the Federal Electric Power Commission. The prices themselves are adjusted quarterly.

Currently a division can be made between two kinds of consumers: final consumers, the prices for whom are fixed by regional commissions, and consumers that are served on the basis of prices fixed by the Federal Power Commission (e.g. major industrial enterprises).

At the moment the consumers from seven different fields fall into two groups: four of them (industrial enterprises, all transport enterprises and other commercial consumers) pay above-average prices, whereas the remaining three (agriculture, resellers and residential customers) pay below-average prices with residential customers paying the lowest prices.[19]

Figure K8: Electricity Prices

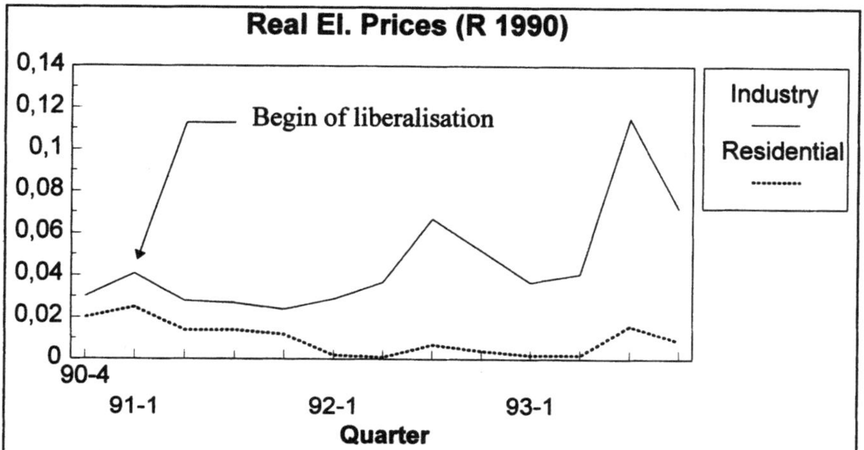

Source: OECD and own calculation

Figure K8 shows that after prices have been liberalised prices for residential customers did not really increase whereas prices for industrial consumers went up considerably in real terms. This clearly shows the special conditions for residential consumption. In the meantime a decision has been made generally to remove such subsidies. But it will be a long way until a cost-oriented pricing system will have been realised. In the Western systems potential competition through auto production of large customers is an important corrective for pricing policies. It is very unlikely that under the present macro-economic conditions such a potential competition will play a major role in Russia.[20] Regional energy systems have to pay transport costs to RAO that have been fixed by the Federal Power Commission. Regional customers fix their tariffs for themselves. Regional tariffs are based on the regional cost of production plus the cost of buying electricity from RAO. Often RAO prices are below the level of cost of regional production.

4. Analysis of the Reform Process

4.1 General Problems of the Present Reform

Highly vertically integrated monopolistic structures were typical of all sectors of production of the former Soviet Union. The technical structure of the electricity sector gives further incentives for a centralised form of organisation. And the process of transformation into a market-oriented structure of the electricity industry is not a necessary outcome, as developments in Western countries have shown. In general, the Russian government has decided to change the market order in the electricity sector by introducing privatisation. This is in contrast to plans for other industries as for example nuclear power or the postal service and telecommunication.

Theoretically, different models are conceivable for opening up the market for electricity in Russia. Such models have to consider the enormous size of the country and the different conditions prevailing in the six former regional electricity districts.

One possibility would have been to set up a number of regional monopolies by keeping the vertical integration intact within the six former electricity districts. Such a development would be comparable to the German system. As it is necessary to balance electricity production and consumption in the regions such a model of regionalisation could have some advantages, for within those regions it would have been possible to maintain a centralist regime.

In a second step one could think of subdividing the regional monopolies vertically by their functional subdivisions. The various parts of the system could be privatised in order to set up independent companies. Thus it would be possible to develop consumer and producer markets.

A different approach would be vertically to subdivide the former unified system by setting up companies on the horizontal level, similarly to the development in Great Britain. To achieve this it would be necessary to split up the regional monopolies which could result in competition among producers. Considering the great distances in Russia the transaction cost of such a model might be quite high. Therefore it seems to be more realistic to develop markets on a regional level.

The instability of the political system and the weakness of central power boards in Russia are on the one hand a good precondition for a removal of the state monopoly. On the other hand a decentralisation might also be risky because it might result in regional separatism rather than regional autonomy. Under the present conditions, i.e. after the breakdown of the central power system, regional interests which are based on large differences in regional resources have a chance of being considered. However, there is a danger that the pendulum will swing too far towards regional interests. The present concept on which the RAO EES is based aims at a unified system under the roof of a monopolistic structure. Certainly such a decision mainly has traditional causes. The political decisionmakers have preferred centralist privatisation to regionalisation.

The RAO was founded in order to achieve the following targets:

- transferring responsibility and risk to the private sector by keeping up governmental control and maintaining a unified sector;
- establishment of market-oriented electricity pricing and gradual decrease of cross subsidies for certain groups of the population

The withdrawal of the state in this sector may, however, also result in a loss of governmental control. The monopolistic RAO EES may develop a dynamic of its own, the degree of which depends on its profit orientation. General interests may require to set up a system of regulation outside of RAO. At present government seems not to be able to enforce such a regulation. Instead, it may be observed that new rules are being set up on the basis of different interests within the structure of RAO. Additional actors try to influence pending fundamental decisions. These are the operators of regional electricity systems (RES) but also the local and regional administrative bodies. Their interests differ from those of the RES.

The Present Function of Different Actors

The foundation of RAO and the formation of regional companies created a kind of wholesale market for electricity. In this market RAO retains a monopoly for the sale of electricity to those RES in need as well as for buying nuclear power and sometimes also for the surpluses from regional systems.

According to the foundation decree RAO has to fulfil the following tasks:

- centralising the co-ordination of the unified energy systems;
- implementation of a unified research and investment policy;
- reinvestment of profits;
- personnel development;
- creation of a market-oriented control system and co-ordination of the process of privatisation in the sector. [21]

The RES are producers and distributors and sell electricity to final users. Most regional systems have to buy electricity from RAO to balance regional demand and supply. This particularly applies to many Siberian regional systems because the share of large power stations which were centralised under RAO was very high in this region. Up to 70 % of all capacities had thus to be transferred to RAO from regional systems in Siberia. [22] Quite frequently the RES are forced to re-buy electricity from RAO, produced by a power station in their territory, at a price exceeding the regional cost of production of the particular power station.

To date the role the Ministry plays is unclear. The 50 % shares of RAO retained by the central government were transferred to RAO itself. It appears that the Ministry intends to take a back seat and that the driving forces for any new regulation lies with RAO and the regional systems.

The diverging interests of the different actors are likely to result in conflicts.

Centralisation and Decentralisation

The major question here is who should be the subjects of the electricity market. This is important for the division of the assets, the right to invest and the methodology applied in price fixing. The interests of regional and local administrative bodies and the operators of regional energy systems have begun to surface. They are still dominated by administrative rather than economic conceptions. At the same time it has become clear that the current division of assets will determine the future policy in this sector.

The RES normally consider the existing capacity to be regional "property" of which they intend to keep control. This refers to the regional systems as well as large power stations in their territory. The transfer of large power stations to the control of the centralised system will increase regional production costs and have immediate effects on regional price fixing (a more detailed analysis of regional conflicts will be given in section 4.2.).

RAO and the Ministry of Energy argue historically. The former investment that has created the present assets came from the state budget. It was financed by society as a whole and the decision on the siting of generators was made on the basis of macro-economic considerations. Therefore it does not seem legitimate to regionalize those capacities. When RAO was established 21 large power stations were transferred to RAO. At the end of 1994 this figure went up to 23 and it is planned to transfer another 25 power stations to RAO from the regional systems.[23]

Also, a number of large generators who formed independent companies take part in the debate about the subject status. In their view the creation of a producer market will strengthen their position in relation to RAO. The present concept does not provide such a possibility.

Investment

There are also discussions about investment decision-making. When RAO was set up it had been planned to stop state investment. Instead, special investment funds were created on the federal and regional level. It was legally fixed that all RES have to contribute a 6 % tax on their total power volume to these funds. [24] RAO itself is supposed to include an investment charge into its tariff in order to be able to accumulate financial means for investment. In 1993 this investment share amounted to 50 % of the cost basis and for 1994 RAO demanded that this share be increased to 50-60 %. [25] The investment fund is a kind of state investment tax. It remains unclear what the responsibilities for the administration of these funds are and what criteria are being applied for expenditure financed from these funds. At present there are no clear rules either regarding the investment activities of RAO.

Competence for investment decision-making and the possibility to form one's own investment funds are being reduced by this centralised rule. There is a strong contradiction between these rules and the general idea of setting up private enterprises as independent agents. We have been told quite frequently that RES refuse to pay any money into the investment funds so that *de facto* the central rules are not being implemented properly.

Subjects of Pricing

Another area of conflict is price fixing and particularly the competence for decision-making of the central and the regional energy commissions.

Differences in regional cost structures are reflected in regional price differences. Thus the regional average price in Siberia with 26.99 rubels/kWh was far below the regional average for the whole of Russia (33.48 rubels/kWh). [26] The transfer of large power stations to RAO raised regional producer prices and at the same time reduced the wholesale prices offered by RAO. The price for electricity in the wholesale market is still higher than the production price that would have been required if the regional power station would have remained in the regional system. The transfer of companies to RAO also increases the transport cost which RES is billed for by RAO. Currently prices in the different tariff classes and cost are inversely proportional. It is planned to level the differences between the different groups of customers by the end of 1995. In general the prices for the tariff classes have come closer to each other since 1993 although the prices for residential customers remained very low (see Figure K9).[27]

Figure K9: Prices by Type of Tariff

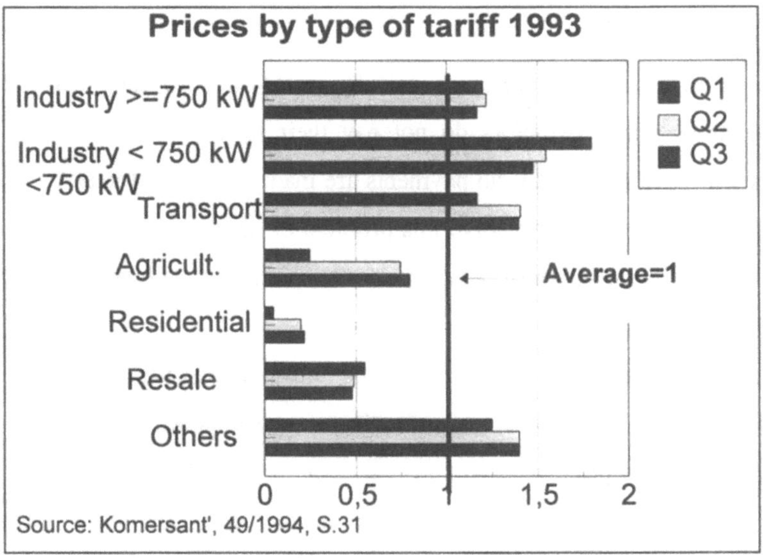

On the one hand the RES may themselves fix the prices for the ultimate consumers. On the other hand there are central parameters for subsidised tariffs for residential customers. According to government decree no. 352 (April 22, 1994) they must not exceed production cost by more than 5 %. This results in real losses because of the high value added tax. This central directive makes it difficult to bridge the gap between the different tariff groups. Losses from subsidised tariffs have to be financed by the RES.

The tariffs of the various regions differ widely. Although government subsidies are not granted any more, the RES, due to their closeness to the ultimate consumer, try to avoid price increases for residential customers and use the possibility of delegating responsibility to central bodies.

Financial Conditions

It is a typical feature of the present economic situation in Russia that the government and many enterprises are unable to meet their financial obligations. This also applies to the ESI.

A discussion about the reform of pricing seems inappropriate because frequently consumers and distributors do not pay their bills at all. Even cost-oriented tariffs will have no effect if no payments are made by the customers.

Attempts have been made to solve this problem of mutual indebtedness. In August of 1994 some basic sectors (metal, coal, oil, electricity, railways) agreed to form a kind of cartel. This contract was aimed at:

- stabilising prices on the basis of indexing mutual deliveries and services,
- a reduction of mutual debts on the basis of a clearing system and
- an agreement on export prices.

Additionally the government was supposed to reduce taxes or wave taxes entirely.[28] The Russian government agreed to the establishment of this cartel and legalised its operation. Another method to pay for debts is to issue short-term papers. Different companies agreed mutually to accept those papers instead of insisting on payment. The Russian commodity exchange trades those papers now. In practice, this results in the creation of an additional form of money supply. In general all those special rules are an obstacle to the proper functioning of newly emerging market structures.

4.2 Regional Interests - Siberia

So far in this paper the term "region" has been used to comprise all those interests differing from those of the RAO and the Ministry (Mintopenergo). However, a distinction must be made between different types of regions - districts (Oblast) and regions in the sense of the former economic territories. Regional interests can become more important on both levels. With the exception of the Tjumen district which is connected to the Ural energy system all Siberian regions belonged to the Siberian Energy Complex in the past.[29] This was part of the former unified energy system of the USSR. At the same time it was practically autonomous because the exchange of electricity with other parts of the system (except for Kazakhstan) only accounted for 2-3 % and was thus very small.

The resource endowment in the different Siberian districts is quite different. The regional systems of Krasnojarsk, Irkutsk and Kemerowo produced electricity surpluses and at the same time have the highest regional concentration as regards industry. There is only one exception, namely the Tomsk region, which is a highly industrialised area with electricity deficits. The other deficit regions, Novosibirsk, Omsk, Tschitinsk, Burjatia and the Altai region consume more heat than electricity.

The present system does not allow a direct transfer of electricity between deficit and surplus regions. Any exchange has to be done through the RAO monopoly. The price policy of RAO, however, is determined by criteria that are applied throughout the country and the specific situation of RAO itself. Power surpluses in the area of Irkutsk and Krasnojarsk are produced by large and cheap hydropower facilities. Since they have been in operation production costs have been low in all of Siberia. Because a transfer of those generators to RAO would have turned the RES of those two regions into deficit systems the RES refused to transfer the right to these capacities to RAO.

With deficit regions in Siberia the same observation was made: "Novosibirsk Energo", for instance, in charge of the Novosibirsk region has to buy 40 % of its consumption from RAO. However, it only transferred 19 % of its shares to RAO instead of the 50 % RAO had demanded. A consequence of this transfer was that Novosibirsk Energo had to lease back a power station it used to possess from RAO. It is difficult to say what service is provided in turn for the amount paid. Other RES renamed power generating stations and termed them "heat generators" in order to avoid a transfer to RAO. The conditions for the monopolistic wholesale market dominated by RAO are not the same throughout the country. RAO offers electricity at regionally varying prices so that actually there are several wholesale markets. It is unclear on what basis the borderlines of those markets were fixed. Regional price differences have increased since early 1994. At that time Siberian deficit regions had to pay 24.6 rubels/kWh, RES of the Ural region paid 14 - 22 rubels, in the central region 11 - 13 rubels were paid and in the Volga region the price was 9 rubels. [30]

In 1990 the regions of Siberia formed a union called "Sibirskoje Soglaschenie". It was the aim of this union to co-ordinate the different policies pursued in the different areas. Among other things the union attempts to gain influence on the price policy of RAO. A major result that the union has achieved to date is that a representative of Siberia has become a permanent member of the Federal Energy Commission. This lobbying resulted in a reduction of the RAO prices for Siberia. Within Sibirskoje Soglaschenie an energy commission has been set up. There are ideas of establishing closed connections among the various Siberian RES. The aims are to reach a price policy which is based on regional cost as well as more autonomy with respect to the use of resources. According to currently valid rules Siberian RES have to use domestic coal whereas natural gas and oil required for power generation have to be taken to European regions. A first step towards unification will be to turn shares of RAO that are assets in Siberian power stations, networks, etc., into regional property. Currently the setting-up of a kind of Siberian holding company is under discussion, however, it is not yet clear how this can be done.

RAO and the RES have some interests in common; RES, however, are dominated more by local interests. There are no sector-specific unions which try to bundle producer or consumer interests in relation to other actors. In general, agreements between RAO and the RES are reached in bilateral negotiation. In areas where surpluses are produced like in Siberia, the position of RES is, of course, very strong. Novosibirsk Energo told us that they refused to accept the average RAO price and instead insisted on being charged a price which was based on regional surplus production cost (hydropower). In order to achieve this they threatened to create additional capacities of their own. On the one hand it is an important step towards a transformation process to take action on the basis of local interests; on the other hand, however, this will not necessarily result in a reasonable optimal operation of the system as a whole. Most regions, for instance, prefer the use of coal-fired power stations (in order to be able to protect the employment of their miners) to buying cheap surplus electricity from hydropower resources available on the wholesale market. This would not only be cheaper but also ecologically beneficial.

4.3 The Concept of Privatisation in the Discussion

It appears to be too early to expect basic conceptual changes that might result from the experiences described in the above paragraph. At the beginning of 1994 a draft version of the presidential decree was discussed. It stated that a redistribution of the RAO shares to the various regions was necessary and should be a function of the respective deficit situations. This proposal was rejected by the majority of the RES. [31]

The Ministry of Economics made similar proposals which were also aimed at strengthening the position of the regional energy systems and at separating the large generators from the RAO and turning them into independent private enterprises. This might be a first step towards a producer market. Power prices could be decreased if large generators and nuclear power stations could supply the regions directly. This would help to promote the setting-up of new industrial enterprises, particular those in the regions which are energy-intensive.

The Ministry of Economics made the following proposals:

1. Turning RAO into a limited financial holding company which would enable a control of investment; investment should not be financed through indirect tax collected from all consumers but through profits and capital contributions made by interested consumers.
2. The central dispatching unit (ZDU) is to be reorganised and turned into an independent joint stock company. It should comprise all networks of 330 kV. This company should be run as a non-profit organisation and its fees are supposed to cover cost and modernisation expenses. Because ZDU is a monopolist the size of its fees should be regulated by government.
3. The government shares of all RAO companies should be transferred to the various regions. Regions with poorly developed energy systems should be enabled to gain a larger share in the power stations of other regions in order to ensure that all regions have a good starting position and to avoid that richer regions reduce the prices. [32]

These proposals of the Ministry of Economics certainly reach much further than the currently on-going process of privatisation. They are aimed at removing the vertical structure of the present RAO monopoly. Some formulations in these proposals which do not seem logical, hint at the political nature of these suggestions. To date these proposals have not been realised and constitute a sharp contrast to the interests of RAO as well as of Mintopenergo, both of which - according to this concept - could be removed.

Development up to now has shown that the political agents on the governmental level do not have enough power to implement their ideas. State property rights provided the possibility to make a political decision on the future organisation of the ESI which could be set up after all by differently redistributing former state-control property rights.

Decisions that have been made so far have reduced the possibilities of future implementation of clear-cut models. New interests will develop as a result of new property rights. They will develop dynamics of their own and entail real world organisations. RAO as a central monopolist has a very powerful position, also in relation to government. It opposes the demand for change put forward by the RES and the Ministry of Economics with the inertia that is typical of historically grown structures. A redistribution of property rights did therefore not take place. Instead, more flexibility in pricing was introduced. Subsidised tariffs can now also be granted by some of the regional energy commissions. This kind of reaction that can best be described as fear of fundamental decisions does not improve the general conditions of competition. The transparency of decision-making and general rules decreases and voluntary action becomes increasingly important.

However, it is far too early, to draw any general conclusions. The fact that the number of new and independent actors, especially at the regional level, has increased, has revealed the different interests and has started to challenge the new monolithic structure. It can be anticipated that new concepts will be implemented at the regional level. The strive for autonomy and entrepreneurial independence is linked to the consciousness of the functioning of the electricity system. A new bargaining model for determining various interests can only develop step by step from the grassroots. It will be determined by sector-specific interests as well as by general economic conditions in the regions.

5. Cooperation with Western Europe

In Russia there is a high need for investment in the electricity sector. This refers to environmental retrofitting, modernisation of existing plant and completion of investment projects under construction. There also is a need for consulting and training-programmes. The second aspect is partly being served by the EC TACIS-programme. In Moscow and St. Petersburg consulting centres for the energy industry have been opened that include management training programmes in western countries. The opening up of another centre in Novosibirsk is being considered.

As far as investment goes there are a number of obstacles for successful involvement of western companies which generally would be interested in a long term involvement in the Russian ESI. These obstacles are the general economical and political instability, the problem of investment security and the problem of finance.

The possibility of direct compensation for investment is only given in industries with high foreign exchange earnings. In other cases investment projects are financed through barter or the exchange of shares for Russian joint stock companies. Western companies hold shares of RAO EES and some of the regional energy companies (Moscow, St. Petersburg, Irkutsk). The value of shares of regional companies lies far above those of RAO. This shows that there is a high interest in the regional companies.

In the capital market there are some activities to increase the security of foreign investment and to open this market to a large a number of foreign investors. For example the Russian bank „Rossijskij kredit“ in this context has negotiated agreements with American and German banks about custody-deals.

A number of joint ventures in the electromechanical sector are on the way comprising different products for the production of power plants.

In the ESI itself the Russian side is offering to open up joint companies that could result in independent producer companies (comparable to the IPPs in the US). At the moment there is no legal basis for such projects so that for each project an individual detailed contract has to be set up. Several such projects are being planned right now. Some large German companies from the electricity sector and the electromechanical industry are willing to get involved. None of these projects is however already on the way.

If this IPP model is being developed further a necessary implication will be to establish some kind of regulation referring to the access rights to the electricity network.

Endnotes:

1 We thank the Volkswagenstiftung for supporting this research.

2 Brunekreeft concludes form a model study that it might not be optimal to exclude the distributor from trade. Brunekreeft, G, (1994): Open access versus common carriage, in electricity dtstribution, Diskussionsbeiträge, Insitut für Verkehrswissenschaft und Regionalpolitik, Universität Freiburg, Nr. 14.

3 This is a general rule in the law of competition: Gesetz gegen Wettbewerbsbeschränkungen.

4 Gesetzentwurf der Bundesregierung zur Neuregelung des Energierechts vom 28.1.1994.

5 Public Utilities Regulatary Policies Act (PURPA).

6 Klopfer, T. und Schulz, W.,1993: Märkte für Strom, München, p. 355. (Translated by WP).

7 According to information from Dr. F.P. Sioshansi of Electric Power Research Institute (EPRI) it is uncertain when this retail-wheeling will be introduced; however the enterprises intend to implement it and currently carry out restructurings in order to be able to cope with it.

8 Member countries can exempt sectors of public importance from competition according to section 90,2 of the Treaty.

9 Recently published research on the development of prices for industrial customers tried to answer the question of whether this is actually trus. The researchers found that contract-oriented prices have led to pool prices (in contrast to what was expected). If the difference between pool (spot market) and contract prices is considered the safety premium of consumers, the result seems quite plausible. Cf. Lucas N.and Taylor, P., An analysis of electricity contract prices to industrial consumers in the United Kingdom, in: Utilities Policy 1994, pp. 309-313.

10 According to information from the Institute for Economic Organization of Industry of the Siberian Academy of Sciences, Novosibirsk.

11 International Energy Agency (IEA) 1994: Electricity in European Economies in Transition, Paris, p. 194.

12 Merekov A. P. and Saneev, G. (1994): Sibiriens Brennstoff- und Energiewirtschaft, in: Energiewirtschaftliche Tagesfragen, p. 798.

13 Merenkov, l.c., p. 799.

14 Merenkov l.c. , p. 799

15 Nekrasov et al. (1993): Al'ternativy razvitia rossijskoj energetiki, in: Problemy prognozirovania, Nr.6, S. 10

16 IEA, l.c., p. 200.

17 IEA, l.c., p. 194.

18 A description of the privatization in Russia can be found in Leipold, H. (1993): Alternative Privatisierungs- und Sanierungsmethoden in Mittel- und Osteuropa, In: Thieme, H.J. (Hrsg.): Privatisierungsstrategien im Systemvergleich, Berlin; Schröder, K. und Kazmin, A. (1993): Der Privatisierungsprozeß in Russland, in: Osteuropa-Wirtschaft.

19 Kommersant, 31/1994, p. 31.

20 To date we had no opportunity to do empirical research on this question.

[21] Pravednikov, N.K. (1994): Upravlentcheskije i organizacionnye aspekty v energetike Rossii, problemy finansirovania i zarubezhnych kapitalovlozhenij, in: Problemy Toplivno-energetitcheskogo Kompleksa, Nr.4/1994, S.8.

[22] According to information from the Institute for Economic Organization of the Siberian Academy of Sciences, Novosibirsk.

[23] Kommersant', 36/1994, S.23.

[24] IEA, a.a.O., S. 191.

[25] Kommersant', 36/1994, S.22.

[26] Kommersant', 36/1994, S.22.

[27] Kommersant', 49/1994, S.31.

[28] Kommersant', 35/1994, S.46.

[29] These are the regions of Novosibirsk, Irkutsk, Krasnojarsk, Tomsk, Omsk, Kemerowo, Tschitinsk, Burjatien and the Altai region.

[30] Kitushin, V.G. u.a. (1995): Sibirskije tarify na elektroenergiu v usloviach stanovlenia rynka, in: Region - Ekonomika i sociologia, 1/1995, S. 128.

[31] Kitushin, V.G. u.a. (1995): Sibirskije tarify na elektroenergiu v usloviach stanovlenia rynka, in: Region - Ekonomika i sociologia, 1/1995, S. 129.

[32] Kommersant', 7/1994, S. 21-23.

References:

BRUNEKREEFT, G. (1994), Open access versus common carriage, in electricity dtstribution, Diskussionsbeiträge, Insitut für Verkehrswissenschaft und Regionalpolitik, Universität Freiburg, Nr. 14.

INTERNATIONALE ENERGIE AGENTUR *(IEA)* (1994), Electricity in Eurpean Economies in Transition, Paris.

KARTTE, W. (1995), Aufbruch zum Markt - als Wirtschaftsberater in Rußland, Hamburg.

KITUSHIN, V.G. u.a. (1995), Sibirskije tarify na elektroenergiu v usloviach stanovlenia rynka, in: Region - Ekonomika i sociologia, 1/1995, S. 128.

KLOPFER, T. and W. SCHULZ (1993): Märkte für Strom, München.

LEIPOLD, H. (1993), Alternative Privatisierungs- und Sanierngsmethoden in Mittel- und Osteuropa, in : Thieme, H. J. (Hrsg.): Privatisierungsstrategieen im Systemvergleich, Berlin

LUCAS, N. and P. TAYLOR (1994), An Analysis of Electricity Contract Prices to Industrial Consumers in the United Kingdom, in: Utilities Policy , S. 309 - 313.

MERENKOV, A. P. and G. SANEEV (1994), Sibiriens Brennstoff- und Energiewirtschaft, in: Energiewirtschaftliche Tagesfragen, S. 798 - 802.

NEKRASOV, u.a. (1993), Al'ternativy razvitia rossijskoj energetiki, in: Problemy prognozirovania, Nr.6

PFAFFENBERGER, W. (1993): Elektrizitätswirtschaft, München

PRAVEDNIKOV, N.K. (1994): Upravlentcheskije i organizacionnye aspekty v energetike Rossii, problemy finansirovania i zarubezhnych kapitalovlozhenij, in: Problemy Toplivno-energetitcheskogo Kompleksa, Nr.4/1994, S.8.

SCHRÖDER, K. and A. KAZMIN (1993): Der Privatisierungsprozeß in Rußland, in: Osteuropa-Wirtschaft

L. Co-operation in Energy Policies: European Union (EU) - Eastern Europe

Peter Palinkas

1. Introduction

The revolutionary upheavals in Eastern Europe have made increased co-operation between Eastern and Western Europe possible (in the energy sector too). The conditions are particularly favourable and promising in the energy sector. Western Europe has a large requirement for imported energy (approximately 43 % covered by imports in the EU). Eastern Europe (and especially the Community of Independent States, CIS) has large energy reserves (natural gas, oil and coal). The security of supply to Western Europe could be improved in the long term by increasing the exchange of energy, i.e. energy exports from Eastern to Western Europe. The first precondition for this, however, is that it proves possible to establish political structures which are secure over a long period in the Eastern European countries. Increased exchange of energy between Eastern and Western Europe could also provide the financial resources necessary for restructuring the energy sector itself (energy saving, improvement of energy efficiency, encouragement of environmental protection) and for generally reconstructing the whole economic system. Western Europe could in this way make a contribution to economic stability in the Eastern European countries, at the same time securing and strongly reinforcing the democratisation process. In addition damage to the environment, which also affects Western Europe, could be reduced as a result. Finally, it is to be expected that increased co-operation in the energy sector between East and West will tend to have a marked stabilising influence on world energy markets.

It would be desirable if important initiatives for further East-West energy co-operation were to be adopted and implemented by the West, especially by the European Union (EU). East-West co-operation in the energy and environmental sectors is a particularly appropriate way for the west to provide the countries of Eastern Europe with the help they so urgently need as they tread the thorny path of economic reconstruction and restructuring.

The countries in Central and Eastern Europe (CEE, including the CIS) face a variety of adjustment processes: economic, institutional and political (WELFENS/WELFENS, 1994). Burdened by severe economic problems, inefficient industrial structures (the dominance of monopolistic large-scale undertakings etc), a shortage of commercial resources, such as human capital, and a lack of technology and technological know-how, all these countries are seeking ways of achieving stability and prosperity. Forging a loose or closer link with the European Community would appear to be an interesting political option. An implicit or explicit extension of the Community eastward that was economically successful and underpinned political stability throughout Europe could make a vital contribution to preventing the continent from sinking back into problems reminiscent of the inter-war period. The sensitive areas of energy and science and research are particularly suited to providing a link between Western and Eastern Europe and thus preparing the ground for long-term and successful East-West co-operation based on equal partnership. The European Community has taken the first steps here with the initiative on the European Energy Charter (PALINKAS, 1992) and the support programmes set in place for restructuring in Central and Eastern Europe (see, for example, the PHARE and TACIS Programmes).

Reliable framework conditions need to be established if the energy industry are to flourish and develop further. Such conditions are particular vital for the energy industry which - unlike almost all other sectors of the economy - requires extremely lengthy period of planning and capital tie-up. Inflationary processes in some of the countries in Central and Eastern Europe are making long-term decisions on planning and investment more or less impossible. High levels of inflation make it impossible to have a properly functioning money economy and mean that a system of barter trade and a short-term economic strategy are almost inevitable. The uncertainties surrounding institutional restructuring in the economy in some of the countries in CEE also add to the general feeling of insecurity.

The framework conditions, that have been described only in brief here and largely remain shaky, are a source of dangers and problems which could critically impede the process of restructuring that is in fact so urgently needed. The establishment of reliable framework conditions of this kind has therefore to be a '*sine qua non*' for further reorganisation, especially in the energy sector.

The following paper tries to describe the existing co-operation between Western and Eastern Europe and will concentrate mainly on a description and analysis of the interconnected systems between East and West in the electricity and gas sector (including prospects for further co-operation and some remarks on the actual Transeuropean-Network-proposals).

2. East-West Co-operation in the Electricity and Gas Sector

2.1 The Electricity Sector

While the interconnected system in Western Europe is largely run through the UCPTE network (Union pour la Co-ordination de la Production et du Transport d'Electricité), the interconnected system is Eastern Europe is provided through the network of what was the Council for Mutual Economic Assistance (CMEA). Because of differing frequency stabilities (level of frequency variations) in the Western European and Eastern European interconnected systems, current can only be supplied between the two separate networks via the appropriate direct current network coupling. Until recently, a link of that kind existed only between Austria and the former CSSR, and since the beginning of 1993 between Czechoslovakia and Bavaria (the network of the former GDR is linked since 1994 to the UCPTE network). Limited exchanges of current between the two networks in Western and Eastern Europe are otherwise possible through so called directional operation (the power station supplying the current is disconnected from the network of its own supply operation) or through isolated operation (the region being supplied is disconnected from the existing network).

The technical, organisational, economic and legal aspects of operating the CMEA and UCPTE systems in parallel are currently under review. It has to be borne in mind here that even now the UCPTE system is the largest, parallel-operated, interconnected system in the world. Extending it to include in the CMEA system would be likely therefore to involve many problems. Any such extension of the UCPTE system will have, unquestionably, to be based on ability to comply with the institutional (power station plant and equipment and network) and operating rules. There will be a particular need for economic assistance from the UCPTE countries here. The countries of Western Europe will have therefore to transfer both human and non-monetary capital to the countries of the former CMEA.

On the basis of the European Energy Charter, a number of Western European electricity companies (eg Preußen-Elektra, RWE and Bayernwerke in Germany and the EDF in France) are offering support for the modernisation of the energy industry in Eastern Europe. The first stage here has been for consultants to enter into contact with those responsible for the energy sector in the Republics of the former USSR and in Poland, the former CSSR and Hungary. An important part of this work involves linking up the existing interconnected systems of Western, Northern and Eastern Europe. These interconnections are essential if a commitment to modernise the energy industry in the countries of Eastern Europe is to be of benefit to that industry in local terms also. The possibility of subsequently ensuring that the investments in Eastern Europe to be financed by the West (production plant and networks) pay for themselves through the supply of current from East to West is now being considered. The European Energy Charter could provide essential assistance and safeguards for forward-looking co-operation of this kind in the electricity industry.

In the short and medium term, there is, of course, little possibility of setting in place an interconnected system for Europe as a whole (from Biscay to the Urals - or further still, from the North Cape to Sicily). We shall, however, look briefly below at how the European interconnected system has developed historically, at how it is currently organised and the possibilities and problems of perhaps one day setting in place an interconnected system for Europe (including the CIS).

Interconnected electricity networks have been set up throughout the world, prompted by the desire for large-scale exchanges of electricity, going beyond national borders, for optimum use of the different energy sources, and for an opportunity of making reserve capacity available, on a reciprocal basis, in the event of power station failure. The UCPTE network, which now encompasses twelve Western European countries and has power station capacity of more than 387 000 megawatts (MW), was first set up in 1958 and has now become one of the largest interconnected networks.

The development of electrical energy supplies in Europe led, after the Second World War, to the establishment of five large interconnected networks which, with the exception of IPS/UPS-system in the former Soviet-Union, are not as yet connected or are connected only via direct current bridges.

Five interconnected systems in Europe

The five interconnected systems in Europe are:

- UCPTE (Union for the Co-ordination of the Production and Transport of Electric Power; twelve Central European countries)
- NORDEL (Nordelektrik; Sweden, Norway, Finland and Denmark)

- CEGB (Central Electricity Generating Board; United Kingdom)
- IPS or VED (Interconnected Power System/United Energy System; Poland, the former CSSR, Hungary, the former GDR, Southern/Western network of the former USSR; the former GDR, Poland, Czech and Slovak Republic are now connected with the UCPTE-system, see below)
- UPS or EES (Unified Power System/Unified Energy System; remaining parts of the former USSR).

In addition to geographical factors (eg the United Kingdom and Scandinavia) and economic considerations, there are also technical features which prevent direct synchronous connection of these networks. Interconnected networks are generally set up with two main aims:

- to improve the security of supply
- to increase economic efficiency.

The reasons for the development of several interconnected systems were partly political but partly also technical. Thus the separation of the high voltage systems in continental Europe into the UCPTE network and the IPS/UPS network was a reflection of the - until recently still existing - rigid separation of the political and economic systems in Western and Eastern Europe. The separation of the NORDEL interconnected system and the interconnected system in the United Kingdom from the UCPTE network is in contrast a technical issue. Those systems are able to be connected to the continent only via cable under the sea, and, because of their high reactive power requirement, they cannot be operated with three-phase current. In this respect, only a direct current link is technically possible, and so a three phase current interconnected network with the continent is not technically feasible.

Table L1 provides an overview based on a number of representative indicators for interconnected systems.

Table L1: Representative Indicators for Interconnected Systems, 1989

	UCPTE	NORDEL	IPS(1)	UPS(1)
Countries	12	5	7	15(2)
area supplied/1,000 km^2	2,286	1,257	1,620	ca10,000
peak load/GW	225	50	120	230
installed capacity/GW	380	85	160	285
capability/TWh	1,440	340	760	1,500
exchange/TWh	125	65	80	

(some of them rounded up),
(1) including southern Russia
(2) Republics

Source: SIEMENS (1992) Standpunkt, Zeitschrift für Energie- und Umweltfragen, No 2/1993

Special attention should be paid here, by way of illustration, to the values given for the UCPTE and the UPS systems for peak demand and installed capacity. While the UCPTE system has peak demand of 225 GW and an installed capacity of 380 GW, in the UPS system 285 GW are installed for a peak demand of 230 GW. That means that to cover more or less the same level of peak demand approximately 100 GW of additional capacity is installed in the UCPTE system. That means that in the UPS system less provision is made for reserve capacity in the event of a breakdown etc. That implies that there is a difference in quality in the various systems which has of course to be taken into consideration when setting up a large interconnected system.

Co-operation in interconnected systems

An interconnected system involves the physical connection of many individual components - power stations, electric lines, switching stations etc - to the general system. As a result of this, the connected networks of the individual partners in the system affect each other. That means, for example, that if only one partner in the system experiences a loss of capacity, the frequency level in the whole of the interconnected system falls. What is important here is that all are affected by this fall in frequency, and every partner is therefore dependent on each of the others. Further examples of this mutual dependency are that changes in power station use in one of the partners in the system have a knock-on effect throughout the network in the form of changes in load flow or that, in principle, all partners in the system are affected by the transit of current between only two partners.

Given that the network results in all the partners being dependent on each other in this way, it is clear that there have to be certain "ground rules" as well as co-operation, if there is to be a properly functioning interconnected operation. The most important principle here is that every partner in the system has to take care of security in its own network. Examples of the co-operation or arrangements required are the rules on frequency maintenance in the UCPTE system and briefing of partners on proposed current transits or electric line shutdowns as well as exchanges of information for the purpose of network security assessment. Co-operation of this kind only works where common ground rules are accepted and recognised by all sides. Were this is not the case, a few might perhaps take advantage of the rest. An important feature of connection to an interconnected system is that everyone has to abide by the agreed ground rules.

In a large interconnected system, the question, of course, immediately arises of whether the requisite co-operation can in fact be secured and will function properly in such a system.

Advantages of co-operation

As mentioned at the outset, the objectives of interconnected operations have always been to improve security of supply and increase the economic efficiency. Both of these objectives are achieved through co-operation in interconnected systems.

Co-operation offers three main advantages:

- Assistance in the event of breakdown

Mutual assistance in the event of disruption to supply or the provision of immediate reserves in the event of power station breakdown are important features of an international, technical, interconnected operation. A power station failure, somewhere in the network, results in a fall in frequency from the target value (50 Hz). That fall in frequency is perceptible throughout the system. All of the partners help rectify it, by making available part of their power station capacity via what is called the system of primary control. The rev controllers of all the power plants operated under primary control react to the change in frequency in the network and temporarily alter turbine capacity. To keep the contributions of the individual partners as low as possible and the period of adjustment as short as possible, many blocks have to be involved in primary control. In a large interconnected system like the UCPTE system in Western Europe, experience has shown that it is sufficient for each of the partners to hold ready for this some 2.5% of its network load in their power stations. In practical terms, that means that in our interconnected network, a fall in capacity of up to 2500 MW can be dealt with without affecting consumers and without further disruption. Thus, the partners help each other, and where there is power station failure, everyone is helped out by all the partners, thereby minimising the reserve capacity needing to be held ready.

- Optimising power supply

As well as minimising reserves, a further advantage of the interconnected system is that it makes it possible to optimise the power supply of every individual undertaking and of those participating in the interconnected system generally. Where a partner has to contend with what are temporarily excessively expensive production options, it can instead take advantage of the more favourable option of procurement from another partner. A classic example of this is the use of Alpine surplus water during the summer to replace, at a cost advantage, thermal production elsewhere, in the Netherlands for example. This has, however, to be done in compliance with the above-mentioned ground rules for co-operation in the interconnected system, as these alone will ensure that the interconnected system functions properly.

- The operation of large blocks of power stations

In addition, with their capacity and network reserves, the interconnected systems alone facilitate the operation of large blocks of power stations, and it is only through interconnected systems that smaller suppliers are able to benefit from economies of scale in large power station units. Specific power station costs - eg cost per kW - are known to fall sharply as unit capacity increases. Furthermore, if there is to be a return on investment, then, for example, a base load power station needs to be operated as far as possible with consistently high capacity. To ensure that a new power station block is used appropriately from the outset, larger production units are often set up by two or more partners to take advantage of the above-mentioned economies of scale. In those circumstances, the share of current of the partners involved is integrated into the appropriate network by the interconnected system.

A large interconnected system also raises the question whether the advantages of co-operation continue to exist in the large system which then develops, eg the advantages of optimisation in regard to exchanges of electricity where this has to be transported over large distances. Are there then further advantages to be gained from co-operation where an interconnected system is extended? This issue, on which further comment follows, is currently a matter of dispute and subject to closer review.

Technical features which prevent connection

The Eastern European IPS and UPS interconnected networks, for example, are operated with a network frequency of 50 Hz, but with a very different frequency control system than is standard in Western networks.

SYNCHRONOUS CONNECTION OF THE ELECTRICITY SYSTEMS OF THE FOUR CENTREL COUNTRIES

To facilitate co-operation between the Eastern European partners in implementing the objectives set, the UCPTE recommended the setting-up of an organising body. In response, on 16 December 1992, CENTREL was established in Prague (member countries today: Czech Republic, Slovakia, Hungary and Poland). On the following day, in Vienna, the historically important list of measures for the integration into the UCPTE-system was signed by all of the participants.

After a longer period of co-operation on 18 October 1995 the synchronous connection of the electricity systems with the countries interconnected in the UCPTE-system took place. With this interconnection in October 1995 a one-year-trial synchronous AC operation between CENTREL and UCTE started. For June 1996 the operational service of the regulation and accounting centre in Warsaw is foreseen (until then this function is carried out by the Dispatch Centre of VEAG in Berlin).

This interconnection with the CENTREL countries will reveal options and possibilities of the further integration of electricity systems in other countries in Central and Eastern Europe in the future.

As a next possible step, UCPTE considers a parallel connection with the Bulgarian and the Romanian networks in the framework of an adequate catalogue of Measures (like in the case of the CENTREL countries).

The consequent 'frequency asynchronism' of the networks has only been able to be overcome in recent times at acceptable cost using advanced technology direct current coupling. In addition, exchanges of electricity between Western and Eastern Europe were for a long time not possible on ideological grounds. It was not until 1983, when the first direct current bridge between the then CSSR and neutral Austria in Dürnrohr, north-west of Vienna, with a capacity of 550 MW, was able to be put into operation, that the 'Iron Curtain' became more permeable as far as electricity was concerned.

2.2 The Gas Sector

2.2.1 Natural Gas from the Former Soviet Union for Western Europe

For more than a quarter of a century (1968) natural gas has been piped from the former Soviet Union to Western Europe. The first country to import it at the time was Austria. Then came Western Germany (since 1973), Italy and Finland (1974), France (1976) and - via Germany - Switzerland (1988).

From the start and right up to the present, the volume of gas supplied from Eastern to Western Europe has constantly expanded. While in 1968, only 0.2 billion m^3 were involved, the natural gas imports of the countries of Western Europe from the CIS states amounts to more than 60 billion m^3 today.

The former USSR has thus become a major supplier of energy to those countries of Western Europe which rely on energy imports:

- supplies of natural gas from the former USSR currently account for 4% of Western Europe's total primary energy consumption;
- 20% of Western Europe's total natural gas supply is currently met through procurement from the CIS states.

Co-operation between the gas industries of the former USSR and Western Europe would be incomplete and unbalanced if - looking at the other side of the coin - we failed to mention the importance of natural gas supplies to Western Europe for the national economies of the supplier countries. Since it first began exporting natural gas to Western Europe, the former USSR has obtained a total between 60 and 70 billion US$ (around 7 billion US$ per year today) in foreign currency earnings.

It has at the same time to be pointed out here that trade in energy - and, as part of that, trade in natural gas - was the most reliable element in East-West relations during the political upheavals of the past. It survived periods of political tension and pointed the way to mutually beneficial co-operation.

2.2.2 The Outlook: Natural Gas from the CIS in the Context of Meeting Future Requirements

Past experience leads us to expect that, in the future also, relations between the gas industries will be particularly important in East-West co-operation, and that it will be in the interest of both sides to continue and extend co-operation in this area (CZERNIE, 1993a, 1993b, 1993c).

It is generally accepted that demand for natural gas in Western Europe will increase substantially. Recent estimates assume that Western Europe's current natural gas consumption of around 350 million t of hard coal units will rise to about 480 million t in the year 2005, that is by some 35% - as compared with total energy consumption which will increase to only a limited extent. One reason for this is the increasing value being placed on natural gas from the point of view of protecting the environment and the climate.

A similarly positive trend in demand is to be anticipated in the countries of Central and Eastern Europe. Gas is being assigned an important role in those areas as part of the process of modernising and restructuring the energy industry.

The possibilities that exist will be able to be used, however, only if, as in the past, natural gas comes on to the market steadily, reliably and subject to proper competition. Those conditions apply to purchases already agreed, whatever their provenance, but they apply also to additional sales still to be agreed.

The procurement prospects of the Western European gas industry must not, however, be viewed in isolation but in the international context. It is apparent here that by the year 2005, for example, for Western Europe alone, further quantities will have to be obtained equivalent to implementation of two or three mega-projects on the scale of the first stage of the Troll project or the Algeria-Mediterranean pipeline, each with a capacity of around 20 billion m^3/year.

Alongside the potential still available in Norway, North and West Africa and the Middle East, the CIS states are also a possible source of this. With some 50,000 billion m^3, the CIS possesses the largest reserves of confirmed winnable natural gas supplies in the world, some 85% of them in the Russian Federation. At the current rate of production of some 800 billion m^3 a year, of which only one eighth is currently exported, the confirmed winnable reserves of the CIS will last for a good 60 years, and that basically opens up the prospect of long-term exports over and above the levels previously agreed.

New projects and further sales will, however, be agreed only if they are in the long term attractive and economic for the exporter, they fit into the supply pattern of the importing country - that is contribute to its policy of diversification - and meet the appropriate framework conditions.

2.2.3 New Areas of Co-operation

The political and economic opening-up of the former USSR has also opened up new fields of co-operation between the gas industries, which in future will cease to be confined to the supply or purchase of gas.

In the light of the political, social and economic problems in the CIS, we have to ask whether and to what extent the capacity to restructure and modernise the energy industry exists. Even the previously successful gas industry is increasingly facing problems in exploration, production and transport. The gas companies of the CIS, and particularly Gazprom, are facing huge tasks here, some of which can be resolved in co-operation with Western companies.

A range of areas of technical or economic co-operation are available:

- The development, renovation and modernisation of gas infrastructures in the areas of prospecting, production, transport and storage. Large sums need to be invested here in the coming years in order to create the economic and technical conditions for maintaining or increasing production and exports.
- Energy-saving as a vital challenge and at the same time an opportunity for using or exporting the quantities released. The application of modern gas technology in regard to production, transport and especially use is particularly appropriate here. It is worth noting in this context the agreement in principle of technical and commercial co-operation in the field of gas distribution signed by Ruhrgas AG, heading a consortium of German gas suppliers, and Rosstrojgazifikacija, Russia's largest gas distributor. This co-operation agreement covers in particular the efficient use of natural gas, employing modern technologies, and measures to encourage consumers to save gas.
- A further possibility consists in the marketing and transport of natural gas by Western companies, working together with partners from the CIS prospecting for and producing natural gas in those countries.
- In addition, there is general co-operation in the form of scientific and technical support on managerial and commercial issues and the implementation of training measures for employees of the gas industry.

There is therefore a huge potential for co-operation. Western companies in the natural gas industry are in close contact with the companies/organisations of the CIS and are looking to extend co-operation in these areas in the coming years.

2.2.4 Stable Framework Conditions as a Prerequisite for East-West-Co-operation

Given the scale and complexity of the reforms needed in the gas industry and the energy sector generally in the CIS (as well as in Central and Eastern Europe) Western input will have a significant effect only if provided on a broad commercial basis. Private sector companies must be the sponsors and driving force behind the process of integration in Europe's energy industry. They need to introduce capital, know-how and commercial experience on the basis of strict economic criteria and develop, if need be, new forms of co-operation between undertakings or of financing. If there is to be a broad-based commercial commitment, there have, however, to exist reliable framework conditions as well as suitable prospects for economic success. Reliable framework conditions are a matter for the policy-makers and can be achieved only on the basis of political measures.

The European Energy Charter seeks and needs to meet this requirement. The specific agreements between undertakings - in the context of the bilateral or international political framework agreements - are able to make a significant contribution to stabilising and thus also maintaining the export capacity of the CIS gas industry. The whole area of technical and economic co-operation is therefore particularly important as well as the fulfilment of the natural gas contracts.

2.2.5 Examples of Co-operation in the Gas Sector

Despite the many opportunities for profitable co-operation so far only a few specific projects have been analysed in detail and considered with a view to implementation. A brief description of a number of these here.

New gas pipelines

Because of the uncertainties and risks of the current route of the gas pipeline supplying Western Europe - namely via the Ukraine, Slovakia and the Czech Republic, a new northern pipeline from the Yamal field via the Baltic and Poland to Germany (supply terminal in the vicinity of Frankfurt-an-der-Oder) is being planned. Intensive work is currently being done on this (expected capacity of the pipeline: 32 billion m^3 per annum). Various other pipelines are currently being planned.

Rational use of energy and energy-saving

Rosstrojgazifikacija (Moscow) and Ruhrgas AG (Essen) started in 1992 a technical and commercial co-operation in the field of gas distribution in Russia. This co-operation covers *inter alia* the efficient use of natural gas by employing modern technologies and encouraging consumers to save gas. This involves general co-operation to promote the rational and environment-friendly use of natural gas which is becoming increasingly important for Russia.

One focus of the co-operation currently agreed between Rosstrojgazifikacija and Ruhrgas is the specific development of a gas-marketing concept. On this basis, Rosstrojgazifikacija will be able to advise gas consumers on the newest technologies and methods of making the best use of natural gas, including the use of combined heat and power stations. Another point of emphasis involves installing gas meters in households. The supply network in the towns will be modernised where necessary. The idea is to begin co-operation in two selected Russian cities and develop a model that can be applied in other towns and cities (PALINKAS, 1993).

A comprehensive training programme for the employees of Rosstrojgazifikacija will effectively supplement the implementation of these measures. In addition, the possibilities of setting up joint ventures in Russia for the manufacture of gas meters are being considered. The development of a manufacturing facility for polythene piping is also under review.

Within the framework of agreements still to be negotiated, the proposed co-operation agreements are seeking *inter alia* to:

- modernise gas distribution in selected towns and cities (reorganisation of the pipe network, equipping households with meters, development of gas-marketing to advise customers on the best use of energy in industry as well as in combined heat and power stations);
- optimise commercial and technical organisational structures;

- train employees in energy-saving, marketing, the use of modern gas equipment and management;
- set up joint ventures for the manufacture of gas meters.

Associated gas

In Western Siberia large quantities of associated gas - an inevitable product of oil extraction - are still being burnt off. Russian statistics put the annual losses in the region of 7-10 billion m^3, and the situation is likely to get worse as projects designed to make use of the gas have been halted or were never embarked upon because the Russian state withdrew the project financing.

A major Western gas company is currently working with Purneftgaz, Gazprom and other Russian partners in a project designed to use the associated gas from the Purneftegaz fields. Those fields are located some 200 km south of Urengoy.

Currently, 4 billion m^3 are obtained by Purneftegaz in the course of oil production, only half of which is used. With the planned development of new oil fields, the volume of associated gas obtained will increase still further.

If the project is viable and the basic conditions for financing are met, Ruhrgas is prepared to invest in Western Siberia within the framework of that project. A gas collection system, a gas processing plant and connections to the existing transport systems will have to be constructed. Both natural gas and liquid hydrocarbons are obtained from processing the associated gas. Economic support for the project will have to come in the form of hard currency earnings from the export of the products obtained. The natural gas is to be marketed through Ruhrgas AG.

A pre-feasibility study, which will provide initial data on economic efficiency and project structure, is currently being made by those involved in the project.

3. Transeuropean Networks (TEN)

3.1 Short Overview

In accordance with Title XII of the Maastricht Treaty and the framework for action laid down in the White Paper on Growth, Competitiveness and Employment the Commission has submitted (starting in 1993) proposals (COM(93) 0685) for decisions whose basic aim is to encourage the development of electricity and natural gas transmission networks (these proposals are part of the general EU-strategy for Transeuropean-Networks, TEN, not only covering the energy sector, but - above all - the transport and telecommunication sector). The proposals put particular stress on insular, isolated and outlying regions of the Community, and the development of interconnections with third countries to help complete the internal market, achieve social and economic cohesion and ensure reliability and security in the Community's energy supply.

As far as TEN-project in the energy sector are concerned, the European Summit in Essen (9-10 December 1994) discussed two projects concerning energy-networks between the EU and Central and Eastern Europe:

- Gas pipeline between Russia and the EU
- "Baltic Ring": electricity grid connection between the countries in the Baltic Sea Region.

These two projects should be discussed briefly.

3.2 Gas Pipeline between Russia and the EU

This project is of particular importance for purposes of diversification and for increasing the natural gas supply capacity of the EU. The project will feature the construction of a new transport line to the north of the current line (Ukraine, Slovakia, Czech Republic) for natural gas from Russia.

The first phase of the project is to be finished by the end of 1998 and consists of an interconnection between the existing gas pipeline running from Siberia to Belarus (200 km), Poland (700 km) and Eastern Germany. The additional phase will start if and when the Russian natural gas production projects in the Yamal Peninsular are completed.

3.3 The East-West High Power Transmission System as an Essential Part of the Baltic Ring

3.3.1 Introduction

The political and economic changes in the East European countries open up the chance to intensify economic co-operation in the region around the Baltic Sea. As a major part of this future co-operation a project called "Baltic Ring" has been launched. The aim of this project is to provide the technical prerequisites for an efficient electric power exchange between the Scandinavian countries Denmark, Norway, Sweden and Finland, the Baltic States, Germany and Poland including Russia and Belarus. Thus, the Baltic Ring represents the interconnection scheme of all high-voltage grids around the Baltic Sea, the NORDEL, UCPTE, IPS and CENTREL grids. This scheme will be formed by a considerable number of high-voltage direct current (DC) links (e. g. submarine cables, overhead lines) and alternating current (AC) grids.

While major parts of the interconnection scheme do already exist other links like the East-West High Power Transmission System which could be the southern flank of the Baltic Ring providing a powerful interconnection between Germany, Poland, the Baltic States, Russia and Belarus still have to be constructed.

Because of its importance for the development of the countries in Central and Eastern Europe in December 1994 the European Commission admitted the Baltic Ring Project to the infrastructure program "Transeuropean Networks".

3.3.2 Rough Investment Evaluation for the Baltic Ring

The estimated cost of the transmission systems already in operation or under construction roughly amounts to 7 billion US$.

This includes the 7500 km AC grids in the countries concerned of 7500 km as well as the DC links between Germany and Sweden, Denmark and Sweden, Denmark and Norway, Sweden and Finland and between Finland and Russia.

The estimated investment costs of the transmission systems under consideration with regard to the idea of the Baltic Ring amounts to 2 billion US$. Three projects are included:

1. Multiterminal High Voltage Direct Current (HVDC) double bipole line between Germany, Poland, the Baltic States, Russia and Belarus (line to be split in Warsaw); Capacity: 4000 MW; length: 1800 km; investment: ca. 1.5 bill US$;
2. HVDC cable link between Finland and Estonia; capacity: 600 MW; length: 80 km; investment: ca. 0.2 bill US$;
3. HVDC cable link between Sweden and Poland; capacity: 600 MW; length: 240 km; investment: ca. 0.3 bill US$.

3.3.3 East-West High Power Transmission System

In general the East-West High Power Transmission System contributes to an optimisation of the generation and transmission infrastructure in Europe and thus, leads to more efficient electricity generation.

The expected economic and technical benefits of the East-West High Power Transmission System can be summarised as follows:

- reduction of installed power capacities by pooling of power plant reserve: a transport channel of e. g. 400 MW connecting all partners could replace the future investment for reserve power stations for a total capacity of 1600 MW. This could counterbalance about 20 - 30% of the investment costs of the High Power Transmission System.
- exchange of peak power: due to the different time zones energy deliveries from Russia in future could cover peak load in the Western countries. This would lead to a reduction of the installed peak generating capacities needed. It is considered that this peak load delivery could pay for a substantial part of the investment.
- short-term exchanges on the spot market
- long-term energy delivery between all partners
- use of available telecommunication channels of fibre optical links installed for control purposes.

The East-West High Power Transmission System has also positive environmental effects. The optimisation of the generating structure leads to a significant reduction of emissions of the thermal power plants. Moreover, the link would contribute to a more extensive exploitation of hydro power resources in Russia that could partly compensate the usage of fossil primary energy resources, especially in Western Europe.

Russia and with it the Russian utility RAO EES Rossij has a strong interest in the realisation of the East-West High Power Transmission System. In the framework of an overall economic rapprochement between East and West Europe the project is qualified to provide Russia direct access to the West European electricity market. This could be a significant benefit for the development of the Russian industry and finally, might even be a contribution to the aim of stabilising the political situation in Russia.

As a countermove to the Western engagement the link can play an important role in the urgent question of future utilisation of nuclear power in Russia and the Baltic States. Exchange of electric power between these countries and Western Europe would help to reduce the power generating capacities in the respective countries which could allow nuclear power plants to be switched off permanently or temporarily for upgrading of safety standards.

3.3.4 Conclusions

The East-West High Power Transmission System as an essential part of the Baltic Ring perfectly meets the aim to establish an efficient electrical infrastructure for a regional energy market around the Baltic Sea including Russia and Belarus. The investment for the link would not lead to a profit in the short and medium term; but in the long run the possible political and economic benefits of a future realisation of the project - especially with regard to the nuclear power plants in Eastern Europe - are so striking that the High Power Transmission System is worthy of substantial support by all relevant international organisations.

Membership of the UCPTE meant to offer still greater advantages

Direct current coupling technology does bridge the separate energy systems in the East and the West, but the operators of the neighbouring networks in Poland, the Czech Republic, Slovakia and Hungary are hoping for further and still greater advantages from direct membership of the Western UCPTE network.

They are therefore keen to leave the IPS/UPS network and join the UCPTE network. Both down-to-earth technical and economic cost-benefit analyses and the emotional draw of moving closer to the West play a significant part in this.

Preparatory studies for network connection

Many studies and investigations were carried out before, finally, towards the beginning of 1993, immediate neighbours Austria and Germany (the latter represented by VEAG Vereinigte Energiewerk AG and Bayernwerk AG) on the Western side and Poland, the CSSR and Hungary on the Eastern side decided to combine their initially uncoordinated individual activities and pursue a common objective.

Direct current bridges will subsequently have new functions

After the linkage of the electricity systems of CENTREL and UCTE, the direct current coupling facilities will largely cease to have a role at their current locations. Trials are currently in place to establish to what extent transferring those facilities or individual components of them to the Eastern boundaries of the CENTREL networks offers a feasible and sensible option.

It is quite possible that this direct current technology will acquire a new role as part of the CENTREL networks' exchange of electricity with the previously directly connected networks of the adjacent states of Eastern Europe. That applies all the more since it is in the longer term unlikely that - with the possible exceptions of Romania and Bulgaria - all of the states of Eastern Europe including the CIS will follow the example of the CENTREL countries and seek to become connected to the UCPTE network under the conditions that currently apply. It will be a very long time before an interconnected operation stretching from the Atlantic to the Pacific - if indeed technically sensible and organisationally feasible - can be put in place.

Concept of an interconnected system for Europe

We can conclude from the above that the installation of direct current coupling between different three phase current interconnected systems would make it possible:

- to take advantage of the - albeit at present from the economic point of view apparently doubtful - benefits of a large interconnected system, but at the same time
- avoid the operational and organisation problems of a large interconnected system.

The concept of a large European interconnected system should therefore comprise not an extended three phase current system but direct current connection between a number of 'reasonably sized' three phase current systems which are in other respects operated in isolation.

That concept has already been put into effect between UCPTE, NORDEL and the interconnected system in the United Kingdom - in this instance, as mentioned, for technical reasons. Assuming that the UCPTE system has already reached a 'reasonable size', direct current connection between the IPS system and the UCPTE network would be the logical next step. Furthermore, that concept of a large interconnected system would mean the possibility of further high-voltage direct current transmissions between the IPS and UPS networks.

It is not, of course, possible to determine on the basis of objective criteria the 'boundaries for interconnected systems' arising out of the concept advanced here. Retaining existing and reliable structures will be an important criterion. On the other hand, this poses a 'psychological problem' for the countries of Eastern Europe in particular: membership of the Western interconnected system will be seen as a prerequisite for membership of the Western European economic area which, as mentioned at the outset, the countries of Eastern Europe wish to join. This fails to take account of the fact that it is possible to participate in full in the electricity industry via direct current coupling technology. We cannot therefore determine here where, in the final analysis, the desirable 'boundaries for interconnected systems' should drawn. It is to be hoped that a plan for an interconnected system for Europe is adopted of the kind implemented in North America, where four asynchronous interconnected systems, connected via high-voltage direct current transmission, have formed the interconnected power supply system for North America.

4. Summary

The revolutionary upheavals in Eastern Europe have made increased co-operation between Eastern and Western Europe possible in the energy sector too. The conditions are particularly favourable and promising of success in the energy sector.

Western European security of supply could be improved substantially in the long run by increased energy co-operation between Eastern and Western Europe.

This paper tries to describe the existing co-operation between Western and Eastern Europe - mainly in the field of interconnected electricity and gas systems between East and West. Also some prospects for further East-West co-operation in the electricity and gas sector are described (including some remarks on the Trans-Europe-Networks - proposals in the electricity and gas sector).

An interconnected electricity system for Europe

The development of the electrical energy supply in Europe resulted, after the Second World War, in the creation of five large networks which are even now, with the exception of the IPS/UPS, connected only in part or via direct current bridges. The five interconnected systems in Europe are:

- the UCPTE network interconnecting the countries of Western Europe;
- NORDEL (Sweden, Norway, Finland and Denmark);
- CEGB (United Kingdom);
- IPS (Interconnected Power System; Poland former CSSR, Hungary, former GDR, southwestern network of the former USSR);
- UPS (Unified Power System; remainder of the former USSR).

As well as geographical factors (e.g. United Kingdom and Scandinavia) and economic considerations, there are technical features which prevent direct synchronous connection of these systems. Generally speaking, there are two main reasons for setting up interconnected systems:

- to improve security of supply and
- to increase economic efficiency.

The issue of extending energy co-operation between Western and Eastern Europe raises in particular the question whether there could not be greater levels of electricity exchange between East and West. There are, however, a number of technical features which stand in the way of this. For example: the Eastern European IPS and UPS interconnected systems are operated like the Western European networks with a network frequency of 50 Hz but with a frequency control system completely different from that usually found in the Western European networks. The consequent network "frequency asynchronism" could only recently be overcome at acceptable cost using advanced technology direct current coupling. In this way, the direct current coupling technology provides bridges linking the separate energy systems in Eastern and Western Europe. Since 18 October 1995 the Czech Republic, Slovakia, Hungary and Poland (forming the CENTREL system) are - after a long period of co-operation - interconnected with the UCPTE-system. But a direct link between the IPS/UPS network and the Western European UCPTE network seems, at the moment, a long way off. Initial technical consideration is, however, being given to a development of this kind involving an interconnected operation for the supply of electricity throughout Europe (an interconnected system from Biscay to the Urals). The possibility of installing high voltage direct current transmission in this type of large interconnected system is also being considered (a project of the Transeuropean Network (TEN)-proposals). However, the proposed large-scale expansion of the UCPTE network to the Urals is at present confined to the drawing-board and unlikely to be implemented in even the medium term, as the economic and technical difficulties will probably not be resolved in the foreseeable future.

The gas sector

Over the past 25 years, the former USSR has become one of the most important suppliers of energy to countries of Western Europe which rely on energy imports:

- Supplies of natural gas from the former USSR currently account for 4% of Western Europe's total primary energy consumption.
- 20% of Western Europe's total natural gas supply is currently met through procurement from the CIS states.

For the supplier country too, namely the former USSR, sales of natural gas to Western Europe are hugely important: since it began exporting natural gas to Western Europe, the former USSR has obtained a total between 60 and 70 billion US$ in foreign currency earnings, approximately 7 billion US$ per year today. If oil export earnings are added to that, the paramount importance of energy exports for the former USSR's foreign trade is clearer still.

In addition to being economically significant, trade in natural gas and energy between East and West is also politically extremely important. During the political upheaval of the past twenty years, that trade was the most reliable element in East-West relations and, even at times of political tension, the path to mutually beneficial co-operation remained open.

The general view is that there will be a substantial increase in demand for natural gas in Western Europe in the coming years. The CIS states in particular offer a potential source of these additional gas supplies to Western Europe, alongside the potential still available in Norway, North and West Africa and the Middle East. With some 50 000 billion m^3, the CIS has the world's largest confirmed winnable reserves of natural gas, and some 85% of these are to be found in the Russian Federation. At the current rate of production of approximately 800 billion m^3 annually, of which only 1/8 is currently exported, the exploitable reserves of the CIS will last a good 60 years. In principle, this opens up the prospect of longer-term exports in excess of the levels previously agreed.

The political and economic opening-up of the former USSR has brought with it new possibilities for co-operation between the gas industries which, in the future, will no longer be confined simply to gas supply or procurement.

Despite the many and potentially profitable opportunities for co-operation, so far only a few specific projects have been analysed in greater detail with a view to implementation. One project is planned in the framework of the European Union's Transeuropean-Networks, i.e. the construction of a new transport line running from Siberia to Belarus, Poland and Eastern Germany.

Bibliography

CZERNIE, W. (1993a), Ost-West Kooperation im Erdgasbereich, in: gwf Gas; Erdgas, No 2, 1993, p 67-71.

CZERNIE, W. (1993b), East-West Cooperation in the Natural Gas Sector, Ruhrgas, Essen.

CZERNIE, W (1993c), East-West Cooperation in the Natural Gas Sector, Presentation at the World Bank, March 17, 1993, Washington DC.

EUROPEAN PARLIAMENT (1995), Esther-News No 2, The East-West High Power Transmission System as an Essential Part of the Baltic Ring, PE 165.505, Luxembourg.

PALINKAS, P. (1992), European Energy Charter and East-West Cooperation on Energy and Environment, in: European Parliament, 'Energy and Research' series No W5, Luxembourg.

PALINKAS, P. (1993), The Russian Federation - Energy and Research: The Current Situation, Problems, Outlook and Prospects for East-West Cooperation, in: European Parliament, 'Energy and Research' series No W9, Luxembourg.

SIEMENS AG (1993), Standpunkt, Zeitschrift für Energie- und Umweltfragen, No 2/1993

WELFENS, P.J.J. and M. WELFENS (1994), Environmental Cleaning-up in Systematic Transformation: Costs, Investment Effects and Policy Options, in: CARRARO, C.; HAURIE, A.; ZACCOUR, G., eds., Environmental Management in a Transition to a Market Economy, Paris: Editions Technique, 323-360.

M. Regulation and Systemic Transformation

*Piotr Jasinski**

1. Introduction

Economic regulation, understood both as regulatory activities of authorities and as regulatory institutions, exists in any economy and is not a constitutive characteristic of any economic (or political) system; its form and methods are, however, and therefore must change in the process of systemic transformation. The form that regulation, defined as an element and tool of public policy (towards different sectors of the economy), takes depends also on economic (and political) systems and their internal structures, on the level of economic development, and on the economic policies actually pursued, as well as on the specific characteristics of various regulatees. Taking into account this multidimensional, dynamic reality we shall start this paper with some general remarks on transition from Soviet-type economies (STEs) to market ones (Section 2). In Section 3 we shall present equally general observations on various aspects of regulation in market economies. The next section will be devoted to regulation in the strict sense of the word, *i.e.* to regulation of public utilities (network industries). Returning our attention in Section 5 to post-STEs we shall analyse and discuss above all the issue of regulation, both *sensu stricto* and *sensu largo*, on the way back towards capitalism. The last section of this paper will try to present some normative remarks and suggestions regarding the regulatory structure to be created and policy to be pursued in the medium to longer run. All of this will be illustrated using as examples the electricity supply and telecommunications industries in the UK and Poland.

2. The Process of Systemic Transformation and its Constitutive Characteristics

If we agree that the process of systemic transformation, *i.e.* the passage from a centrally planned STE to a market one or, perhaps better, from (real) socialism to capitalism started, or considerably accelerated, in the Summer and Autumn of

1989, two main characteristics of this unique enterprise have to be stressed. Firstly, it was a process started by a process: there was no single cause, and no single event can be identified as its beginning or as its exclusively decisive turning point. Thus in the search for relevant factors we have to go back in time perhaps almost as far as the moment of Stalin's death, because, for example, in Poland that proved to be a precondition for the decollectivisation of agriculture and in consequence for an increase in the share of privately owned (productive) assets in the Polish economy[2]. Paradoxically, one may reach the conclusion that there is no such a thing as systemic transformation. What we actually observe are market economies returning to their capitalist equilibria after having suffered a prolonged external metaeconomic shock caused by the October Revolution and the Soviet imperialist policy towards Central and Eastern Europe as well as towards the Third World[3]. In other words, the processes of building and dismantling socialism not only immediately followed each other but perhaps even overlapped!

Secondly, in this process politics is inseparably linked with economics. This characteristic of systemic transformation has its roots in the fact that under the old regime in a sense everything was politics. Equally important, however, was the feedback effect between developments in spheres traditionally understood as typically economic or political. In the first systemic transformation the economy was transformed for political and/or ideological reasons, but the failure of the new economic system to produce the expected results eventually forced political changes which in turn made possible retransformation of the economy. This simple fact is all the more important as regulation represents the political in an economy in general and in the realm of the determinants of economic efficiency in particular.

Whatever our opinion on whether STEs constituted a separate (meta)economic system, real socialism was based on central planning, *i.e.* on *ex ante* coordination of economic activities and on public ownership of the means of production. In consequence, systemic transformation has to change both these defining or constitutive characteristics. In the most general terms, therefore, it consists of marketisation and privatisation. Marketisation, in turn, embraces liberalisation, *i.e.* extending the sphere of economic freedom and reconstruction of the legal and institutional framework within which economic agents exchange goods, services, and assets. Privatisation of post-STEs, instead, means above all increasing the share of the GDP produced in the private sector, which can be achieved not only by transferring assets and enterprises from the public sector to the private (the so-called privatisation from above), but also by the setting up of new private enterprises and their growth (the so-called privatisation from below).

These two, or rather three, types of actions, *i.e.* privatisation, liberalisation and, so to say, reinstitutionalisation, differ from each other with respect to their internal dynamics and degree of complexity. It was relatively easy to remove restrictions but in doing so without putting in place a new, consistent legal framework one risked compromising the whole undertaking. Institutional structure does not consist exclusively of written law, but also of certain behavioural patterns. Both were missing, and developing them had to take time. What is more, the knowledge level of economic agents regarding the operation of market mechanisms was and still is far from sufficient, let alone satisfactory. Political instability and various vested interests, often based on exploiting existing legal gaps, on the one hand did not encourage farsighted law making, and, on the other, were instrumental in creating behavioural pattern which soon or later will have be eradicated. Last but not least, privatisation of post-STEs has proven to be much more complex than expected. Although privatisation from below has proceeded almost everywhere very rapidly, privatisation from above has already taken too long and caused too much excitement. This in turn has created a dual economy with most large industrial enterprises still hopelessly remaining in the public sector, putting additional strain on the regulatory structures of post-STEs, and most private enterprises remaining small and operating above all in retail trade and services. What is more, in some STEs - in 1989 it was Poland, and in 1992-93 the republics of the former Soviet Union - systemic transformation had to be accompanied or even preceded by macroeconomic stabilisation policy, with all the undesired and dysfunctional trade-offs that result.

3. Regulation in Market Economies

The very existence - and necessity - of an institutional structure in any economy (economic activities are never conducted in a legal, customary and institutional vacuum) shows that some form of explicit, external regulation is inseparable from all economic activities and therefore also from market self-regulation. What distinguishes economic systems from each other is the way in which these regulatory problems are solved. Since at the most general level explicit regulation and the official exercise of regulatory activities are part of the legal framework of economic activities[4] and, normatively speaking, can be looked at from the point of view of the rule of law, (re)introducing this kind of regulation is a necessary element in moving towards a market economy, or returning to capitalism, which in the medium to long run seems to require a stable (liberal) democracy. This kind of regulation will be distinguished by its degree of generality and universality from regulation in the strict sense of the word, *i.e.* from public policy towards certain industries, which will be discussed in Section 4. This section will end with a look at various aspects of the relationship between regulation and competition.

3.1 The Rule of Law and the Law of Contract as Regulation: an Institutional Approach

In any society there are institutions and organisations, to use North's terminology, which both facilitate, or even make possible, the exchange of goods and services, whilst at the same time imposing restrictions on what kinds of actions (taken by economic agents) in any particular sphere are possible and allowed. It is far beyond the scope of this paper to discuss the role of social rules and law in society, but there is no doubt that in the most general sense they simply regulate human behaviour and therefore also economic activity, both production and exchange. Just as at the individual level human beings cannot enjoy absolute freedom because they live in history, and matter is part of their nature, there cannot be a regulation-free society, although the existing regulatory systems will differ from each other in terms of their formalisation, scope, and degrees of oppressiveness and of commanded consensus. It is also in this context that one could mention, again without entering into any details, the link between morality, regardless of whether or not embodied in any legal code, and economic activities.

The rule of law is a more general concept than simply what follows from requirements regarding the process of preparing and enacting legal acts. The procedure, that is its legality, legitimacy, and the authority behind it, are of course important aspects of any legal and therefore socio-political order.

Nevertheless, liberal democracies, the corner stone of which is the rule of law, differ from all types of authoritarianism not simply because they have freely and democratically elected parliaments as their legislatures or because, in most, of them these legislatures are separate from the executive and the judiciary (not to mention the various in-built checks and balances), but also because of which areas of social and individual life are regulated at all and, if so, to what extent. On the other hand, while one has to distinguish between Rechtsstaat and Gesetzstaat, what is regulated is regulated and not open to almost totally arbitrary interpretations and applications of the law, and that regardless of the legal tradition of the given country. Stability is also an important element of so generally understood regulation.

More specifically, the law of contract, together with private property, constitutes the very foundation of decentralised operation of markets in a market economy. Its general principles specify rights and obligations of dealing parties, involved in exchange of goods and services. Its importance stems from both providing a structure within which economic freedom can be expressed and brought to fruition and from its ability to reduce transaction costs.

To the extent that parties are free to enter this kind of relationship, the law of contract and the content of contracts made in accordance with the law is part of market self-regulation, *i.e.* of the process in which individual agents interact and thus influence each other's behaviour, which by analogy could also be called regulation. At the same time, the law of contract is part of the explicit, even though general, regulatory structure of any economy because it is the state through its legislative powers and judicial interpretation that determines the formal requirements for a valid contract, adjudicates conflicts between parties, and enforces obligations on those who do not want to honour their commitments. In other words, it is in all the arrangements concerning the law of contract that the state finds its, some would say only, indispensable economic role. If this role is performed effectively, *i.e.* if administrative and, in particular, judicial structures are able smoothly to fulfil their tasks and duties, transaction costs are reduced, just as they are when the state takes care of delineating and assigning property rights. For those involved in production and exchange of goods and services such a reduction is a source of security in producing and trading and an encouragement to entrepreneurship which in this way does not have to go hand in hand with sheer power. The law of contract is, therefore, a necessary condition for economic freedom, but limiting the economic role of the state to precisely this sphere is also the best guarantee for economic freedom, *i.e.* a sufficient condition for exploiting all possible gains from trade as well as for making the fullest possible use of productive skills of individuals and groups.

3.2 Regulation and Competition

In economic terms, regulation (understood as, firstly, a general system of law and analogous "regulators" and, secondly, as the mutual influence exercised by dealing parties) prepares the ground - ideally, a level playing field - for competition which, being inseparable from economic freedom, is the most important market mechanism. Although the theoretical construct of perfect competition has nothing to do with rivalry, market coordination of economic activities consists in firms competing with each other for buyers willing to purchase their products and services. It is ultimately competition, dependent as it is on private ownership of the means of production, that leads to economic efficiency, both allocative and productive, and therefore assures the best possible use of resources as well as responsiveness to consumers' needs and preferences.

One could say that the "invisible hand" regulates markets from within. There is no regulatory body, just as in the real world there is no auctioneer, but within the framework of the law in a sense everybody is everybody else's regulator. In this way a decentralised, self-regulatory system is created. Safeguarding competition in any economy as a whole should therefore be considered as a part of universal regulation. On the other hand, it is precisely this kind of regulation that also prepares ground for a trade-off between competition and regulation in the strict sense of the word, to be discussed in the next section.

Protection and even promotion of competition is necessary because, very roughly speaking, each and every economic agent would like to be a monopolist. He or she competes with others in various markets but ultimately for these markets. In this sense, economic agents compete with each other not for the sake of competing but to win. These anticompetitive or, perhaps better, anti-competition inclinations of all economic agents, understandable as they are, may be realised either by entrepreneurship or by collusion and abuse of market power.

While the former leads to the best possible use of information in a given society, the latter runs against its interest and therefore is usually considered as not acceptable. In other words, a dominant position or a huge market share may mean that a victory has been achieved, a victory which should rewarded. But at the same time the way in which this victory manifests itself is not unequivocal enough to eliminate the possibility of punishing the most efficient and innovative firms. This possibility has its source in the difficulty of distinguishing and classifying forms of behaviour and is one of the paradoxes of much antitrust legislation. The whole situation is made additionally complex and even ambiguous by the fact that almost everybody is at the same time a consumer and a producer, either a capitalist or an employer or an employee. Competition seems to be in the interest of all (consumers) but at the same time of nobody in particular, *i.e.* of no producer - which is another of the apparent internal contradictions of market economies. And taking into account different degrees of easiness with which the two groups can organise themselves, it is usually the state that becomes the main arena of their unequal struggle.

Competition policy as part of public policy is meant to remedy the difficulties following from the anti-competitive inclinations of all economic agents, *i.e.* to foster entrepreneurship and combat collusion and abuse of market power. This policy is institutionalised in, and exercised by, a diverse set of antitrust institutions, like, just to give a few examples, the Office of Fair Trading and the Monopolies and Mergers Commission in the United Kingdom, Bundeskartellamt in Germany, the Antimonopoly Office and Court in Poland and many other institutions in other countries, including bi- and multilateral agreements as well as supranational bodies like the European Commission and the GATT. These institutions constitute a part of the regulatory framework of the economies in which they operate and which they supervise, and their activities may be rightly described as regulatory ones.

Assessing competition is part of the brief of these institutions. At least two different points of view may be (and usually are) adopted, *i.e.* those concentrating on the structure of individual markets and those concentrating on the behaviour of individual (or groups of) agents. While markets are sometimes difficult to delineate, behaviour can be evaluated only through its results, a given market structure being one of them. Multidimensionality of the relationships between structure, conduct and performance is an additional source of theoretical and practical difficulties with broadly understood competition policy. However, even the most vigorous pro-competition policy is in some cases unable to create effective competition, the very possibility of which is in these cases thought of as either impossible because of the technology predominantly used in the given industry, or undesirable from the point of view of public interest, the doctrine which underlies, for example, British antimonopoly legislation.

4. Regulation in the Strict Sense of the Word

The main objective of the brief considerations, presented above, was to remind us that what is usually understood as economic regulation has to be seen in a broader context, especially when systemic transformation of an economy of a given country is accompanied by a transformation of its (legal and) political system. As we have already pointed out, phenomena like regulation and competition are not system specific, although the way in which they appear and work is. In what follows we shall adopt a more standard understanding of regulation and consider it as, among other things, public policy towards public utilities (network industries, natural monopolies)[5].

4.1 Regulation as an Institution of Market Economies

Before we discuss regulation (in the strict sense of the word) of public utilities, two aspects of regulatory structures and activities in market economies have to be pointed out. The first one may be labelled as actual: regulatory institutions are simply part of the economies of industrialised countries. Regulatory agencies do exist; regulatory actions are taken and, in consequence, some economic activities are regulated explicitly and specifically. But from this simple fact one is not allowed to draw normative conclusions, and that is why we have to consider a second aspect of regulatory institutions in market economies, that is we have to look at them from the normative point of view: Is it really in the public interest (and/or does it maximise social welfare) to have regulation (in the strict sense of the word)? Does it make sense from the purely economic point of view, and if so, how much and what kind of regulation?

4.1.1 Regulation as an Actually Existing Institution in Market Economies

Any economy consists, among other things, of a hierarchy of institutions. In a decentralised market economy at the very bottom of such a hierarchy one should place private property, which is why privatisation of post-STEs is a necessary element of their transformation as well as a part of the institutional reform therein. The second layer of this hierarchy may be said to be various markets: labour, capital, and goods and services markets. Each of these markets, in turn, requires systems for them to function, not only properly but also at all. For example, an integral part of the capital market is the banking system which channels households' savings to investors. Similarly, in the labour market various job centres facilitate the flow of information between those who look for new employees and those who look for jobs. And the operation of all of them is regulated by laws of various importance, ranging from acts of legislature to codes of practice. At the top of this structure come economic agents, either firms, households or simply individuals, and their single actions. The relationship between lower and higher layers consists, among other things, in that the lower ones provide a framework for what is above.

Of course, in the real world the structure of a market economy is much more complex and perhaps even not fully susceptible to such a formalisation. The point, however, is that apart from civil courts supervising the law, and apart from antitrust institutions protecting and promoting competition, in almost all industrialised countries there are also special institutions, generally speaking, taking care of (the activities and behaviour of) public utilities, especially if they are in the private sector. Sometimes these "caring" institutions are non ministerial government bodies like in the UK, or semi-autonomous agencies like American commissions, or form parts of other governmental bodies or even of public administration. Their relationships and links with other branches of government also differ. Sometimes, instead, it is easier to identify strictly regulatory functions which are exercised by institutions whose brief is much broader or even has directly nothing to do with public utilities or with the economy as such.

Whatever the model, and regardless of whether it evolved or was simply introduced at one stage purposefully, it is a matter of fact that this kind of regulation, defined as exercising public policy towards public utilities (network industries, natural monopolies), is also a part of the institutional framework necessary for the exchange of goods and services and therefore for the smooth functioning of any market economy. The importance of public utilities' regulation stems, among other things, from the role that these industries play in all economies, being part of their infrastructure. Regulation as such may be also considered as part and outcome of a more general evolutionary process in the result of which industrialised countries achieved their present standard of living and position in the world. Therefore at any given moment regulation has to be seen, together with competition and ownership, as one of three key determinants of economic efficiency (YARROW, 1986)[6].

Here it suffices to recall that private ownership ultimately provides incentives for the best possible use of resources, from material ones to such ones as information, and, as the Austrian school has reminded us over and over again, together with free-market pricing of goods and rational economic calculation is a *condition sine qua non* of proper functioning of an economy. Competition is, in this context, nothing else but a consequence of the pursuit of similar objectives by many agents simultaneously, and regulation, as we have already seen, is both a necessary condition for private property being able to fulfil its promises of economic efficiency and, because of the ongoing interaction between the spheres of the economic and the political, a threat of political failures possibly replacing and overriding market imperfections.

These three determinants of economic efficiency are closely interwoven and changes in any one of them will have an effect on how the other two work. This is above all true about privatisation, the direct effects of which change the way in which the agency problem is solved. But the indirect effects of privatisation are able, *ceteris paribus*, to change market structure, and therefore to affect the level of competition[7], and to increase efficacy of regulation[8]. Similarly with regulation. Even if no changes with respect to the other two are intended, there are unavoidable trade-offs between the way in which the right to private property (and to compete) is exercised and both the framework in which it is exercised and the public policy pursued, and everything is reduced to the necessity to strike the right balance. Too little regulation may be just as damaging as over-regulation.

4.1.2 Why Regulation? Regulation and Economic Efficiency

The very existence and operation of various regulatory bodies, indicated and briefly interpreted above, does not by itself prove that the example of industrialised countries should be followed by the governments in post-STEs which, at least from this point of view, start with a clean sheet. That is why the issue of theoretical and practical justification of such institutions has to be at least briefly touched upon.

A textbook approach to regulation suggests that an intervention (assumedly of the state or its organs) is necessary whenever there appear considerable market failures (externalities, public goods, etc.). According to this approach goodwilled intervention aimed at maximising social welfare is usually able to correct distortions and remedy failures, even though the theorem of the second best sometimes has to be evoked.

Coase theorem questions this by trying to show that if we have well defined property rights, competition, and negligible transaction costs, individuals or firms involved will find an optimum solution without government's intervention. The problem, however, is that transaction costs not only exist but may also be quite considerable[9]. Yet other approaches to regulation point out that instead of remedying the consequences of market failures, regulation and regulatory agencies serve the interests of those regulated and in this way become a cause of yet another market failure. One must not forget either that the public interest theory of government, underlying many textbook discussions, is seriously challenged by its public choice competitor. The role played by public opinion, often simply demanding from politicians that something be done in response to more or less unfortunate current developments, does not help to achieve, from the economic point of view, the best possible result either.

All in all, if one does not let a single theory or approach monopolise one's views on regulation, and especially if one takes into account the multiple and multidimensional interactions between the spheres of the economic and the political as well as the possibility - a very realistic one, to say the least - of there being market failures in the latter, the issue of whether - and if so, what kind of - regulation is desirable will not have a definitive solution.

4.2 Regulation of Public Utilities

We have already said that thanks to competition - or the process of competing - economic agents in a sense regulate each others' behaviour. This is not, however, always the case, and in the real world of industrialised countries there is a trade-off or inverse relationship between (economic) regulation and competition, with regulation co-present with so-called market failures. Whatever the reason for this state of affairs, usually the more regulation there is, the less competition and *vice versa*[10]. The word "regulation" is, therefore, much more often used in the sense of specific institutional arrangements designed to control and/or supervise behaviour of economic agents in certain industries and their mutual relationships, mostly but not exclusively in those industries classified as public utilities.

Public utilities (network industries, natural monopolies) are industries which, because of their cost structure and/or technological requirements, are quite distant from the theoretical construct of perfect competition. This, in itself, is not an argument against attempts to introduce and/or promote competition, but nevertheless such specifics of these industries have to be taken into account in the process of (re)creating relevant markets and designing - or letting evolve - their structure. In market economies such a situation will usually coincide with either the given industry being in the public sector or, as for example in Britain and in the U.S., with existence of special regulatory arrangements, some of them introduced at the moment of privatisation[11].

4.2.1 Public Ownership as Regulation

The apparently easiest solution to the problem of public utilities is to have them in the public sector, especially if one believes that civil servants do (or can) play the role of *the highest custodians of the public interest*. This can (and did) happen in two different ways: either public utilities are from the moment of their inception in the public sector, like for example many telecommunications companies in the Third World countries, or they are nationalised at one stage of their development. Electricity supply industry (ESI) in the UK is just one example of the latter, much more important from our point of view, and its history in England and Wales well illustrates the trade-offs involved[12].

In the interwar period the ESI in the UK consisted of literally hundreds of small companies generating and distributing electricity. In 1926 the Electricity (Supply) Act set up the Central Electricity Board to construct and operate a national system of interconnected generating stations. That marked the beginning of efforts to centralise production and control (CURWEN, 1986), but was not much of a success. At the moment of nationalisation there were still 369 municipal and 200 private electricity undertakings[13].

The decision to nationalise the ESI in the UK could be seen as following from the recognition that the then arrangements failed and did not augur well for the future. And fail they really did, because, as Foster writes, *by the 1930s it could be said that the British electricity industry was high cost and the most economically inefficient in the world* (FOSTER, 1992, p. 64). But one could hardly blame the free market or the market forces for the failure of the ESI if their operation was stifled by so many legislative and other regulatory restrictions and definitely not helped by the given ownership structure of the industry. The ESI was in fact yet another example that *rescue from the complications that had grown up through successive layers of regulatory legislation and inconsistency in government policy was an important motive for nationalisation* (FOSTER, 1992, p. 73).

In the process of nationalisation the Attlee government radically changed the structure of the ESI in the UK. The Electricity Act 1947 established the Central Electricity Authority (CEA), a public corporation responsible for the generation and supply of bulk electricity, and 14 Area Boards, each of them a public corporation which was to take care of the electricity distribution in its own region[14]. This vertically separated structure survived until the moment of privatisation and the potentially prone to conflicts monopoly/monopsony structure learnt to solve its internal disputes in a corporatist way[15]. The only major legal change, the Energy Act 1983, which was supposed to reduce entry barriers and in this way introduce some competition, failed completely.

Any assessment of the more than 40 years of history of the ESI in the public sector has to take into account the shortcomings that this industry shared with the rest of the public sector and indeed with the rest of the British industry. There were inherent conflicts in the Morrisonian model: no clear dividing line was drawn between decisions to be taken at ministerial level and those to be taken by management and both groups of decision makers tried to pursue commercial and non-market objectives at the same time (ROBERTS *et al.*, 1991). The celebrated NEDO report, published in 1976, painted, among other things, a picture of *ad hoc*ery, confusion, and blatant political manipulation. Among the listed deficiencies were lack of trust and mutual understanding between government and management, confusion about roles, and the absence of an effective system for measuring performance and managerial competence. Last but not least, capture by trade unions and politicians was all too visible (VELJANOVSKI, 1991a).

The CEGB, as well as the distribution Boards, was instructed to set prices proportional to marginal cost, but since the CEGB possessed a monopoly of cost information, this amounted to a cost plus contract[16]. Furthermore, various groups within the industry were strongly linked with each other and *created a powerful interest in favour of the* status quo, *while new technologies or ideas which threatened the interest of those within, such as renewable generating technologies, combined heat and power, and energy efficiency, remained out in the cold* (ROBERTS *et al.*, 1991, p. 43).

Despite the fact that the ESI in the UK did not do badly by international standards on a number of measures of efficiency (POLLITT, 1992), as time went on became more and more evident deficiencies of the public sector ESI. But it was only the process of privatisation that fully revealed some of the problems, the case of nuclear electric being the best example of the surprises in stock[17]. The most serious consequence of this state of affairs was that the CEGB appeared to underestimate greatly the costs of decommissioning nuclear power stations at the end of their active life, which made them unsaleable to the private sector and strongly reduced the scope for pro-competitive restructuring.

This lack of cost transparency and difficulties with liberalising third party access to transmission and distribution networks, made in theory possible by the Energy Act 1983, is not unrelated to attempts to regulate public utilities by taking them into public ownership[18]. On the one hand, regulation without access to all necessary information is a tall order. On the other, although the main lesson to be learnt from this failure is not that the liberalisation of the ESI is unviable but that regulation for competition is needed, *i.e.* that new entrants have to be protected against the anti-competitive behaviour of incumbent firms; increasing economic efficiency by competing with a publicly owned monopolist was never going to be easy.

Obviously, it is difficult unequivocally to distinguish which features of this grim picture can be directly traced back to public ownership and which appeared simply because things developed as they did. But as in the case of the unending dispute about whether the Soviet Union truly implemented socialist principles, this is, by and large, irrelevant. The picture is as it is, and should discourage further experiments.

4.2.2 Regulation and Private Sector

As we have said, public ownership itself is a form of regulation, and in the case of public utilities, especially if we accept that they always require some regulation, this regulatory aspect of public ownership is combined with regulation in the strict sense of the word. Although this was the state of affairs in most industrialised countries, this solution was by no means universal and is becoming even less so. It is therefore appropriate to discuss separately two cases: regulation of public utilities which were always in the private sector and regulation of privatised utilities. The former refers above all to the situation in the United States, while the best illustration, and even a paradigm, of the latter is to be found in the United Kingdom.

4.2.2.1 Regulation of Private Public Utilities (Network Industries)

Although from the point of view of the present state of affairs in Western Europe, regulation of privately owned public utilities may appear as something special or unusual, when we broaden our perspective, and especially when our investigation goes back in time, we may reach a reverse conclusion. Public ownership, followed by privatisation, is a relatively recent phenomenon which appeared - and became rather widespread - in some countries, mostly after the Second World War. Be that as it may, we shall briefly consider the main institutional features of the attempts to regulate the industries under consideration, using as examples the U.S., because of continuity, and the pre-WWII U.K., because later on this country experienced all three regimes.

As FOSTER (1992) reminds us, *Britain before nationalisation had more than 100 years of economic regulation of natural monopoly [that] is now almost forgotten* (p. 17)[19]. In the first part of this period Parliament tried to regulate, mainly through the so-called private bills, which addressed each case separately, and with growing intervention from the courts. In the second one, starting in the 1870s, the main instrument of state regulatory activities became various commissions.

Despite the fact that quite a few industries were regulated in this way or another, according to FOSTER (1992) it was the regulation of railways that raised the most fundamental issues. The Railway and Canal Traffic Act, 1888, was one of the turning points because it started the only period in the history of economic regulation in the UK *during which the form of regulation became as legalistic as it has remained in the United States* (FOSTER, 1992, p. 48)[20]. Namely, the commission, established as a result of that Act, had the authority of a high court and adopted legal procedures and rules of evidence, with the possibility of appeal on points of law to the High Court. But a few years later it was asked:

> "Can anyone imagine the strict technical procedure of a Law Court - with its preliminary proceedings and precise issues, its sworn testimony and its rigid exclusion of whole classes of evidence which, though logically irrelevant, are often precisely the considerations which would influence a businessman in deciding a business proposition - applied to a question as to whether the passenger services of a particular company are reasonably adequate, taking account broadly of the sufficiency of trains, their speed, their punctuality, their comfort, their cross-country connections, the volume and class of the traffic catered for, the financial position of the company and a dozen things more which ought to be taken into account if a fair all round conclusion is to be arrived at?" (ACWORTH, 1905, p. 8, quoted in FOSTER, 1992, p. 58).

The Railway Rates Tribunal, which replaced the Railway and Canal Commission, was an attempt to set up a less judicial body. Renamed the Transport Tribunal after 1947, its proceedings remained, nevertheless, intensely legalistic. Not only were parties represented by lawyers, but formidable preparations had to be made - of the same magnitude as before US regulatory bodies. That is why in the 1930s as a minister supported by civil servants Herbert Morrison believed that he would be in a far better position to regulate the railways and match their expertise than any tribunal could be when confronted with the expertise and knowledge of the railways on the one hand and the poverty of the information available both to the railways themselves and to objectors on the other (FOSTER, 1992, p. 60). Although the fall of the Ramsay MacDonald Labour government in 1931 frustrated those plans, in 1945-51 ministers sponsoring nationalised industries took over most regulatory functions not only for the railways but also for the other industries, and that with the support of many people who, with no ideological predispositions, were led to believe that nationalisation was the only way to solve the problems of the railways.

Unlike in the UK, the institutional framework used for regulating privately owned public utilities in the United States of America has not changed much. At the federal level it is a century-old phenomenon whose beginning is generally traced to the establishment of the Interstate Commerce Commission (ICC) in 1887[21] and it still consists of administrative agencies, created by Congress to carry out broadly defined policies (SPULBER, 1989, p. 71).

British constitutional doctrine, according to which *the sovereignty of Parliament was almost defined by its ability at any time to pass any legislation on any domestic topic* (FOSTER, 1992, p. 17) simplified the solution of the problems caused by the apparent contradiction between regulation and the right to private property. In the United States it was the Supreme Court that in *Munn* v. *Illinois* in 1877 held that the state of Illinois had a right to fix prices since the business was *clothed with a public interest* (SWANN, 1988, p. 91). After having to solve from time to time similar cases, in 1934 in the case of *Nebbia* v. *New York*, the Supreme Court decided that *there was no clearly defined collection of activities which were clothed with a public interest. It was up to each state to decide what policies would promote the public welfare and in principle there was no limit to the range of industries which would be brought within the ambit of regulatory control. It was not the job of the courts to set limits* (SWANN, 1988, p. 92). Also the difficulties following from the federal structure of the United States were solved by the same institution. In *Wabash, S. Louis and Pacific Railway* v. *Illinois*, the judges, drawing on the commerce clause of the US constitution, defined the realm of federal regulation as regarding interstate operations, in this case - interstate railroads operations.

However, the worry about the sacrosanct character of private property was soon replaced by questions about the extent to which congressional agencies violate, at least in principle, the objective of separation and delegation of powers[22]. The main difficulty consisted in the almost self-contradictory nature of regulatory agencies. On the one hand, since they were established by the Congress and represented an exercise of congressional and executive power, they had to be accountable to the Congress or the Executive. On the other, it was desired that they accord individuals the due process of law, and that their decision be consistent with judicial review.

To sum up, although similar approaches to regulating economic activities had completely different results in Britain and in the United States, one can nonetheless say that issues like the sacrosanct character of the right to private property and the principle of the separation of powers constitute inherent problems associated with any attempt to regulate private economic activities, not least of privately owned public utilities. Whether their solutions in individual countries result in a regulatory failure, as in the case of many public utilities in Britain, or in excessively legalistic, expensive and prolonged procedures, as in the United States is of secondary importance.

4.2.2.2 Regulation and Privatisation

In the 1980s in the United Kingdom the consecutive Conservative governments pioneered a new approach or policy towards public utilities: privatisation. More specifically, shares in these utilities were sold to a very broad public, the whole process becoming not only almost a synonym of Thatcherism but also a well publicised part of an attempt to construct popular capitalism or a property owning democracy. Whatever the reasons for, and the objectives of, this initiative, there is no doubt that it caught the imagination of politicians all over the world, from corporatist Germany (Deutsche Telekom is to be privatised in 1995) to the rapidly growing "small dragon" of Singapore (Singapore Telecommunications was privatised in 1993) and to Argentina, firmly on its course away from the decades wasted from the economic point of view. The post-STEs are only the latest to join the worldwide process which will certainly go on for many years to come.

One consequence of privatisation of public utilities is that it forces the government initiating and implementing this kind of policy to rethink its regulatory role. The regulatory aspect of public ownership obviously disappears[23], but there remains a very practical problem of the potentially damaging consequences of transforming public monopolies into private ones, equally protected by the law. That is why privatisation was usually accompanied by more or less procompetitive restructuring of the given industry as well as by creation of special governmental agencies to regulate it. Different countries went for various institutional solutions, the British experience being the most complete and longest in operation and therefore deserving most attention and already enabling a critical review. Only very few countries decided that the general antitrust law will suffice to do the job. The two kinds of actions, restructuring and creating regulatory agencies, will be analysed separately.

4.2.2.2.1 Procompetitive Restructuring

First of all, one has to be reminded that although it is whole industries that are described as public utilities or even, less correctly, natural monopolies, they are not monolithical entities and it is usually only some parts of them that display these cost characteristics. Again the most recent history of the ESI in England and Wales provides a good illustration of the possibilities, dangers and trade-offs present.

The most obvious characteristic that ESIs share with other utility industries is that most of them are usually described, somewhat uncritically and imprecisely, as natural monopolies. And what makes them natural monopolies is that they all comprise networks or grids which, firstly, give a first-mover advantage over potential competitors and, secondly, require substantial investment. In consequence, it is claimed that one producer can serve the whole market most efficiently. Public utilities are also renowned for their complexity and multidimensionality.

It is thus almost self-evident that any attempt to treat the ESI, as well as any other public utility, as a monolithic, naturally monopolistic structure is simply wrong, although naturally monopolistic cost patterns of wholly vertically integrated structure remain sometimes an open issue[24].

In the case of the ESI one usually distinguishes between generation, transmission (high voltage grid), distribution and supply (retail sales)[25]. Competition can be relatively easily introduced into generation, which in terms of costs is by far the largest sector of the ESI. Also supply of electricity can be relatively easily developed into a competitive industry. This leaves us with only two potentially naturally monopolistic elements of the ESI: transmission and distribution[26].

The UK privatisation of the ESI followed the division of the supply chain into four stages and therefore did represent a revolution in the economics of the industry when judged by international standards. In the process of breaking the CEGB power stations into new groups, apart from the still publicly owned Nuclear Electric, two new companies were created: National Power and PowerGen, the market shares of which were, in the moment of privatisation, equal to, respectively, 70 and 30 per cent[27]. Taking into account the already existing possibilities to import electricity[28] and the number of power stations under construction, one can expect that the relative importance of the negative consequences of insufficient break-up of the CEGB will most likely diminish over time (NGC report, FINANCIAL TIMES, 19.05.1993).

In the process of abolishing the CEGB one more company was created, namely the National Grid Company (NGC), which was also privatised but not by public offer: it is now jointly owned by the privatised Regional Electricity Companies and run at arm's length. In this way generation has been vertically separated from transmission. The NGC is responsible for the high voltage transmission of electricity from generators to distributors and some consumers, its wires being open to common carriage by third parties. Tramsmission charges are subject to regulation. The NGC also runs the spot market for electricity, acting as an auctioneer and load dispatcher with authority over the operation of generating stations (WEYMAN-JONES, 1991). The merit order dispatch is operated on the basis of bids made by generators, but buyers and sellers are linked not only by the short-term spot market but also by various longer-term financial contracts[29].

The next link of the ESI chain is distribution, in the process of which electricity transmitted through the high voltage grid gets to final consumers. The 12 Regional Electricity Companies (RECs) are direct heirs of the same number of Area Boards and, in order to avoid allegedly wasteful duplication of network, they enjoy monopoly rights in their own respective territories. They are also responsible for system reliability.

Distribution with its common carriage provisions should not, however, be confused with supply, which essentially consists of *the contract between final consumer and supplier in which the supplier delivers electricity which it may have bought from a generator, or another regional distribution company, or which it may have generated itself* (WEYMAN-JONES, 1991, p. 12).

Although at the moment for buyers with maximum demand less than 100 kWs (previously: 1 MW), the RECs are also necessarily their suppliers, the two roles are nevertheless separated in these companies' accounts[30]. In this way, regulated and unregulated activities are separated from each other, or, if we assume that market failures and regulation coexist, separation is introduced between activities which have different degrees of (potential) competitiveness. In consequence, the supply business of any REC must purchase electricity transportation services from its own distribution business, and that on terms similar to those available to other suppliers.

Who else then can be a supplier? Subject to regulation, the answer is: anybody who signs a contract with a final user (and a producer, if necessary)[31]. The potential list includes, of course, all generators in England and Wales as well as in countries linked with the national grid, but in principle a supplier does not even have to be a generator. A company supplying electricity may, for example, itself buy either directly from a generator or from the pool, as was, and by and large still is, the case, for example, with the 12 RECs. What is more, the list of generators and, therefore, of potential suppliers is to grow rapidly in the near future as new projects are completed, which in turn is a direct consequence of freeing entry into the ESI[32].

In other words, the restructuring of the ESI in England and Wales introduced competition not only into generation of electricity but also into one part of final supply, *i.e.* into the contract market. This competition has still to be overseen by the authorities and protected against the abuse of market power and thus is not yet self-sustaining, but the process of introducing it, envisaged as lasting for eight years, is still under way. At the same time, there is convincing evidence that anybody learning from British experiences can go even further, especially as far as the number of companies in electricity generation is concerned[33]. It will not, however, be easy, because this ESI's management were not exceptional in their attitude (read: opposition) to any pro-competitive restructuring[34]. What is more, they were not alone in displaying this kind of attitude: *The management of the corporation will want to retain its market power, so will its unions which can conclude much cosier arrangements with monopolies than with firms in competitive markets, and so will the City which finds it easier and more profitable to bring monopolies to market. Not least, politicians will want to raise large sums quickly, without management obstruction, from the sale of corporations which face little competition and therefore appear to have good earning prospects* (ROBINSON, 1989, p. 114).

This in turn shows how widespread the anti-competitive inclinations of almost all economic (and political) agents are. Very few governments, staffed by politicians as they are, are able to act against what seems to be in their best interest. However, from the point of view of regulation of privately owned public utilities, pro-competitive restructuring consisting, as we have seen, in separating naturally monopolistic and potentially competitive elements of a given industry and introducing competition where it is possible, has much to offer. On the one hand, it limits the scope of regulation to the spheres where the case in favour of it is strongest. On the other, it allows, in the process of competition, various economic agents to regulate each other in the way almost automatically guaranteeing accordance with the public interest and/or maximising social welfare.

Privatisation was, in practice, an almost necessary condition for this kind of pro-competitive restructuring of the ESI in England and Wales, which at least to some extent limited the scope for regulation and arbitrary intervention. Unfortunately, what happened with other British public utilities in the process of their privatisation is much less encouraging, although, as we have seen, quite understandable and easy to explain from the political point of view[35]. First, in 1984, it was British Telecom, which in the moment of sale to the public of 50.1 per cent of its shares, not only was kept intact but also promised that until 1990 it would have only one competitor, Mercury Communication, which had got its licence before BT's privatisation anyway, but whose market share at that time was close to zero. In other words, there was no pro-competitive, nor any other for that matter, restructuring either of the company under consideration or of the industry.

BT was sold with its market power and market share, and further developments depended on how the Director General of Oftel would fulfil his duty to promote competition[36].

Equally important, if not more, was, however, rapid technological progress which ultimately undermined the naturally monopolistic character of telecommunications, if there ever was any. This in turn allowed for an acceleration in the process of increasing the number of competitors as well as in the intensification of their rivalry once this market was liberalised in 1990. Whether and in what way this newly and rapidly developing competition will change the role of Oftel it is still too early to predict. Looking at regulation from the point of view of supply and demand it is by no means certain that the demand elasticity of Oftel's services is low.

The second example, which also comes from the days when Mrs. Thatcher was firmly in control of the British economy, is that of gas industry. Even more than in the case of telecoms, this example should serve as a warning. At the same time the whole story throws new light on the relationship between - and relative importance of - ownership, competition and regulation, as well as on the political economy of the three.

British Gas was not only a vertically integrated monopoly combining both transport and distribution of natural gas. It also enjoyed a monopsony power as far as purchasing gas from the North Sea was concerned. In the process of preparing for privatisation, its management even contrived to keep its high street shops in the company. Competition, therefore, seemed to be all but impossible. However, although the list of the Director General of Ofgas's duties mentioned only *enabling competition*, it was the concerted, even though not coordinated, action of Ofgas, the OFT, the MMC and the DTI, that eventually managed to change the pre/post-privatisation *status quo*. Yet, the very structure of British Gas, despite many suggestions and threats to the contrary, has been unchanged since its transfer into the private sector.

4.2.2.2.2 Regulatory Agencies and Procompetitive Regulation

In the moment of privatisation of public utilities, (network industries, natural monopolies), ownership is by definition separated from regulation. That is, ownership as a form of regulation disappears and it is decided whether new institutional arrangements, *i.e.* either new institutions should be created or new prerogatives given to the existing ones. It was when ownership rights were to be transferred from the public sector to the private one that such quintessentially regulatory bodies as Oftel, Ofgas, Ofwat and Offer were created, even though one has to add immediately that it was perfectly conceivable - and in fact considered - for the Office of Fair Trading to have its brief appropriately extended. In other words, some changes in the arrangements inherited from the old regime are necessary but the necessity of specialised regulatory agencies is by no means self-evident and has to be decided upon[37].

The creating of separate regulatory agencies for various utilities was ultimately a pragmatic decision, justified by highly specialised expertise required as well as by the fact that when British Telecom was to be privatised, the OFT declined to take care of it[38]. These agencies are non-ministerial government departments headed by a Director General, appointed by the relevant Secretary of State (in most cases it is the President of the Board of Trade, *i.e.* the Secretary of State for Trade and Industry), for not more than five years. He may be removed only *on the ground of incapacity or misbehaviour*. The regulator appoints his staff within a budget provided by the Treasury, the size of which is beyond his control (FOSTER, 1992).

The duties of the regulator are specified in the respective privatisation acts as well as, in much more detail, in the licences issued to individual utilities by the Secretary of State. All regulators have a primary duty to ensure some form of universal service and to guard that the regulated utility has the financial resources to meet the obligations imposed on it. Subject to meeting these objectives the regulator has a secondary duty to take steps to increase competition (although the wording differs substantially from one Act to another), to control prices, quality and service levels, to promote efficiency and economy and other duties relating to health, safety and R and D.

The cornerstone of post-privatisation economic regulation UK-style is the price cap "RPI-X". This is enforced by the regulator and modified in agreement with the industry, or in absence of this by referral to the MMC. The MMC acts as an "appeal court" for licence revision and re-setting of the price cap[39]. Apart from the possibility (or the threat) of a referral to the MMC, regardless of who and on what grounds does it, and apart from pressure exercised by media, sometimes called "sunshine regulations" (FOSTER, 1992), judicial review or rather its possibility is yet another important element of the regulatory game in the UK. However, the principles of judicial review require that a regulator's decision is not tainted with "illegality, irrationality or procedural impropriety", which is rather difficult to establish if, as is the case in the UK, the decision makers are not obliged - and usually decline - to present any reasons for the decisions reached[40].

The whole system is sometimes accused of imposing insufficient legal constraints on the regulator: weak accountability, weak judicial review, and the absence of procedural safeguards (VELJANOVSKI, 1991a)[41]. What is perhaps worst is that almost everybody expected the whole system to work in a completely different way. Regulators wanted to work themselves out of a job and it took them some time to discover that permanent regulation is much more complicated than temporary; regulatees seem to have expected "regulation with a light rein", as promised or at least suggested by some government officials. However, before we despair about the state of regulation in the UK and rush to express our solidarity with unjustly oppressed utilities and their managers and shareholders, should not we ask whether, assuming that the picture drawn is accurate (which is by no means self evident) it is not the case that such an arrangement is also cheaper, more flexible, quicker in addressing new issues and in solving them, especially when compared with regulation US-style? In the last analysis conformity with formal, legal and/or constitutional requirements, important as they are, should be weighed against the results achieved, and courtrooms certainly are not the only places in which regulatory decisions can legitimately be taken. The border line between advantages of pragmatism and dangers of arbitrariness may be very fine indeed and *the delegation of discretion to regulators has obvious costs and benefits* (VELJANOVSKI, 1991a, p. 22).

In a sense regulation, whatever form it takes, is as old as economic activities and a state's attempts to affect it. Forms, procedures, and goals of regulation change over time, and the most obvious form of regulation for competition consists of general competition or antitrust law. Instead, regulation for competition, as distinguished from general competition and consumer protection laws, is a new invention[42] and the enabling Acts of Parliament on the basis of which British regulatory agencies were created, all contain references to competition. For example, the 1989 Electricity Act talks about the Director General of Offer having a duty to *promote effective competition* and the earlier Gas Act of 1986 mentions *enabling competition*. The concept itself, however, seems to be, in the light of most recent debates, a very controversial one. Regulation for competition is usually considered as complementary with respect to pro-competitive restructuring[43].

Why does one need regulation for competition? Why does regulation have to go any further than establishing a level playing field for competition between the incumbent firms and new entrants (HELM and YARROW, 1988)? At the most general level one could point out the very complex relationship between restructuring, competition, deregulation, regulation, and liberalisation. This relationship is both path- and environment dependent, and politics, especially politics understood as a process of interaction among various pressure groups and vested interests, which differ from each other in their relative strength, has considerable influence on specific outcomes. More specifically, the duty to promote competition reflects in part the practical difficulties of moving in one go from a nationalised monopoly to a competitive industry (BEESLEY and LITTLECHILD, 1991). Whether or not regulators' activism in this field is a direct consequence of the fact that restructuring did not go far enough, or whether it simply takes time for competition to start to work, is a different matter, but most likely it is not an either/or issue.

Promoting competition involves facilitating the entry of new competitors, including the entry of existing competitors into new parts of the market, which obviously requires that the regulator is aware of - and able to assess - the relationship between his actions and the probability of entry (BEESLEY and LITTLECHILD, 1991). Whether specific measures taken to fulfil this duty go any further than setting terms of access to networks for potential competitors and involve changes in the structure or boundaries of the industry, varies from industry to industry[44].

The fact that regulation for competition, as put into practice by the British regulators, contradicts the idea of a level playing field constitutes one line of attack. Veljanovski, an outspoken critic of the way in which the present system works, points out that *the corollary of promoting competition is overt entry assistance by the regulator. In practice this means subsidising entry by extracting favourable interconnection and other terms for the new entrant presumably to overcome the initial hurdles it faces. However, this is a marked departure from the ideal of regulation based on a level playing field* (VELJANOVSKI, 1993, p. 55). What is more he interprets this departure as seeing competition as an end in itself, manufactured by the regulator, rather than something which should be initiated by robust entrepreneurs who win market shares (VELJANOVSKI, 1993).

Other arguments against regulation for competition point out that the promotion of competition as a distinct objective of the economic regulator creates a potential conflict of interest with consumer protection: competition can lead to reductions in benefits to some consumers, especially if they were previously receiving preferential treatment; it is also hardly compatible with many social objectives. The very concept of cream skimming shows the difficulties of squaring rivalry and universal service (VELJANOVSKI, 1993).

5. Regulation on the Way Back towards Capitalism

Sections 2 - 4 have shown that, variously understood, regulation is an integral part of any economy and market economies are no exception. The main problem of regulating economic activities is therefore the way in which it is done and and the rules governing it. In industrialised countries their legal systems and their adherence to the rule of law, the importance that they ascribe to promotion and protection of competition, as well as the institutional frameworks within which, just to give one example, their public utilities operate, all exemplify regulation. Their existence is a matter of fact, even if the reasons given to justify it are not always beyond doubt. That is why introducing these kinds of regulation, or rather changing the structure, character and activities of the inherited regulatory institutions is part of the process of systemic transformation, of turning post-STEs into market economies. Transformation of the economic system will have to be accompanied by transformation of the political system, in the process of which transformation the rule of law will have to become a real corner stone of the new order. Competition will have to be appreciated, protected and promoted. Perhaps even *par excellence* regulatory agencies taking care of, for example, public utilities (network industries, natural monopolies) will have to be established and their briefs defined. To these ends institutional arrangements of market economies of industrialised countries will have to be both imitated and adjusted to local needs.

5.1 The Rule of Law and Competition

From the purely (meta)economic point of view, the constitutive characteristics of STEs were, as we all know, central planning and public ownership of the means of production (see Section 2, above). But equally important was the political system within which these economies operated. This authoritarian system, called by those in charge of it "socialist democracy", was characterised by the unquestionable dominance of communist parties, arbitrariness of its rule, and, in the case of countries other than the Soviet Union, dependence on the motherland of socialism. Politics was the main reason for which STEs were established, but their performance eventually forced political changes which in turn allowed countries of Central and Eastern Europe to begin their return to capitalism.

To change the political system it was apparently enough to call a free general election and let the newly elected MPs to gather together and form a government. To reestablish the rule of law it would then suffice to ask the legislative, executive and judiciary branches of government to follow some procedures as well as to consider properly the scope for their activities. But the rules first had to be set up and the laws written, to say the least. In some countries pre-communist democratic tradition and western culture seemed to be on the side of the newcomers, but there were no political parties which usually mediate between voters and candidates and structure political processes. The social groups, the actions of which made the breakthrough possible, perceived the world around them as a black and white cliché of us and them and it was naïve to expect that forty or so years of propaganda would leave the minds of the public intact. What is more, even in those Central and East European countries where the opposition was best organised, things moved so quickly that programmes had to be written only after the new equipe was already in power. On the one hand, the implementation of these programmes had to be entrusted to the old nomenklatura, who were too large a group to be sacked and replaced by new people. On the other, the ideas that some members of the "new nomenklatura" had about what market economies are all about also left much to be desired.

In this context a fully fledged, western-style liberal democracy could be easily confused with its external appearances, but its taking roots and gaining support from universally acknowledged social values had to take time. Corner stones are not laid down from one day to another and the same is true about the relationship between the rule of law and the political system built upon it. Obviously, things could have moved in the right direction and they were moving in the right direction, but the tasks faced by those now in charge of the polity and of the economy were not only enormous but also very urgent.

It was precisely this sense of urgency that was one of the main threats to developing democracy and reestablishing the rule of law. Creating a political vacuum had to avoided at any cost and new structures had to be in place and operating as soon as possible in order to, paradoxically, smooth the transition. Furthermore, most of the problems which had to be solved as soon as possible regarded the economy, and the economic situation in some countries was quite dramatic, again calling rather for a strong, decisive government than for procedures slowing down or even making necessary changes impossible. The temptation to cut the corners was - and still is - strong.

Against this background of political systemic transformation one has to look at more specific problems of the law in force. Whatever one's views on the legitimacy of the old regime and therefore on the legal acts introduced by it, one has to acknowledge that declaring all the inherited laws as null and void has never been a realistic option. From the theoretical point of view it would have been a rather controversial option and from the practical point of view one that would have been almost impossible to follow. The revolutionary changes of 1989 had their source in a rejection of the (particular form of the) communist rule in Central and Eastern Europe, but no revolution is able to break all the links with the past, least of all such a peaceful revolution as the one witnessed by most parts of the former Soviet empire in the late 1980s. Some continuity is unavoidable and the legal system of these countries is just one example, even though their legislatures have been very busy over the last few years. Eventually any parliament may decide to overturn almost all the laws passed by its predecessors, but this pragmatic attitude was potentially able to undermine the legitimacy of the new regimes, not least because of the limitations of the productive capabilities of any legislature. What ordinary people often saw were simply old bureaucrats implementing old laws.

But the problem of legitimacy, although important in itself, was not the only one connected with regulation *sensu largo*. Leaving aside the problem of whether and to what extent the introduction of the rule of law can be simply declared, one could not close one's eyes to the fact that piecemeal reform of the legal system was creating the risk of serious inconsistencies and gaps, regardless of whether or not introduced or overlooked purposefully. This was particularly dangerous in the sphere of what we called above the law of contract (see Section 3.1), in which sphere the inherited was simply incompatible with the newly established. Obviously, (economic) systemic transformation was started and is being brought about above all by changing existing laws and introducing new ones and quite a lot had to be done at the very beginning, as it were, in one go. But reworking the whole of the legal system, however long it was to take, was one thing, and making the new one operational, quite another. And all of this runs against one of the most important requirements with respect to economic laws, namely their stability.

Another context in which the issue of regulation appears is that of competition. Competition understood as rivalry in and for market places not only was all but impossible under the old regime but also thought of as highly wasteful and therefore undesirable. This did not mean that SOEs did not compete with each other. In the world of scarce, administratively allocated resources this was the only way to survive. State-owned enterprises (SOEs) had to compete with each other, most often in the process of increasingly multilateral bargaining, to force central planners to give them as much as possible (and demand as little as possible). However, this competition for investment funds as well as raw materials and other inputs had nothing to do with the competition in and for markets which can be observed in market economies, where it is the most effective mechanism to increase economic efficiency. From this point of view the degree of monopolisation as well as that of openness of STEs did not really matter, and their importance was further diminished by persistent shortages. Freedom of entry was excluded, and the concept of anticompetitive practices (*e.g.* abuse of market power) was known only from Western economic literature.

The situation did not change much in reformed STEs which tried to give autonomy to enterprises and introduce better incentives. These attempts, considered audacious at that time but later on labelled as naïve, were called marketisation. But even much bolder liberalisation and creation of the necessary institutional framework, although both are preconditions of competition, would have been unable to transform economic system of STEs. As long as most assets belonged to the state and money played only a passive role, competition in STEs and in market economies were too distant cousins to use the same name. In other words, it is privatisation and monetisation that have to accompany marketisation for the economic system to be transformed and, in consequence, for competition to develop and bring about improvements with respect to economic efficiency[45].

It was only after the break of 1989 that the possibility - and, as time passes by, the necessity - to compete in and for markets arose; now even casual observation suggests that such competition is a relatively widespread phenomenon in former STEs. What is, however, most important is that, compared with *status quo ante*, competition in (post-)STEs changed its nature, methods and objects. Only one thing remained as it had been: success was still necessary to survive.

At the most general level such competition in and for markets, in itself a necessary attribute of any market economy, is a consequence of systemic transformation, *i.e.* of marketisation and privatisation. Opening the borders, reducing tariffs, demonopolising foreign trade and, last but not least, introducing (internal) convertibility, all these factors played important roles. However when one looks at the behavioural and structural inheritance with which post-STEs started to rely on competition in allocating their resources, there arises a problem of what kinds of measures are necessary to protect what has already been achieved as well as to encourage future spontaneous developments. Equally important is the question regarding more positive actions aiming, for example, at changing the structure of previously monopolised industries.

In Section 3.2 we argued that more or less institutionalised competition policy belongs to the regulatory infrastructure of any market economy. Trying to translate these general remarks and considerations into what is required for post-STEs on their way back to capitalism, one should point out that sooner or later, and rather sooner than later, antitrust laws will have to be passed, specialised institutions created, and practices established. All of this is an integral part of systemic transformation, of creating an institutional structure suitable for market self-coordination of economic activities. Yet the difference between transforming economies and mature market economies consists in that although, as we have said, all economic agents would like to be monopolists, in post-STEs there are many enterprises entering the new brave world as monopolists which never won anything apart from the arbitrary favours of central planners interacting with local party bosses. This makes the dilemma regarding not punishing the most competitive enterprises even more acute, because available criteria are blurred by unreliable data. What is more, the usual conflict between a vocal minority claiming that their industry is special and the silent majority in favour of (fruits of) competition, *i.e.* of lower prices and larger variety of goods, is not going to go away, and those trying to solve it in the public interest cannot as yet rely on the support of the public at large.

Nonetheless, the relevance of this kind of approach to regulation *sensu largo* in the process of systemic transformation is additionally strengthened by the desire of all Central and East European countries to join the European Union, which necessitates some kind of harmonisation of existing legal arrangements and institutional structures[46]. Despite there existing a variety of particular institutional solutions and a pluralism of political traditions, democracy and its corner stone, the rule of law, are both an important common denominator among existing country members and a necessary requirement for those who want to apply. More specifically, although there are differences as far as the level of enthusiasm and of being convinced is concerned, all country members, both separately and in sharing their responsibility for the single market, try at least to protect the existing level of competition, and Articles 85 and 86 of the Treaty of Rome are used more and more to this end.

But are these kinds of common denominators sufficient to help post-STEs to make the right decisions? All the choices that have to be made in this transition period are tainted with indeterminacy, because although the direction of necessary changes is clear enough from it one cannot deduce the exact institutional form that the political system including protection and promotion of competition, ought to take, not to mention regulation *sensu stricto*. The solutions applied elsewhere are a result of more or less smooth evolution of market institutions, with all the limitations involved, and therefore of limited applicability in the countries in which capitalist continuity was interrupted. And there is much more to it than only differences in traditions; evolution of economic institutions means also that there are solutions suitable for the given stage of economic development, which is one more argument against simply copying the laws from elsewhere.

5.2 Public Utilities and Systemic Transformation

Even if one is not convinced by theoretical arguments in favour of regulating economic activities in certain industries, *i.e.* even if one doubts the ability of politicians, operating in highly imperfect markets, to correct market failures, there is no doubt that various institutions regulating public utilities exist in most industrialised countries, whose economic systems Poland and other Central and East European countries want to emulate. Although it is highly doubtful whether they should do it uncritically, the debate is more about what, to what extent and how, and not about whether at all.

The ongoing changes of the systemic characteristics of the post-STEs have already been, quite obviously, forcing the authorities not only to change their policy towards public utilities but also the policy tools they are using in its implementation. Since liberalisation, undertaken to bring about systemic transformation, consists in the last analysis in separating those parts of the economy that can be regulated by applying general rules such as the law governing contracts or the antimonopoly legislation from those which have to be taken care of individually, the old regulatory structure of STEs has to undergo a radical change. As long as such (considered as) special industries remain in the public sector, regulatory functions can be exercised, as in most mature market economies, by the same state organs which are supposed to play the role of an owner. Public ownership will therefore provide a form of regulation, but regardless of how we interpret the given state of affairs there is no doubt that state administration will have to rethink its role and change the policy tools that it uses. This is forced on civil servants and politicians both by changes in the economic environment in which these industries operate and by the desire to (re)introduce the rule of law.

What is more, as most Central and East European countries want to join the European Union rather sooner than later, in the process of preparation for their entry they will be subject to strong pressures to harmonise not only their laws but also their institutional structures, including institutions regulating public utilities[47]. Since in the European Union there appears a tendency, and rightly so, to create single regulatory agencies for the whole Union, without which agencies a single market in network industries is difficult to imagine and even more difficult to organise, the need to create similar regulatory structures in post-STEs is, quite understandably, ever greater.

But there is more in it than mere blind imitation or harmonisation required by the prospects of future integration with the European Union. One could perhaps say that central planning was nothing else but an all embracing regulation of the kind exercised towards publicly owned public utilities. Obviously, it went much further than even the most restrictive regulations in market economies, because specific decisions were taken on behalf of or imposed on the regulatees. What is more, in trying to regulate too much central planners were unable to solve informational and incentive problems unavoidably arising when the agency problem is tackled by regulation alone. This was one of the reasons for which STEs could not work and eventually collapsed, the poor state of these economies precipitating such political changes as made possible systemic transformation.

Be that as it may, one seems to have to do something about public utilities, once central planning has been abandoned. One of the decisive factors in shaping institutional choice is the degree of progress in privatising public utilities in these countries. It is therefore from this point of view that we are going to look at regulation in the strict sense of the word on the way back towards capitalism.

5.2.1 Regulation within the Public Sector: Institutional Reform and Restructuring

Regardless of whether and when more or less complete privatisation becomes a realistic, implementable option, something has to be done with public utilities as they are at present, not least because of the importance that public utilities have for any economy. The question of whether it is done in view of privatisation is in this context of secondary importance, and this "something to be done" has as it were two dimensions: institutional reform and restructuring.

One aspect of the necessity to do something with public utilities once issuing commands about the level of production and the distribution of inputs and outputs has started to run against the logic of the economic system being (re)established follows from the institutional vacuum created by abandoning the old system. This was accompanied by the high level of dissatisfaction with the inherited structure, both within the industries themselves and within the state administration and their advisors, and its all too visible inefficiency. Prices, especially in the energy industry, were hugely distorted and there was pressure from international financial organisations to correct their structure and level, using those in OECD countries, sometimes called "world prices", as a reference point if not a paradigm. Changes in the degree of vertical and horizontal integration seemed to be necessary and, last but not least, environmental policy was to be harmonised with (or even become a part of) the energy policy, and the same is true about water and sewage companies.

In Poland the transfer of hundreds of water and sewage companies to local authorities, *i.e.* to *gminas* or their unions, and the separating Telekomunikacja Polska S.A. from Poczta Polska, *i.e.* telecomms from post office, were relatively straightforward and did not involve any restructuring worth its name. The changes being implemented in the energy sector and in particular in the Polish ESI deserve a little more attention. PSE (Polskie Sieci Elektorenergetyczne S.A.), the Polish National Grid Company, was created, and after having taken over the whole of the high voltage grid and the dispatch system as well as the pump and storage power stations it became a monopsony buyer of almost all electricity produced and a monopoly seller to Zak ady Energetyczne, 33 regional distribution companies. At the same time the 15 most important hard coal power stations gained full autonomy; what is now being discussed is the number of (holding) companies that are to be created out of them[48]. Coal prices were finally liberalised, although not without difficulties, but prices of electricity for final users are still decided by the Ministry of Finance. When PSE buys electricity from power stations, each of them has its own price formula. Whether and to what extent competition is going to be introduced remains to be seen[49].

Although many think that competition is excluded in public utilities almost by definition, it is surely not true that all these industries are naturally monopolistic[50]. For example, dividing the ESI in England and Wales into generation, transmission, distribution and supply created, as we have seen in Section 4.2.2.2.1, ample space for competition in generation and supply, while the other two, standing on their own either as separate companies (National Grid Company) or at least as accounting units, preserved their monopolistic character and they are subject to strict regulation.

Pro-competitive restructuring creates space for competition and delimits the really naturally monopolistic elements of these industries. In the context of post-STEs the question to be answered runs, however, as follows: is such a restructuring enough? Will SOEs compete with each other? Western experience as well as that of the reformed STEs are not very encouraging as far as the possibility of competition within the public sector is concerned. If these experiences can be generalised, preserving public utilities in the public sector will almost certainly exclude not only actual competition but also the very possibility of competition, and therefore will put excessive burdens on regulatory structures and, if these fail, on consumers, both domestic and business. In the last analysis such a failure may seriously cripple a given country's international competitiveness and comparative advantage.

The fact that the economic and institutional environment in which public utilities operate is different in industrialised countries and in (post-)STEs is of course very important. Since by and large post-STEs are moving in the right direction as far as creating such a market environment is concerned, if they are successful, some improvement in economic efficiency may be expected. What is more, once the monolithic façade of "moral unity" of these nations has been abandoned and various pressure groups are able to voice their interests, not all problems can be swept under the carpet, which in turn should make acting in the public interest a little easier, at least in theory.

Nevertheless, in market economies publicly owned public utilities were not exactly examples of economic efficiency and responsiveness to consumers' needs. Solving the problem of how to increase economic efficiency of publicly owned public utilities was not helped by the fact that in those OECD countries which did not privatise their public utilities, ownership and regulatory functions were exercised by the same ministers. In theory this could be remedied by independent regulatory agencies but of whom should these agencies be independent? There are certainly ways of solving the problem of regulators' accountability and incentives, but the lack of precedents, of actually operating institutional solutions, constitutes a warning.

In post-STEs the situation is even worse because of still not very stable and underdeveloped political systems and because of the history of the relationship between public utilities and other enterprises and ministries. In other words, the necessity to switch to a new type of regulation of publicly owned public utilities in post-STEs provides more questions than answers and even a superficial attempt to make use of the experience of other countries provides plenty of reasons for scepticism about possible efficacy of such an arrangement. All in all, there is, therefore, a strong presumption in favour of privatisation.

5.2.2 Regulation and Privatisation

Western experience suggests that regulatory problems have also - or perhaps, in particular - to be solved when the governments of Central and Eastern Europe are deciding about what - but the "when" and "how" are not less important - should be privatised. Although it is often pointed out that these economies were highly monopolised, most of the industries there are at least potentially competitive, just as in any other market economy, especially if free trade prevails and borders remain relatively open. In consequence most of privatisation - even though, as in Poland, individual decisions have to be approved by the Antimonopoly Office - will be, so to say, regulation-free. Or perhaps better said, the regulatory aspect of state ownership will disappear in the moment of privatisation; the activities of given enterprises will be subject only to general legal regulation, and no industry specific regulatory structures will be created.

So far privatisation has been proceeding very slowly even in structurally simple cases and most of the growth of the private sector has come from setting up new private enterprises, and not from transferring assets and enterprises from the public sector to the private one. The present speed of privatisation in post-STEs suggests that it will take a long time before all assets in non-public utilities sectors of the economy are transferred to the private sector, which would put them in more or less the same position as the British government was in the early 1980s. But should they wait so long? Should they start with at least potentially competitive industries and proceed to public utilities only when they are done with them? Are not public utilities so important that they should be taken care of as soon as possible? These are by and large rhetorical questions. Since, as noted above, new structures and ways of proceeding have to be created anyway, and since regulation both in theory and in practice works better if it is separated from ownership, it seems that this kind of privatisation-cum-regulation should not be postponed for too long.

Whatever the answers to the questions listed above, there are also important *caveats*. If the British experience has any relevance for post-STEs, it can certainly tell them how complex privatisation of public utilities is. For all these reasons in the case of post-STEs the possibility of having public utilities in the public or private sector is not an either/or alternative describing the dilemma that politicians have to solve here and now, but rather refers to a sequence of future events. In other words, in the immediate future one can at best expect a mixed ownership structure of public utilities.

This is precisely the case with, for example, the ESI in Poland, where privatisation does not seem possible in the near future. Apart from the resistance within the industry and criticisms from other defenders of national interests, Western advisors, especially from other than the UK member countries of the European Union, are in no hurry either to advise such a complex and time consuming change. That is why in this section we shall consider the case of mixed ownership structure of public utilities in post-STEs, which in turn assumes piecemeal privatisation of these industries, with all its consequences for the best possible regulatory structure[51].

Such privatisation may happen in two ways, either by privatising some enterprises as in the case of power station Opole and combined heat and power station e.g. near Cracow, Poland, or by selling part of shares in a monopolist to a private investor, usually foreign, as in the case of Hungarian telecommunication company Matav[52]. Each of these options has its own specific consequences and implications as far as regulatory structures and their efficacy is concerned.

Whatever the share and form of the private sector's participation, being regulated by the owner of potential competitors is not a very attractive option and may result in the desire by private investors to acquire some monopoly rights, *e.g.* for the given territory. While this should be resisted especially in the potentially competitive parts of these industries, mixed ownership structure of the given industry raises the problem of creating and preserving a level playing field for publicly and privately (and perhaps even municipally) owned enterprises. Since it would also be definitely wrong to continue with the arrangements in which ownership and regulation are united in the same institution or even person, *i.e.* with the arrangements in which an (public) owner of some enterprises is also a regulator for all enterprises in the given industry, the creation of an independent regulatory agency seems to be of extreme urgency.

Such an agency, which is in this context a much less problematic solution than in the case of exclusively publicly owned utilities, would have to combine a few different tasks. It would oversee, perhaps with veto powers, future restructuring of the given industry as well as ownership changes within it. It would regulate the behaviour of incumbent firms but at the same time it should facilitate new entries. However, its most difficult task will be, as already mentioned, to assure a level playing field for existing enterprises, regardless of their ownership status, and new entrants.

Some post-STEs have a long history of declaring equal treatment of various sectors of their economies but they rarely went beyond declaration and - surprise, surprise! - there was always one sector that felt discriminated against, and most often both of them claimed the same. The problems to be solved in this area range from old (nomenklatura) boys' networks uniting the industry, the state administration and the banking system, to obvious differences between the public and private sectors' enterprises, which question the rightness of treating them equally. Since the sectors differ from each other from the point of view of how the agency problem is solved within them, of the degree of risk aversion in general and towards bankruptcy in particular, of their access to capital markets and of their exposure to markets, all these factors will have to be taken into account by the regulators, who, at the same time, will have to be independent and accountable.

Once this is done mixed ownership structure in the potentially competitive subsectors of public utilities has much to be recommended, at least from the theoretical point of view. Assuming that all enterprises are subject to the same constraints imposed on their behaviour, equal access to capital and labour markets including, and that the publicly owned firms will maximise social welfare, *i.e.* the sum of consumers' and producers' surpluses, and the privately owned firms will maximise profits, it follows that the very presence of SOEs in any industry will act pro-competitively and therefore exercise additional pressure towards economic efficiency, which is the most important rational for advocating competition (DE FRAJA, 1991). This result, however, is strongly dependent on actually establishing a level playing field for all enterprises, regardless of their ownership, which in practice may be quite difficult because the general difficulty of state authorities to commit themselves is in a country like Poland made worse by the persistence of the soft budget constraint (JASINSKI, 1993).

The second model of partial privatisation of public utilities involves selling some shares in a monopoly to a private (foreign) investor, while preserving its monopoly position, seems to be less controversial from the point of view of potential buyers of these shares. Regulator who owns the remaining shares should not be expected to wish to hurt both shareholders. But their objectives may nevertheless diverge. This effect is well known from the Western experience and was even quantified by BOARDMAN and VINING (1989), who drew our attention to the possibility of what they call *cognitive dissonance* (p. 26), caused by the conflict between public and private shareholders, *i.e.* their divergent or even contradictory objectives. One more *caveat* follows from Section 5.2.1 where we were very sceptical about governmental regulation of publicly owned public utilities. The arguments presented there seem to hold here too.

Obviously, things will change over time, especially if restructuring and privatisation progresses as planned. One cannot even exclude the possibility that the scope for regulation will diminish over time. That is why free entry and competition as well as pro-competitive commitment of regulators are perhaps much more important than privatisation understood as transfer of existing assets from the public sector to the private one. In other words, privatisation which was, for political reasons, almost a *conditio sine qua non* of pro-competitive restructuring and regulation in the UK, gives its priority to other actions, and although it preserves its long-term desirability it losses much of its short-term necessity.

The situation will differ from one public utility to another. We have already mentioned the undergoing changes in the Polish ESI. To give one more example, in telecommunications in Poland, Telekomunikacja Polska S.A. is still a typical public sector/public utility monopoly exposed to competition, unlike

Western telecomms, only in local networks and with a statutory monopoly for international calls[53]. Theoretically competitive, but in fact rather monopolistically complementary, networks[54] can be established - and appropriate licences are being issued in relatively large numbers - almost exclusively in rural areas with a very small number of telephones, but usually cover the territory of only few *gminas*[55]. But even in these attempts they do in fact work for TP S.A. and the monopolist is able to protect and preserve its dominant position by being allowed to dictate the conditions of interconnection[56]. The only considerable attempts at entry, Telekom Pi a and Silesia (with participation of the local government and, among others, of an American company Sprint, where new networks wer to cover one vojvodship each - one of 49 - in Western and South-Western Poland, *i.e.* more than 100,000 and about 1 million lines, respectively) are still struggling to get off the ground. TP S.A. is also a major stakeholder in a cellular phone network and at least one cable TV company.

On the other hand, there is no Oftel, and apart from public opinion it is only the Antimonopoly Office that is able to check the activities of TP S.A. and that to a very limited extent. There is therefore a strong case for developing both regulation and competition. What is more, as the example of interconnection shows clearly, the two are in fact inseparable and therefore Polish telecommunication seems desperately to need an independent regulator, committed to promote competition, *i.e.* a Polish Bryan Carsberg[57].

All in all, changes in the regulatory structure regarding public utilities will demand a change of mentality ranging from the engineers who used to run public utilities to politicians whose understanding of economics even in much simpler cases still leaves much to be desired. New, different sets of information/accounting data are also required for the commercial success to become possible at all. At the same time there certainly are factors which will work in favour of necessary reforms and systemic transformation, and the necessity to do something with public utilities is only one of them. Distorted prices, for instance, cannot be really corrected only by calculating their shadow values, and letting some degree of competition in may certainly facilitate achieving their market clearing and allocativelly efficient levels. Public utilities are also part of the post-STEs' infrastructure, the modernisation of which is a *conditio sine qua non* of future economic growth and of attracting foreign investors into these countries. Backwardness of this infrastructure may even undermine, for example in telecomms, standard assumptions about economic sense of duplicating networks and therefore, together with rapid technological progress in this industry, point in favour of more competition and openness. What is more, when we compare autonomy of power stations in Poland with the monolithic structure of the CEGB before splitting and privatising it, even the existing internal structure of public utilities in post-STEs may provide a promising starting point. Politics, however, will almost certainly remain as the most important obstacle on the road towards regulatory reforms.

6. Regulation as an Element of the System Being Created: Some Normative Remarks and Suggestions

Changes in the regulatory structure of post-STEs are to a large extent forced upon the authorities because of the undertaken task to transform former STEs into market economies.

However, on the one hand, these changes consist above all in discontinuing certain forms of control, supervision and decision making and may be labelled as abandoning central planning. On the other hand, systemic changes only to some extent predetermine institutional choices and the structures actually existing in industrialised countries show how many and how different solutions are possible. Of course, all of them are a result not only of search for the best possible solutions but also of legal traditions of given countries and of actually operating interest groups and their relative strength. Nevertheless, it seems appropriate to finish this paper with a few remarks regarding the desired regulatory structure in the countries of Central and Eastern Europe once they successfully completed transformation of their political and economic systems.

Following the assumptions on which we worked in the previous section, these final remarks and suggestions will refer to a system to be aimed at rather than to arrangements which have to and can be introduced in the relatively short period of time to answer the most urgent needs of the transitory period. This period will have come to an end - or will have been considered as finished - once economic and political stability have been achieved and sound economic growth resumed. The rule of law should be fully operational and market institutions in place. Privately owned assets should be producing much more than half of the GDP, and the private sector employing majority of the workforce[58]. Money should play an active role; capital markets should complement markets for labour and for goods and services, and economic agents should compete with each other. Private public utilities - and their possible regulators - only complete this idyllic picture.

To the extent that post-STEs are going to - ideally not uncritically - imitate the institutional structure of industrialised countries, one may realistically expect that, as we have analysed this above, the new regulatory structures will go beyond the rule of law and promotion and protection of competition, even though this state of affairs does not seem absolutely necessary and alternative solutions are feasible. In other words, there will be regulation in the strict sense of the word. But how should it look like? Should each industry be regulated separately or should there be just one regulatory body, for instance for all subsectors of the energy industry? Should they take the form of a committee or be headed by a director general who on his or her own takes decisions and is responsible for them? What is the best place for regulatory bodies in the government structure and how can such a position be squared with their independence and accountability? Who is going to have the final say in conflicts between regulators and regulatees? Is sunshine regulation to be relied upon? What is to be the role of the public opinion in the regulatory process? And, going from the general to the more specific, should it be profits that are regulated or only prices?

As we have said above, apart from the U.S., it is only the United Kingdom that has a separate regulatory system which has been in operation long enough to allow first assessments of its functioning. The ongoing debate shows how difficult it is to provide a disinterested assessment. We presented some of the most salient features of these two regulatory paradigms in Section 4.2. In practice, however, although the jury still seems to be out on regulation UK-style, this institutional model has to compete with its American counterpart, and attempts to create regulatory offices or agencies face resistance from those who prefer (federal) commissions. The choice is far from being straightforward, but at least two things have be taken into account.

Firstly, one similarity between the UK and post-STEs is that, regardless of the speed of privatisation in the latter, sooner or later regulatory agencies will have to deal with industries in which most economic agents recently were or even still are in the public sector. Regulation UK-style was designed for a transitional period and therefore most likely can be used in the process of systemic transformation too.

Secondly, it seems that British regulatory institutions are more modern. Such a statement does not uncritically assume that newer institutions are necessarily better than older ones, but at least in some industries, like for example telecommunications and, perhaps, ESI, technological progress has to have an effect on regulatory structures. And this is precisely what happened in the UK in the 1980s. Institutions designed for and set up in post-STEs have to be forward- and not backward-looking.

Poland, on the other hand, has its first experiences with a regulatory committee responsible for broadcasting called Krajowa Rada Radiofonii i Telewizji, responsible, among other things, for granting licences to prospective broadcasters. The results of the first year of activities of this Rada are rather disappointing, and the degree of controversy it managed to provoke was unexpectedly high. These developments were at least partly due to clashes of personalities, but excessive politicisation of everything that Rada did or tried to do should be taken into account when institutions supposed to regulate the economy are established. Giving the Sejm, the Senate and the President the right to appoint the members of this Rada was almost necessarily a recipe for disaster, and a more corporatist strategy, apparently envisaged for the Energy Agency, seems to augur equally badly.

One can use the two examples already introduced, namely the ESI and telecommunication, for further considerations. According to the letter written by the Polish government to the World Bank, a special agency for the whole of the energy industry was to be created before the end of 1992, but in the moment of writing this paper (April 1994) the birth process is still going on and the change of government after the elections of 19 September 1993 is certainly not helping to accelerate it.

If we were nevertheless to try to formulate some tentative recommendations, based on the British experiences of regulating this industry, the following comes to mind: a separate regulatory agency, in the sense of taking public utilities out of the Antimonopoly Office's domain, seems to be preferable;

the Antimonopoly Court could still play the appeal role. Such a regulatory agency, despite the obvious risk of becoming too powerful and therefore too politicised, would cover the whole of the energy sector, instead of having separate agencies for gas, electricity etc. One must not forget either how important personalities are and therefore how crucial it is that the person heading such a agency - and a single person seems to be preferable to a committee - is really committed to promotion of competition. It would be also advisable to create such an agency as soon as possible so that it could take part or at least give advice when strategic decisions about the future structure of industry are taken.

On the other hand, regardless of the future shape of regulatory institutions, it seems necessary to stress the importance of pro-competitive restructuring of the ESI. Namely, from the history of the ESI in England and Wales in general and from its structure in particular one can draw three kinds of lessons. Firstly, there are dangers of creating a monolithic structure and keeping it in the public sector. Secondly, there are negative experiences with half hearted liberalisation and with leaving unsolved problems of the coal industry. Thirdly, and perhaps most importantly, pro-competitive restructuring not only is possible but also can really work in practice. There is also convincing evidence that anybody learning from British experiences can go even further, especially as far as the number of companies in electricity generation is concerned.

In telecommunications, for the time being, it is the Ministry for Telecommunications that has its loyalties divided between regulation and ownership. It regulates telecomms, among other things, by issuing licences for new operators of private and public networks[59] as well as guidelines for interconnection contracts[60], but at the same time represents the Treasury, the only shareholder of TP S.A., in dealings with and supervision of this company.

The activities of the Antimonopoly Office, the only other player in this regulation game, are bound both by Government's policy and by the laws passed by Parliament, which obviously limits its room for manoeuvre. The Antimonopoly Office's brief certainly covers public utilities, but the results are rather disappointing, partly because of the lack of qualified staff and partly because other acts of parliament excessively restrict the areas open for investigation and intervention. Nevertheless, it is important for the Antimonopoly

Office to have clearer ideas about the institutional form that regulation should take as well as about the extent to which competition should be promoted and enabled.

A useful framework could be as follows:

- Consumer protection handled above all by contractual agreements (and liabilities);
- prices, both final and for interconnection - by the Antimonopoly Office;
- technical issues regarding both equipment and interconnection - by a separate, specialist body, possibly within the Ministry for Telecommunications.

Endnotes:

* Regulatory Policy Research Centre, Hertford College and Campion Hall, Oxford University. Paper written for the ACE project on *Regulation in the process of systemic transformation*. The author wishes to thank W odzimierz Brus, Anna Fornalczyk, Julia Gibert, Tim Keyworth, Helen Lawton Smith, Andrzej Szablewski, Paul J. J. Welfens and George Yarrow for their useful comments. The usual disclaimer applies.

2. This action constituted a step towards privatisation of the Polish economy but it did not result in a systemic transformation because the coordination mechanisms remained unchanged and no critical mass of privately owned assets was achieved.
3. The cases of socialism in Yugoslavia, China and Cuba, although different in their origins, only confirm this general rule.
4. The issue of behavioural patterns, interesting and very relevant as it is, will have to be left aside in the remainder of this paper.
5. Such examples of regulation as that regarding health and safety, various standards or professional associations will not be discussed in this paper. This should not, however, be interpreted as an attempt to question their regulatory character.
6. Obviously, we concentrate our attention on economic determinants, leaving aside such factors as the political system and its stability as well as culture and traditions.
7. The very possibility of competition is listed among direct effects of privatisation.
8. This depends on two scale variables: the unit cost of raising public funds and the quality of the political system and the strength of short-term political pressure (Shapiro and Willig, 1990, Jasinski, 1992, Jasinski and Welfens, 1994).
9. In this sense Coase never advocated the Coase theorem - see Coase (1988).
10. From a more dynamic point of view, well exemplified by the developments in the UK, any attempt to introduce competition in a previously monopolised industry, or any action aiming at separating naturally monopolistic elements from potentially competitive submarkets, sometimes result in a growth of regulation. Whether this is a permanent development remains to be seen. However, one could say that increasing competition, or, more specifically, the number of players in a given industry, increases also demand for regulation, which is all too eagerly supplied by governmental bodies which find self-restraint particularly difficult. In other words, changes of the balance between competition and regulation do not seem to be a zero sum game.
11. Telecoms in New Zealand are most likely an exception in that ordinary antimonopoly legislation is used to oversee a public utility.
12. The history of the gas industry in the UK is equally revealing and illuminating.
13. The latter operated mainly in distribution and were heavily regulated by the Electricity Commission, established in 1919 to promote, regulate, and supervise the supply of electricity on a national scale (Vickers and Yarrow, 1988).
14. Despite the change of government in 1951, the only modifications introduced by the two consecutive Electricity Acts of 1954 and 1957 consisted in setting up two independent, vertically integrated, Scottish Boards, and, south of the border, in creating the Central Electricity Generating Board (CEGB) and giving more autonomy to the Area Boards. The CEA was transformed into the Electricity Council, a federal body entrusted, despite the Herbert Committee's recommendations, with consultative and deliberative function rather that with powers of direction, control and supervision.
15. Leaving aside constitutional problems with any form of corporatism as well as the issue of whether and to what extent this form of governance is compatible with liberal democracies, one has to notice that *corporatist institutions are better able to take a long view (...) However, the long term view taken by such bodies will be very heavily coloured by institutional self interest and, therefore, is likely to be resistant to change. In addition,*

corporatism leads to producer domination and consequent inefficiency, inflexibility and inequity (Roberts *et al.*, 1991, p. 177).

16. The inevitable inefficiency manifested itself in two ways: an excessively reliable system in which interruption costs were not evaluated, and an investment programme that was heavily nationalistic in its technology choice. This inefficiency was further compounded by the use of a risk free social discount rate which grossly underestimated the private capital market's opportunity cost of capital for long-lived high risk projects (Weyman-Jones, 1991).
17. Its story was best summarised by Lord Marshall: *I do not like this form of electricity privatisation, but the broad story of nuclear power in this country is the most powerful argument in favour of privatisation that I have ever seen. Over the last forty years governments have interfered with this business so continuously, and with such appalling effects, that I am thoroughly convinced that it must be best to do everything that can be done in the private sector* (Marshall, 1990, quoted in: Roberts *et al.*, 1991, p. 117).
18. In the latter case a relatively simple readjustment of the pricing structure used by the CEGB, which changed the balance between fixed and variable costs, deprived to a large extent entry of its profitability. But much more important was the fact the CEGB did this and could do it unpunished which suggested that more anti-competitive surprises were in stock for too audacious entrants who took the provisions of the Energy Act at their face value.
19. Toll roads and canals were regulated from the eighteenth century, water - from the beginning of the nineteenth century, railways a little later, gas, electricity and telephony from soon after their commercial exploitation (Foster, 1992, p. 17).
20. On the other side of the Atlantic the Interstate Commerce Act was passed in 1887.
21. State regulatory commissions originated with a Rhode Island law of 1839.
22. *The multiple goals that Congress attaches to the regulatory process has resulted in a broad range of powers for regulatory agencies and diverse instruments for carrying out the agency's mandate. Agencies create laws and standards in the form of administrative rules. Agencies enforce laws by monitoring compliance with laws and imposing legal sanctions. Agencies perform arbitration of conflicting interests and allocate resources as in a court of law. Thus the powers and procedures of regulatory agencies resemble those of the legislative, executive, and judicial branches of government* (Spulber, 1989, p. 72)
23. Some governments use the so-called "golden shares" to have their cake and eat it.
24. In such a case, gains from vertical integration have to be weighted against gains in efficiency which only competition can bring about in the potentially competitive elements of the supply chain.
25. The more complete list of production stages in the ESI could include industries providing inputs as well as the regulator, regardless of the legal form taken by the latter (Yarrow, 1991).
26. It seems that the former is - because of the advantages of having an integrated power system is the real source of economies of scale that in not too big a country give plausibility to the existence of a natural monopoly. As far as the latter is concerned, at first glance *to have more than one set of distribution conductors running down a suburban street [is] both economically and environmentally ridiculous* (Evans, 1989, p. 137), but the evidence gathered and analysed by Primeaux suggests that, because of economics of density, in many situations consumer prices are in fact lower despite duplication of networks. This in turn suggests that from the possibility of minimising costs and prices by having only one producer one should not too easily draw the conclusion that this is precisely what is going to happen when there is only one producer. There are many more determinants of economic efficiency than only the optimal structure of an industry and this also has to be taken into account when actual and potential benefits of competition in the ESI are assessed (Primeaux and Walter, 1989). In other words, the problem of promoting competition in the ESI is not to look for applications of textbook definitions but for an institutional framework which, because of the rules and

incentive structures it implies, can perform most efficiently or simply relatively better (Evans, 1989).

27. *Much further breakdown into county or metropolitan sized areas (the precise boundaries based on technical considerations) would have been possible without unduly compromising economies of scale. Such a breakdown would increase the potential for competition at area boundaries and increase the potential for community based demand management and CHP schemes* (Roberts *et al.*, 1991, p. 181).
28. The pool receives electricity also from Scotland (Scottish Power, Scottish Hydro-Electric, Scottish Nuclear), and from the French EdF.
29. On the one hand, the existence of such contracts is an important safeguard against collusive attempts to raise prices in pool. On the other, *the ability to strike long-term contracts gives new entrants the capacity to make pre-emptive (pricing) moves against incumbents, and may, therefore, serve to increase the competitive pressures that potential competitors exert on established firms* (Vickers and Yarrow, 1991, p. 209).
30. The subdivision of the supply market into a monopoly franchise and an open (contract) sector is supposed to disappear by 1998.
31. From the viewpoint of such a user it is worth noting that signing a supply contract with a firm different from the local REC considerably modifies the duties of the latter. *The Electricity Act put RECs under an obligation to continue supply to existing customers; to offer connection to the distribution system; and, if required, to supply new customers on request in most circumstances. The major exception concerned those customers already being supplied by a company other than the REC with the PES licence for their area. Should such customers wish, for whatever reason, to receive electricity from their REC they had no rights to supply, only a right to connection to the distribution network* (Roberts *et al.*, 1991, p. 93).
32. Unlike in the aftermath of the Energy Act 1983, in the restructured and privatised ESI liberalisation seems to have caught on and instead of feared under-investment it is rather the case that too many projects are under way, due to new technology and, as mentioned above, long-term contracts. This *par excellence* competitive process has also some other positive side-effects, both in terms of security of supply (diversification of inputs) and of limiting environmental damage (in terms of CO_2 emission, gas is much more environmentally friendly than coal), and technological developments, thanks to more sophisticated metering and associated two-way communication, promise even more benefits for customers (Littlechild, 1991).
33. From the point of view of developing effective competition, freedom of entry and of contracting are much more important than the initial number of generating companies.
34. For the managers the ideal privatisation should combine financial independence from the government ministers and civil servants, especially on capital expenditure, but without disadvantages of restructuring and/or liberalisation.
35. On the brighter side one could point out that both telecommunication duopoly and privatised monopoly/monopsony in the gas industry predated the case of the ESI in England and Wales.
36. See Section 4.2.2.2.2, below.
37. This is just a special case of a more general problem of whether antimonopoly legislation should take care of market structure or only of behaviour of individual agents.
38. Since neither privatisation Acts nor individual licences issued by the Secretary of State override UK and EC competition laws, this decision should not have had too much of an impact. Namely, it did not deprive the rest of the legal system of its pro- or anti-anti-competitive bite, even though in practice non-utility specific legislation is rarely used or referred to. As in utilities there is plenty of scope for anti-competitive practices which usually take two forms: direct predatory behaviour (*e.g.* cross subsidisation and predatory pricing), and indirect strategic entry deterrence (*e.g.* pricing for the use of the transmission system) (Helm and Yarrow, 1988), whoever was to be in charge of administering competition law in

the realm of public utilities, there was no danger that that institution would remain without having anything to do.

39. The reference is usually a general public interest reference rather than a narrow determination of the issue in dispute, and the managers of regulated utilities are usually keen on avoiding it because of, as they themselves say, their lost time that a full scale reference would entail. Moreover and far more to the point, after the MMC report has been published, the regulator can change the licenses so as to remedy the detriment to the public interest, but he is not required to accept the recommendations specified by the MMC. That is why implicit regulation (Vickers, 1991), *i.e.* the very threat of - or the risk of triggering - both a regulatory intervention and a referral to the MMC have themselves powerful disciplining effect and play such an important role in the British regulatory system.
40. But this legal framework does not describe exhaustively the politico-economic reality of regulation in the UK. Graham and Prosser (1991, p. 230) wrote that *it is not too extreme to say that negotiation and bargaining are institutionalised in the regulation of privatised enterprises in Britain*. Others put it more bluntly: *British regulation is a system of ambushes backed by blackmail* ("A source close to BT", quoted in Veljanovski, 1993, p. 21). The *Independent on Sunday* was not therefore an exception when it wrote in August 1991 that *over the last couple of months regulators have been cracking down on privatised monopolies with increasing severity as if engaged in some kind of regulatory virility contests - each seems intent on outdoing the other* (quoted in Veljanovski, 1993, p. 26).
41. The list of items intended as accusations includes also the fact that *much of the law is informal, consisting of guidance notes, codes, determinations, and negotiated modification to the utilities' licence and the price cap* (Veljanovski, 1991a, p. 9). High on this list is also the degree to which regulation in the UK became personalised, especially in the case in which a regulator faces just one company, *i.e.* in gas and telecommunication. And if many of these issues matter only for those involved and some specialist, the public face of regulation is sometimes described as even uglier: *secretiveness; deals between utility and regulator; the clash and influence of personalities, all force the conclusion that we are witnessing a challenge to the fundamental principle of a free society - the rule of law - by the rule of men, who are able to exercise considerable discretion* (Veljanovski, 1993, p. 81).
42. Nowhere can it be seen better than in comparison with regulation in the US: *there is nothing in US utility regulation approaching a statutory duty to promote competition* (Beesley and Littlechild, 1991, p. 45).
43. This is not, however, as uncontroversial as it may appear at first sight. For example, the creation of the Office of Electricity Supply (Offer) may be legitimately considered as part of the restructuring of the ESI and the Director General of Electricity Supply (DGES) or rather his activities as another link in the supply chain (Yarrow, 1991). The real difference between the two is that the restructuring was a one-off action, or so at least one hopes, while regulation is a process and it is by no means certain that it is going to wither away.
44. An intermediate approach could consist in the "ring-fencing" of monopolistic activities through either separate ownership or regulated accounting separation (Yarrow, 1991).
45. Regulation, which apart from competition and ownership, is the third most important determinant of economic efficiency, will be changed almost automatically in the course of these changes, but as it is better to shape it more explicitly.
46. This in itself is an attempt to hit a moving target as the official commencement of the single market has so far had almost no effect such industries as telecommunication and energy. George Yarrow's background paper for this project addresses these issues directly and therefore we shall not dwell on them here any longer.
47. European integration aspects of regulation *sensu largo* have been mentioned in Section 5.1.
48. The 3 lignite or brown coal power stations provide base load. Hydro-electric power stations play a negligible role.

49. For more on the electricity supply and gas industries in Poland see papers written by Piotr Jasinski, Andrzej Szablewski and George Yarrow for the same ACE project.
50. Whether vertical integration extends the range of what may be justifiably considered as naturally monopolistic is a debatable issue.
51. The issue of whether and how private public utilities should be regulated will be discussed in the very last section of this paper.
52. Electricity company CEZ taking part in the Czech(o-Slovak) voucher privatisation may be considered as some kind of exception.
53. Getting a licence for provision of long distance voice telephone is restricted to companies in which the share of foreign investors is not larger than 49 per cent.
54. In the short run such a arrangement has nothing to do with promotion of competition as it creates alternative, local monopolists. What is more, their financial viability is, because of the present price structure, debatable, which means that most new entrants are dependent on external financial assistance, like receiving free exchanges.
55. There are about 2500 *gminas* in Poland.
56. Only very recently, i.e. in the Winter of 1994 the Ministry for Telecommunications prepared some guidelines for interconnection contracts, but the effect remains to be seen.
57. For more on telecommunication see in papers written by Leszek Stachów, Paul J. J. Welfens and Cornelius Graack for the same ACE project.
58. By the end of 1993 only Poland fulfilled these criteria. The Polish economy, however, is certainly an economy in transition, however advanced this transition is, and not yet a mature market economy.
59. TP S.A. operates in force of the Telecommunication Law and not of a licence.
60. The legal status of the latter is uncertain.

References

ACWORTH, W.M., (1905), *Elements of Railway Economics*, Clarendon Press, Oxford

BEESLEY, M.E. and S.C. LITTLECHILD (1991), The regulation of privatised monopolies in the United Kingdom, in: VELJANOVSKI, C., ed. (1991b), *Regulators and the Market. An Assessment of the Growth of Regulation in the UK*, IEA Readings 35, IEA, London, p. 29 - 58

BOARDMAN, A. and A. VINING (1989), Ownership and performance in competitive environments: a comparison of the performance of private, mixed and state-owned enterprises, *Journal of Law and Economics*, 19, 1989, p. 1 - 33

COASE, R.H. (1988), *The Firm, the Market and the Law*, The University of Chicago Press, Chicago and London

CURWEN, P. (1986), *Public Enterprise. A Modern Approach*, Wheatsheaf Books, Brighton

DE FRAJA, G. (1991), Efficiency and privatisation in imperfectly competitive industries, *Journal of Industrial Economics*, 39, p. 311 - 321

DEMSETZ, H. (1989), *Efficiency, Competition and Policy*, (*The Organisation of Economic Activity*, vol. II), Basil Blackwell, Oxford

DEMSETZ, H. ([1969], 1989), Information and efficiency: another viewpoint, in: Demsetz (1989), *Efficiency, Competition and Policy*, (*The Organisation of Economic Activity*, vol. II), Basil Blackwell, Oxford, p. 3 - 24

EVANS, R. (1989), Property rights, markets and competition in electricity supply, in: VELJANOVSKI, C., ed. (1989), *Privatisation and Competition. A Market Prospectus*, IEA Hobart Paperback 28, London, p. 135 - 143

FOSTER, C.D. (1992), *Privatisation, Public Ownership and the Regulation of Natural Monopoly*, Blackwell, Oxford

GRAHAM, C. and T. PROSSER (1991), *Privatising Public Enterprises*, Clarendon Press, Oxford

HELM, D. and G. YARROW (1988), The assessment: the regulation of utilities, *Oxford Review of Economic Policy*, vol. 4, no. 2, p. i - xxxi

JASINSKI, P. (1992), The Transfer and Redefinition of Property Rights: theoretical analysis of transferring property rights and transformational privatisation in the post-STEs, *Communist Economies and Economic Transformation*, 4, p. 163 - 189

JASINSKI, P. (1993), *Financial regulation of state-owned enterprises in Central and Eastern Europe. The soft budget constraint under central planning, market socialism and in the process of systemic transformation*, Regulatory Policy Institute, Oxford

JOSKOW, P.L. and R. SCHMALENSEE (1983), *Markets for Power*, MIT Press, Cambridge, Mass.

LITTLECHILD, S.C. (1991), Office of Electricity Regulation: the new regulatory framework for electricity, in: VELJANOVSKI, C., ed. (1991b), *Regulators and the Market. An Assessment of the Growth of Regulation in the UK*, IEA Readings 35, IEA, London, p. 107 - 118

MARSHALL, W. (1990), The future for nuclear power, *Atom*, 400, p. 5 - 13

POLLITT (1992), *Ownership, Scale and Efficiency in the Electricity Generation Industry: A cross-country study*, Paper presented to Industry Group Seminar, Institute of Economics and Statistics, Oxford University, 6 February

PRIMEAUX, Jr. and J. WALTER (1989), Electricity supply: an end to natural monopoly, in: VELJANOVSKI, C., ed. (1989), *Privatisation and Competition. A Market Prospectus*, IEA Hobart Paperback 28, London, p. 129 - 134

ROBERTS, J.; ELLIOTT, D. and T. HOUGHTON (1991), *Privatising Electricity. The Politics of Power*, Belhaven Press, London and New York

ROBINSON, C. (1989), Privatising the energy industry, in: VELJANOVSKI, C., ed. (1989), *Privatisation and Competition. A Market Prospectus*, IEA Hobart Paperback 28, London, p. 113 - 128

SHAPIRO, C. and R.D. WILLIG (1990), Economic Rationales for the Scope of Privatisation, in: SULAIMAN, E.N. and J. WATERBURY, eds. (1990), *The Political Economy of Public Sector Reform and Privatisation*, Boulder, CO, San Francisco, CA and Oxford, Westview Press

SPULBER, D.F., (1989), *Regulation and Markets*, The MIT Press, Cambridge Mass.

SULAIMAN, E.N. and J. WATERBURY, eds. (1990), *The Political Economy of Public Sector Reform and Privatisation*, Boulder, CO, San Francisco, CA and Oxford, Westview Press

SWANN, D. (1988), *The Retreat of the State. Deregulation and Privatisation in the UK and US*, Harvester, Wheatsheaf, New York, London, Toronto, Sydney, Tokyo

VELJANOVSKI, C., ed. (1989), *Privatisation and Competition. A Market Prospectus*, IEA Hobart Paperback 28, London

VELJANOVSKI, C. (1991a), The regulation game, in: VELJANOVSKI, C., ed. (1991), *Regulators and the Market. An Assessment of the Growth of Regulation in the UK*, IEA Readings 35, IEA, London, p. 3 - 28

VELJANOVSKI, C., ed. (1991b), *Regulators and the Market. An Assessment of the Growth of Regulation in the UK*, IEA Readings 35, IEA, London

VELJANOVSKI, C. (1993), *The Future of Industry Regulation in the UK. A Report of an Independent Inquiry*, European Policy Forum for British and European Market Studies, London

VICKERS, J. (1991), Government regulatory policy, *Oxford Review of Economic Policy*, vol. 7, no. 2, p. 13 - 30

VICKERS, J. and G. YARROW (1988), *Privatisation. An Economic Analysis*, MIT Press, Cambridge, Mass.

VICKERS, J. and G. YARROW (1991), The British electricity experiment, *Economic Policy*, 12, pp. 187 - 232

WELFENS, P.J.J. and P. JASINSKI (1994), *Privatisation and Foreign Direct Investment in Transforming Economies. Theoretical and Policy Issues of Systemic Transformation in an International Perspective*, Dartmouth Press, Aldershot

WEYMAN-JONES, T.G. (1991), *Regulating the Privatised Electricity Utilities in the UK*, Paper prepared for the Conference "International privatisation: strategies and practices", University of St. Andrews, 12-14 September

YARROW, G. (1986), Privatisation in theory and practice, *Economic Policy*, 2, p. 323 - 379

YARROW, G. (1991), Vertical supply arrangements issues and applications in the energy industries, *Oxford Review of Economic Policy*, vol. 7, no. 2, p. 35 - 53

List of Tables:

Table A1:	The World's Leading Telecommunication Operators (Ranked by 1992 Revenue)	20
Table A2:	Degree of Digitization of the Telephone Network of European Public Telecommunications Operators	32
Table A3:	Main Lines and Telecommunications Investment for Network Expansion and Network Modernization in EC Countries (1980-90)	34
Table A4:	Development of Optical Fibre Networks of Selected PTOs	36
Table A5:	Planned LEO Satellite Systems	40
Table A6:	Peak Rate Intra-European Telephone Calls between European Member States, December 1991	44
Table A7:	Applications for Terminal Equipment in France 1991	47
Table A8:	World Distribution of Telecommunications Equipment by Type	48
Table A9:	Source of Revenue for Public Telecommunications Operators (1990)	50
Table A10:	Telephone Service Margins in the UK	54
Table A11:	Mainlines per Employee in PTO 1980 - 1994	57
Table A12:	BT Share of UK Public Switched Telephony Market in %	62
Table A13:	Market Shares of UK Telecommunications Carriers (per cent)	63
Table A14:	European Mobile Network Operators: GSM Digital Cellular Tariffs (in ECU)	67
Table B1:	Telecoms Revenue ($R^T/Y=ß$) as Percentage of National Output and Investment per Capita I/P in US \$, 1989	91
Table B2:	Selected International Telecoms Data	92
Table B3:	East European Cellular Telephone Subscribers (January 1993)	93
Table D1:	Telecommunications Administration in CEC	156
Table D2:	Legislative Frameworks	160
Table D3:	Evolution of the Number of Subscribers Connected 1980-1995 in Poland	164
Table D4:	Forecasted Growth of Subscribers Number in 1996-2000 in Poland	165
Table D5:	Development Programs in CEC	166
Table D6a:	Telecommunications Development in the Czech Republic	168
Table D6b:	Telecommunications Development in Hungary	168
Table D6c:	Telecommunications Development in Poland	168
Table D6d:	Telecommunications Development in the Slovak Republic	169

Table D7:	Private Operators in the CEC	176
Table D8:	Privatization Programs	187
Table D9:	Annual Fees Paid by Subscribers	190
Table D10:	Cellular Telephony Subscribers	192
Table E1:	The Process of Increasing the Main Line Density	198
Table E2:	Siemens Public Communication Network Group - Joint Ventures and Subsidiaries in Eastern Europe	209
Table G1:	Comparison of Electricity Prices for Industry in Germany, France and the Czech Republic (1993, DEM per MWh, rounded)	278
Table G2:	Telephone Traffic Domestic PSTN Tariffs in the CR Compared with OECD 1992 Average, USD	285
Table G3:	Main Regulation Features in PSTN in the Czech Republic, Poland and Hungary, 1993	286
Table G4:	Main Regulation Features in PSTN in Selected EU Countries, 1993	287
Table I1:	Shares of Fuel Consumption	312
Table I2:	Electricity Generation Capacity and Production, 1990	314
Table K1:	Industrial Production, Energy Consumption and Power Production in Russia (1990 =100)	366
Table L1:	Representative Indicators for Interconnected Systems, 1989	399

List of Figures:

Figure A1:	Structure of Telecoms Network	15
Figure A2a:	Network Operation - Role of Demand Shifts over Time	16
Figure A2b:	Shift of Cost Curve under Competition vs. Static Scale Economies	16
Figure A3:	Prices under Monopoly, Regulation and Competition	17
Figure A4:	Stages of Privatization in Western Europe	25
Figure A5:	Interconnection of LAN, WAN, MAN and ISDN-B	37
Figure A6:	Development of Existing and Planned Airwave Systems in Germany	39
Figure A7:	Present Situation and Telecommunications Model	53
Figure A8:	Cost-Oriented Tariffing and Welfare Gains	55
Figure A9a:	Cost of International Calls, Average Single Line Business Customer, Average Cost per Call Minute, 1983-1994 (in 1994 UK £)	58
Figure A9b:	Total Cost of Calls, Average Residential Customer, Average Cost per Line, 1983-1994 (in 1994 UK £)	59
Figure A10:	Sunk Costs and Market Entry	61
Figure A11:	Market Entry Barriers and Entry Easing Factors	70
Figure A12:	Market Segments, Likely Competitors and Regulatory Measures	71
Figure A13:	Consequences for Central and Eastern Europe	76
Figure B1:	Structure of the Telecoms Industry	95
Figure B2:	Effects of Transaction Cost	98
Figure B3:	Effect of Reduction of Information Costs in Information-intensive Industry	99
Figure B4:	Network - Competitive Telecoms Market with Different Demand	101
Figure B5:	Telecoms Market with Different Demand under Duopoly	101
Figure B6:	Monopoly Profit and Cost of Reduced Innovativeness	103
Figure B7:	Competition-induced Product Innovation	104
Figure B8:	Average Costs Pricing as a Problem in Industry with Falling Marginal Costs	106
Figure B9:	Monopoly Pricing with Falling Marginal Costs and Tax Credits for Users	107
Figure B10:	Natural Monopoly with Positive External Effects on the Demand Side (Network Effects)	111
Figure B11:	Small Deviation Between Full Cost Recovery and Optimum Supply	112
Figure B12:	Monopoly versus Duopoly	114
Figure B13:	Duopoly: Asymmetric Costs and Competition	116

Figure B14: Effects of Competition in Telecoms Services 128
Figure B15: Organization of Regulatory Agency 129

Figure K1: The Electricity Supply Industry 355
Figure K2: Structure of German ESI 359
Figure K3: Primary Energy Consumption 366
Figure K4: Power Consumption per Capita / GDP 367
Figure K5: Power Production by Source 369
Figure K6: Power Consumption by Sector 370
Figure K7: Power Network 372
Figure K8: Electricity Prices 375
Figure K9: Prices by Type of Tariff 381

List of Contributors

Stephen Economides
Telekom Fachhochschule Berlin

Cornelius Graack
University of Potsdam and
European Institute for
International Economic Relations

Zdenek Hruby
Economics Institute of Academy of
Science, Prague

Piotr Jasinski
Hertford College
Oxford University

Hanna Kontkiewicz-Chachulska
Franco-Polish School of New
Information and Communication
Technologies, Poznan

Petra Opitz
University of Oldenburg

Peter Palinkas
European Parliament, Luxembourg

Wolfgang Pfaffenberger
University of Oldenburg

Anita Ronne
Energy Restructuring Group,
Warsaw

Marius Schwartz
Council of Economic Advisors,
Washington D.C. and
Georgetown University

Marec Béla Steffens
Siemens AG, München

Andrzej Szablewski
Polish Academy of Science, Warsaw

Paul J.J. Welfens
University of Potsdam and
European Institute for
International Economic Relations

George Yarrow
Hertford College
Oxford University

Index

access discrimination 231
access rights 389
accounting separation 317
ACWORTH, W.M. 438,474
additional capacity 399
administration 129
turnkey project 148
access barriers 51
accountability 3
advanced information services 258
advisory board 128
aggregate investment 1
AirTouch 125
airwave services 63
airwaves systems in Germany 39
alternative network 73,171
AMBROSE, W.W. 127,140
American model 304
ANALYSYS 58,59,81
anti-competitive behavior 171
anti-monopoly law 305
Antimonopoly Court 467
Antimonopoly Office 302,304,306, 322,329,463,467
antitrust institutions 429
antitrust laws 453
antitrust tests 225
arbitrage process 86
Argentina 92,440
artificial industry definitions 215
aspiration levels 88
Association of Independent Operators 172
asymmetric costs 115
asymmetrical information 46
AT&T 12
ATTENBOROUGH, N. 54,56,62, 64,81
auction 123,124,223
Australia 91
Austria 395,404
autogeneration 314
average costs 278
average costs pricing 106
awarding exclusive concessions 175

balance of payments 94
Baltic Ring 411,412
BAR, F. 88,140
barriers to entry 113,119,326
barriers to exit 119
barriers to trade 336
BARTH, H. 38,81
BAUER, J.M. 91,141
BAUMOL, W. 29,81
Bavaria 396
Bayernwerk 396,416
BECKMANN, M. 137,140
BEESLEY, M.E. 448,474
Belarus 92
Belgium 24
BELINFANTE, A. 54,82
benchmark 96
benefits of resale 246
BERLAGE, M. 33,77,81,93,140
biased competition 121
bidding 323
bidding procedures 159
big firms 86
Bildschirmtext 51
billing functions 248
BLANKART, C.B. 27,81,140
blocks of power stations 402
BMPT 56,73,81
BOARDMAN, A. 462,474
BOHLIN, E. 73,77,81
BOLCSKEI, I. 179,196
BORRUS, M. 88,140
bottleneck 88,134,204
bottom up 352
breakdown 399,401
breakup of AT&T 253

BRENNON, T.J. 230,239,269
Britain 320,440
British Gas 320,445
British Telecom 444
broadcasting 221
BROCK, G.W. 230,253,256, 265,269
brown coal 312
BROWN, S.J. 251, 252,269
BRUNEKREEFT, G. 390,392
building alliances 86
Bulgaria 91,92
business potential 90
bypass 252

cable rate decontrol 225
cable television provider 61
cable TV 25,85,117,135,175,220
CABY, L. 91,141
California State Commission 311
call rates 87
call-back services 101
capability 399
capacity 87,353
capacity utilization 93,100
capital gains 130
capital markets 90,125,126
capital stock 369
CARSBERG, B. 62,64,81
cartel 382
CAULEY, L. 240,269
CEGB 398,436,442
cellular mobile networks 113,123, 204
cellular telephone subscribers 93
cellular telephony 190,222
Central and Eastern European countries 72,76,153,451, 455,459
central dispatch 363,386
central planning 424,456
central power system 377
centralisation 379
centrally-planned economy 351
CENTREL 412,416
centrel countries 403
CHAPON, J.-P. 127,140
cheap lobby 275
CIS 397,404,405,406
climate 405
CMEA 396
co-operation 245,388,393,400, 406,408
co-operative structure 169
co-ordination 378
coal 272,393
coal and steel 335
coal industry 274
coaled-fired power station 356
COASE, R.H. 469,474
combined heat and power (CHP) 321
COMECON 289,365
commercialization 299
common costs 248,251
common technical protocol 26
communication lines 90
Communications Act of 1934 213
Community action programs 42,78
Community Law 338
comparative advantages 458
compatibility questions 206
competition 17,102,118,135,282, 291,301,305,314,415,319,325,327, 337,339, 428,429
competition for dispatch 324
competition policy 429
competition policy body 113
competition regulation 155
competitive checklist 242
competitive market entry 161
competitive networks 90
competitive safeguards 217,241
competitive tenders 164
competitiveness 88,128
competitors 71

complementary services 112
concept of overlay networks 203
concept of privatization 385
concession fee 93
conciliation procedure 342
CONGRESSIONAL RECORD 213, 269
consequences of delay 329
consistency of telecommunications policy 206
construction units 373
consultants 396
consumer benefits 237
consumer protection 449,468
consumer rent 98
consumer surplus 102,103,104
consumer tax 277
contracts 129
control market entry 171
conventional power plants 272
corporate users 124
cost estimations 315
cost of international calls 58
cost-accounting 155
Cost-of-Service-Regulation (COSR) 230
cost-oriented pricing systems 375
cost-oriented tariffing 55
cost-related tariff structure 155
COSTELLO, N. 30,33,35,39,42,49, 51,83
costs 122,316,336,419
costs and revenues 201
costs of monopoly regulation 215
costs of regulation 18
costs of structural separation 237
Court of Justice 338,340
CRAMTON, P.C. 224,269
CRANDALL, R. 252,256,269
crash-type program 163
cream skimming 250,259
credits 290
Croatia 93
cross-ownership 225
cross-subsidies 249,251,275
cross-subsidization 4,11,27,66,100, 109,118,132,158,217,231,278,286
CSFR 91,92,279
CSIS 140
CSSR 396
current bridges 416
CURWEN, P. 435,474
Czech Republic 153,175,415
CZECH TELECOMMUNICATION OFFICE (CTO) 183,196,283
CZERNIE, W. 405,422

data banks 97
data services 281
data-processing 30
DBP Telekom 35,51,74,81,146
DE FRAJA, G. 462,474
de-monopolizing 337
deadweight loss 103
decentralization 86,379
DECKER, P. 39,81
DECT 135
delegation of powers 327
demand for energy 273
demand side 110
democracy 451
demonopolisation 282
DEMSETZ, H. 474
Denmark 413
Department of Justice 230
DER TAGESSPIEGEL 144,146, 149,151
deregulation 128,200,351
deregulation pressures 26
DEUTSCHE TELECOM E.V. 38, 45,66,67,68,81
DEUTSCHE TELEKOM 145, 146,148,149,150,151,152,440
development programs in CEC 166
development strategy 161
DIE WELT 150,151

different political systems 358
digital cellular tariffs 67
digitization 30,31,87,161
direct competition 56
direct current link 398
direct fiscal revenues 120
directory information 116
discretionary power 124
discriminating against competitors 217
discrimination 241,357
dissemination of information 97
distorted consumer decisions 254
distorted entry decisions 254
distorted prices 272
distorting competition 230
distorting relative prices 236
distribution 318,321,325
divestiture of AT&T 10,62
DOYLE, C. 293
duopoly 62,114,291
Dutch KPN 121
dynamic benefits 215
dynamic economies of scope 108

East Germany 3,73,143
East-West cooperation 5
Eastern Europe 91
Eastern Germany 412
EBRD 120,133,167
EC Commission 9,12,42,43,45, 46,81,82
EC Directives 42
EC Green Paper 282
EC Service Directive 43
EC TACIS-programme 388
EC Treaty 338
economic agents 371,431
economic assistance 396
economic efficiency 251,398,419, 428, 432,433,453
economic growth 88
economic opening up 86,421
economic policy 206
economic regulation 423,430
ECONOMIC REPORT OF THE PRESIDENT 213,269
economic welfare 128
ECONOMIDES, S. 3,143
economies of scale 14,95,105
economies of scope 102,108
ECONOMIST 40,82
ECXC 338
EDF 396
EES 398
effective competition 226
effective price 98
effective price level 98
efficiency gains 126
efficient component pricing 29
efficient economic solution 105
efficient regulation 117
efficient regulatory polic y 127
EIB 167
EINHORN, M.A. 117,140
electric lines 400
electric power 276
electrical power system of Russia 372
electricity 121,346,418
Electricity (Supply) Act 1926 435
Electricity Act 1947 435
Electricity Act 1989 448
Electricity ESI 441
electricity generation 312,313,413
electricity grid company 132
electricity industry 307
electricity networks 397
electricity prices 375
electricity production 369
electricity sector 288
electricity supply industry (ESI) 327,351,355,434
electricity trade 341
electricity transport 314
electricity-oriented products 354

electrification 368
Elektrim 170
ELLIOTT, D. 475
emissions 272
end user prices 286
energy 306
Energy Act 1983 435
Energy Charter 336
energy co-operation 419
energy conservation 368,272
energy distribution 442
energy enterprises 304
energy exports 393
energy generation 442
Energy Law 300,335,343,348
energy lobby 275
energy markets 308
energy policy 335,393,457
Energy Policy Act 361
energy prices 311,347
energy production 311
energy regulation 308
Energy Regulation Authority (ERA) 344
energy reserves 393
energy resources 365
energy sector 274,289,298,299,302, 311,317
energy situation 365
energy supply 442
energy systems 315
energy transmission 442
energy use 311
energy-saving 407
engineering 129
entrepreneurial accountability 163
entry easing factors 70
environment 272,405
environmental damage 311
environmental impact 348
environmental policy 457
equal access 63,234
equipment 468
equipment industry 88,95,120
equipment network 28
equipment prices 100
equipment suppliers 108
ERBE, W. 148,151
ESPRIT 13
established rights 309
Estonia 93
ETSI 63
EU directives 288
Euratom Treaties 338
Europe 19
European Bank for Reconstruction and Development 200
European Commission 307
European Council in Essen 154
European Energy Charter 394,397, 408
European Investment Bank (EIB) 97,133,200
European Mobile Network Operators 67
EUROPEAN PARLIAMENT 78, 82,422
European Union 9,307,335,348,362, 411,455
EVANS, R. 470,474
ex-CMEA countries 11,73
excess demand 85,87
exchange of information 87
exchange rate deviation index (ERDI) 277
expansion 34
expansion of trade 86
external economies 356
external effects 109

failure rate 121
fair profits 274
falling average costs 369
fault rates 131
FAZ 150,151
FCC 131

FDI promotion 110
Federal Energy Regulation Commission 361
Federal Power Commission 374
fibre optics 35
fibre to the home 35
fibre-optical network 102
final energy consumption 312
finance 272
financial conditions 382
financial holding 386
financial obligations 382
financial requirement 278
financial services 86
FINANCIAL TIMES 28,35,60,62, 63,67,68,82
financing 201
financing of telecoms 90
Finland 136,404,413
fiscal aspects 120
fixed-link segments 2
flexibility 3
flexibility in pricing 387
flexible tariff policy 202
foreign consortia 174
foreign currency earnings 404
foreign direct investment 19
foreign exchange 367
foreign investment 86,210,221,287
foreign investors 96,124,288
foreign participation 162
foreign partner 287
FOREMAN-PECK, J. 27,82
Forum on the Information Society 154
fossil resources 370
FOSTER, C.D. 435,438,439,445, 447,470,474
framework conditions 394,406,407
France 91,287,363,404
FRANCE TELECOM 44,51,82
franchises 124
free trade 335
freedom of establishment 339
FREEMAN, C. 154,196
frequency 400
frequency asynchronism 403
frequency management 129
frequency spectrum management 283
frequency variations 395
FROHBERG, W. 36,82
fuel consumption 312
full facilities 245
funding of universal service 257,259

gas 307,312,319
Gas Act of 1986 448
gas distribution 407,409
gas enterprises 346
gas industry 405,445
gas infrastructure 407
gas pipelines 408
gas restructuring 320
gas sector 295,404,420
GATT 338
Gazprom 406
general competition or antitrust law 447
German Democratic Republic 3,73, 143
German unification 146
Germany 91,287,358,404,440
Gesetzstaat 427
GIBBS, D. 212
global business communication 87
global economic competition 197
global price caps 235
GLUCK, G. 180,196
GOELRO plan 368
golden share 136
GORE, A. 225,269
GOSPLAN 371
government 123
government barriers to entry 216
government control 377

government intervention 256,354
government procurement 95
government regulation 360
GOVERNMENT RESOLUTION 196
GRAACK, C. 2,9,41,72,82,83, 141
grace period 112
gradual process 317
gradual strategy 131
GRAHAM, C. 472,474
GRANDE, E. 206,212
GRANSTRAND, O. 73,77,81
granting licences 342
Great Britain 307,310,352,364, 377
Greece 24,287
Green Paper 13
Green Paper on Telecommunications 42
GREENE 332,333
grid 349
grid company 85,321
grid subsector 295
grid supply 314
gross demand 98
ground rules 400
group-specific networks 14
GRUPP, H. 37,82
GSM 38,121

HABUDA, J. 97,140
HANCHER, L. 332,333
HANDELSBLATT 151
hands-off deregulation 261
hard coal 312
hardware bottleneck 94
harmonising policies 5
harmonization 42,337
HARPER, J.M. 37,82
HAUSMAN, J. 54,82
HAY, D. 332,333
HDTP 45,53,82
heat generation 272,384
HELM, D. 448,471,474
HENNEMEYER, P.R. 127,140
HENRY, C. 13,82
HESOUN, F. 293
high definition television 37
high sunk costs 109
high voltage transmission 322
HITCH-SWEEZY demand curve 111
HOPE, E. 332,333
horizontal integration 5
HOUGHTON, T. 475
HRUBY, Z. 4,189,196,271,293
Hungary 92,93,94,121, 153,173,396,415
HUNYA, G. ET AL. 89,140
HVDC 413
hydro 314
hydro plants 272
hydropower 385
access charges 117

IEA 390,391,392
import competition 110
imported energy 272,376,393
imports 94
incentive regulation 46,124
incentives for cost reduction 214
income transfers 123
incumbent's monopoly power 194
incumbent utilities 307
independent actors 387
independent body 171
independent companies 169,379
independent control 171
independent private enterprises 386
independent producer companies 389
independent regulation 304,330
indirect revenue effects 120
individual price agreements 360
industrial consumers 375

industrial production 366
industrial restructuring 5,308,317
industrial structure 325
industrial users 343
industry-specific regulation 6
inefficiencies 311
information 118,316,325
information costs 97,98
information society 87,154
information-intensive industry 98
infrastructure fund 133
infrastructure networks 89
innovation 26,103
innovation dynamics 118
innovative regulation 102,128
innovative use 119,125
innovativeness 1
input prices 272
instability of political system 377
installed capacity 399
institutional choice 351
institutional reform 371,456
institutional set-up 353,359
institutional structure 425,454,465
integrated services digital network33
intellectual property184
inter-war period 394
interconnected system 396,402
interconnection 247,417,468
interconnection agreement 243
interconnection charges 14,63,130, 131
interconnection rules 63
Interexchange Carrier 228
internal energy market 336,341
international alliances 88,133
international competitiveness 1,2, 10,458
international consortia 23
INTERNATIONAL ENERGY AGENCY 332,333
international experience 328,330
international services 131
international telecommunication 87,283
international trade 134
internationalization 86,121
internationalization of business 127
internationalization of the economies 119
Interstate Commerce Commission (ICC) 439
intra-European long distance call 43
intraLATA 228
investment 118,380,396
investment costs 414
investment evaluation 413
investment per capita 90
investment regulation 361
investment tax 380
investors 314
IPS 398,412
IRWIN 131,140
Italy 22,91,287,404
ITU 32,80,82, 138,139,140

Japan 19,91
JASINSKI, P. 6,171,196,462,469, 474
joint ventures 410
joint-stock companies 273,280
JOSKOW, P.L. 332,333,474
JURZYCA, E. 188,196
justified costs 276

KARTTE, W. 392
Kazakhstan 92
KAZMIN, A. 390,392
KELLY, T. 91,141
key elements necessary to attract investor 193
KITUSHIN, V.G. 391,392
KLEIN, M.W. 87,140
KLOPFER, T. 390,392
KNIEPS, G. 14,27,81,82,133,140
knowledge 128

KOMMERSANT 390,391
KONTKIEWICZ-CHACHULSKA, H. 3,153,186,196
KRUSE, J. 123,140
KUBASIK, J. 121,140
KÜNZI, H. 137,140
KWEREL, E.R. 223,269

lack of transparency 255
LAFFONT, J.-J. 29,82,235,251, 252,269
LAMB 332,333
land-based communications 38
large generators 373
Latvia 93
law of contract 427,452
law of one price 44
leading telecommunication operators 20
learning by doing 109
leased lines 134
LEBEOUF 332,333
legal basis 305
legal entry barriers 219
legislative frameworks 160
LEIPOLD, H. 390,392
LEITE, A.P.N. 293
LEWIS, W. 267,269
liberal democracies 427
liberalization 89,312,362,372
liberalizing legislation 307
licence fees 206
licence system 345
licences 123,130,349
lignite 272,314
limit of foreign capital 157
line rentals 126
liquid fuels 277
Lithuania 93
LITTLECHILD, S.C. 448,471,474
local administrative bodies 378
Local Area Networks (LAN) 36
local competition 241
local content 164
Local Exchange Carriers (LECs) 228
local telephone service 228
locational advantage 128
long-distance carrier 228
long-term contracts 324
lottery 123,223
low earth orbit satellite systems 39
LRMC 278
LUCAS, N. 390,392
LUCENA, D. 293
LUTZ, E. 40,82

Maastricht Treaty 336
MACRAE 332,333
macroeconomics stability 154
MagyarCom 178
MAIJ-WEGGEN, H. 63,82
main lines 92
major industrial entreprises 374
majority ownership interest 180
management 186
management buy-outs 187
marginal cost price 364
marginal cost pricing 29
marginal costs 17,98,102,214
market economies 431
market economy 199
market entry 109,116
market entry barriers 61,70
market exit 116
market exit barriers 68
market failures 433
market for energy 343
market for information 109
market potential 94
market power 311,327
market segments 71
market self-regulation 427
market-oriented set-up 356
market-oriented structures 376
marketing 186,318

marketisation 424,453
markets 431
MARSHALL, W. 470,474
MATAV 121,158,460
MERCER 32,82
Mercury Communication 444
MEREKOV, A.P. 390,392
meritorial good 10
Metropolitain Area Networks (MAN) 36
MEYER, J.R. 27,83
micro-electronics 30
Middle East 420
ministerial control 193
Ministry of Economics 395
MINISTRY OF ECONOMY 175,196
Ministry of Energy 372
Ministry of State Security 144
Ministry of Telecommunications 468
Minitel 51
MITCHELL, B.M. 100,140
mixed ownership structure 461
MMC 87,446
mobile network competition 66
mobile services 281,284
mobile telecommunication 2,21,38, 93,123
model of regionalisation 376
modernization 6,98,165,405
monetisation 453
money supply 382
MONITOR 150,151
monopolies 87,337
monopolist 273
monopolist behavior 100
monopolistic competition 216
monopolistic parts 326
monopolistic practices 172,347
monopolistic pricing 109
monopolistic wholesale market 384
monopolization 113,327
MONOPOLKOMMISSION 74,83
monopoly 17,90,102,113,315
monopoly price 105
monopoly pricing 107
monopoly rent 116
MORRIS, D. 332,333
Morrisonian model 435
MÜLLER, J. 27,82
multinational companies 12,19,96, 132,199
multinationalization 96
mutual indebtedness 382

NATA 47,83
NATIONAL AUDIT OFFICE 126, 140
national energy policy 344
National Grid Company (NGC) 320, 442
nationwide standard 220
natural gas 273,276,277,312,393, 404,407,409,410
natural gas consumption 405
natural gas markets 336
natural gas supply 404
natural monopoly 2,4,102,105,214, 299,304,441
NEDO report 436
need for communications 88
NEKRASOV ET AL. 369,390,392
net demand 98
Netherlands 52
network access 98
network competition 95,117,131
network development 85
network digitization 31
network effects 111
network electricity 357
network expansion 90,96,105,108, 130
network externalities 95,105,110
network growth 28

network liberalization 121
network modernization 34
network operation 134
network operator 94
network outsiders 116
network performance 280
network unbundling 246
network-oriented energy sector 353
network-oriented sectors 351
new compensation rules 65
new entrants 116
new projects 406
new technologies 23
new technologies 88,113
new transmission techniques 30
New Zealand 91
Newly Industrialized Countries 88
NGC report 442
NOAM, E. 77,83
nomenklatura 461
non-profit organisation 386
nondiscrimination 261
nonpecuniar costs 98
NORDEL 397,412
North America 19
nuclear energy 371
nuclear plants 272
nuclear power 312
nuclear power station 374
NULTY, T.E. 212
number of licences 66,67
number portability 25,63,136,245
numbering plans 283

OBERPOSTDIREKTION 149,151
obligation to supply 360
obligations 346
OECD 32,34,36,50,51,55,56,77, 83,85,88,91,367,279,308,370
Ofgas 445
OFTEL 287,291,445
oil 393
oil imports 311
oil prices 272,336
oligopolistic networks 112
oligopoly 111
OMSYC 48,83
once-and-for-all subsidy 65
one-stop-shopping 246
ONP principles 46
open access standards 26
open network provision 13,42
opening-up of government procurement 21
operator services 244
OPITZ, P. 5,351
opportunity costs 124
optical networks 35
optimal allocation 109
optimal resource allocation 15
Optimus 170
organisational structures 409
organizational solutions 298
Ostpolitik 144
ownership separation 318

Pactel 125
PALINKAS, P. 5,339,394,409,422
PAP, Z. 199,212
parallel concessions 175
partial facilities 245
partial-facilities entry 246
particular power station 378
pay-on-demand systems 19
pay-per-view 19,37
peak demand 399
peak load 399
peak power 414
peak-time phone calls 10
PEARL, D. 268,269
penetration rate 279
PEREZ, C. 154,196
peripheral regions 98
personal communication services 224

personal communications network (PNC) 38
personnel development 378
petroleum 312
petroleum products 339
PFAFFENBERGER, W. 5,351, 392
PHAN, D. 153,186,196
PHARE 120,394
pipeline networks 321
pipelines 272,408,411
Poland 3,74,91,92,93,94,121,130, 169,296,297,303,308,311,320,322, 329,335,337,343,347,396,413,415, 424,457, 460,462
policy for developing telecommunications 163
policy gains 320
policy of „non-communication" 143
policy options 119
Polish ESI 457,462
Polish Oil and Gas Company (PGPN) 312
Polish Power Grid Company (PPGC) 313
Polish TPSA 121
political constraints 5
political decisionmakers 377
political instability 388
political side-effects 328
political systemic transformation 451
political uncertainties 117
POLLITT 436,474
Polskie Sieci Elektorenergetyczne (PSE) 457
population density 123
Portugal 91,287
position of the community 78
positive externalities 2
positive network externalities 17
POSPISCHIL, R. 47,83
potential competitors 227
power consumption by sector 370
power network 362,372
power plants 401
power pools 323
power production by sources 369
power stations 362,400
power subsector 277
power supply 402
power transmission system 412,415
PRAVEDNIKOV, N.K. 391,392
precise interconnection rules 171
Preußen-Elektra 396
preventing access discrimination 239
price adjustment problem 316
price cap regulation 46,56
price caps 232,446
price control 346,349
price distortion 11,54
price elasticity 90
price fixing 379
price law 274
price liberalisation 275
price reform 316
price regulation 109,117,118,231, 274
price subsidy 275
price transparency 342
price-cutting 323
price-structure problem 315
prices 276,278,285,289,297,316, 323,328,374,381
prices of fuels 275
pricing 116,346
pricing methods 277
pricing policies 126
pricing regulation 123
primary domestic supply 312
primary energy 273
primary energy consumption 366, 404
PRIMEAUX, Jr. 470,475

principles of average tariffing 54
private capital 299,331
private operators in the CEC 176
private ownership 432
private property 427,431
private sector 459
privatisation 22,88,89,119,131, 178,200,280,282,286,297,307,314, 337,372,373,424,425,432,440,444, 453,459
privatization of banks 126
privatization of state enterprises 154
privatization programs 183,187
privatization stages 23
privatization strategy 178,188
pro-competitive commitment 462
pro-competitive regulation 445
pro-competitive regulatory reform 213
pro-competitive restructuring 441, 458,467
problem of deregulation 326
problem of interconnection 172
process of privatization 378
procurement policies 47
producer rent 98
producer surplus 104
product differentiation 114
product function 87
product innovation 104,214
productivity 57,86,107,128,318
productivity gains 291
productivity growth 325
profit margins 324
profit-maximization strategy 3
program of privatisation 373
project financing 410
promote local competition 230
property rights 69,309,352,427
proprietary know-how 116
PROSSER, T. 472,474
PSTN 279,287
PTT 83
PTT employees 127
public bodies 360
public choice theory 433
public enterprises 339
public interest theory of government 433
public intervention 296
public ownership 434,455
public procurement 337
public utilities 113,432,441,455
Public Utilities Regulatory Policy Act of 1978 361
publicly owned public utilities 458
purchasing power parity (PPP) 277
Purneftgaz 410

quality 87,121
quality of services 341

R&D activities 110
R&D costs 21
R&D intensity 110
R&D programs 13
RACE 13
radio communications 124
radio spectrum 222
Railway and Canal Traffic Act, 1888 438
railways 121
Ramsey pricing 251
rate of return regulation 117
rating 290
re-regulation 351
rebalancing 126
rebalancing of prices 54
rebalancing of tariffs 132
Rechtsstaat 427
redistribution 103
redistributive effects 316
reform process 376
reforming universal service 255
reforms 295,297,407

Regional Bell Operating Companies 213
regional cost structures 380
regional electricity administrations 371
Regional Electricity Companies (RECs) 320,442
regional energy commission 380
regional energy systems (RES) 373
regional interests 377
regional market integration 96
regional monopolies 321,354,361
regional price differences 384
regional separation 359
regulated prices 100
regulation 17,113,271,282,283,287, 288,296,298,300,304,306,314,315, 320,325,326,327,344,428,432,433, 440,464
regulation costs 114
regulation for competition 447
regulation of public utilities 434,437
regulation sensu largo 454
regulation versus deregulation 216
regulator 118,286,345
regulatory agency 113,326,440, 445,461,466
regulatory asymmetry 235
regulatory authority 303,304
regulatory body 274,290,344
regulatory commission 329
regulatory developments 40,41
regulatory discretion 242
regulatory framework 11,295, 331,348
regulatory functions 301
regulatory institution 301
regulatory issues 295
regulatory measures 71
regulatory policy 128,276,326
regulatory processes 328
regulatory reform 213,216,271, 307,308,310
regulatory regime 305,349
regulatory structure 302,430
regulatory system 303,348
reinvestment 378
reliability 121
rent-seeking 18,118
reorganisation 96,395
resale of capacity 132
research & development 88
research capacities 373
reseller 245
residential users 124
resistance of employees 120
restructuring 289,307,319,321, 405,456
restructuring in generation 321
retail supply 318
retailing functions 248
returns on investment 201
revenue aspects 26
revenue for public telecommunication operators 50
revenue per main line 130
reversible costs 60
RHEINISCHE POST 145,151
RICKE, H. 150,151
rights-of-way 219
ROA EES Rossij 413
ROBERTS, J. 436,470,471,475
ROBINSON, C. 444,475
Romania 91,92,122
RONNE,A. 5,335
ROSE, N.L. 332,333
Rosstrojgazifikacijy 409
rotation of members 303
RPI-X regulation 48
RUD, L. 332,333
Ruhrgas AG 407,409,410
rule of law 426,464
rule of law and competition 450
Russia 92,93,351,364,411,413,415
Russian bank 388

Russian Company for Electricity 373
Russian Federation 406
RWE 396

safety standards 415
SANEEV, G. 390,392
satellite communications 39
satellite networks 88
satellite systems 102
scale economies 110,214
scarce frequencies 123
scarce resource 222
scarcity-oriented prices 358
SCHMALENSEE, R. 332,333,474
SCHNÖRING, T. 13,33,77,81,83, 93,140
SCHRÖDER, K. 390,392
SCHULZ, W. 390,392
SCHWARTZ, M. 3,218,261,270
screening evaluation 123
SEABRA, C. 293
second-best 330
secondary energy law 340
secondary monopolization 136
sector-specific interests 387
sector-specific unions 385
security of supply 336,341,398,419
SEDA, R. 189,196
separate accounting 134
separate affiliates 242
separation of property rights 363
separation of trade and distribution 357
service price 87
service providers 94,108
service users 102
services sector 279
shadow price 88
SHAPIRO, C. 469,475
Siberia 366
Siberian Energy Complex 383
SIBLEY, D.S. 251,252,269
SIDAK, G. 29,81
SIEMENS 92,140,209,422
signaling and control systems 244
Singapore 440
Singapore Telecommunications 440
SINGH, B. 332,333
single buyer 363
single European market 10,13
sliding scale regulation 46
Slovak Republic 93,153,415
Slovak Telecommunications 186
social acceptability 276
social aspects 123
social policy 300
social problems 126
social utility 109
social welfare 461
socialist command economy 199
socialist democracy 450
SOETE, L. 37,82
soft budget constraints 311
solid fuels 273,312
South East Asia 197
Soviet Union 271,311,368,404,450
Spain 91
special institutions 431
special networks 203
spinning-off 125
SPT Telecom 280
SPULBER, D.F. 439,470,475
stages of privatization 25
stand-alone costs 252
standardization 116
standardization interconnection 132
Stasi 144
state aid 337,339
state budget deficits 23
state control 274
state equipment firms 94
state investment 380
state monopolies 339
state property rights 386

state public utility commission (PUC) 243
State Regulatory Commissions 361
state-control property rights 387
state-owned entreprises (SOEs) 452
static losses 215
station maintenance 313
STE 450
steamless network 26
STEFFENS, M.B. 3,197,205,212
STEINFIELD, C. 91,141
stock market 273
STOCK, W.G. 97,140
strategic alliances 19
strategic foreign investor 120,131
strategic foreign partner 188
strategic industry 97
strategic investors 96,164,203
strategic planing 186
structural decisions 324
structural separation 234,318
structure of prices 301
structure of telecoms network 15
structure of the telecoms industry 94
sub-additive cost function 40
sub-additive costs 14,113
subscriber charge 87
subsidies 109,272,376,318,337
subsidised tariffs 387
subsidization 96
subsidized tariffs 381,382
substitution of electricity 354
SUK, J. 293
SUNDRUM, L. 197,212
sunk costs 60,61,113
supra-normal profits 10
surplus 402
SURREY, R.S. 185,196
SWANN, D. 439,475
Sweden 11,59,91,413
Swiss PTT 121
switches 244
switching stations 400
Switzerland 404
synchronous connection 403
systemic transformation 88,135, 295,423,424,449,455
systemic transition 200
SZABLEWSKI, A. 5,295

TACIS 394
TARDIFF, T. 54,82
tariff adaptations 202
tariff policies 271
tariff policy reform 184
tariff regulation 284
tariff structure 182,284,346
tariffs 271,282
tax 277
tax credit 107,110,133
tax receipts 26,120
tax revenues 128
TAYLOR, L.D. 62,83
TAYLOR, P. 390,392
TAYLOR, W.E. 62,83
technical characteristics 284
technical compatibility 63
technical feature 403
technical issue 398
technical minimum standards 63
technical standards 116
technical tests 47
technological changes 30,215
technological discrimination 239
technological dynamics 30,85
technological economies of scope 238
technological progress 15,100
technologies 311
technology transfer 132
technology-intensive branches 210
telecom modernization 146
Telecommunications Act of 1964 159
Telecommunications Act of 1990 157

Telecommunications Act of 1992 158
Telecommunications Act of 1996 213
telecommunications administration 155,156
telecommunications channels 413
telecommunications equipment 11, 47,242
telecommunications pricing 189
telecoms aid rules 96
telecoms business 129
telecoms equipment industry 94
telecoms network 94
telecoms privatization 120
telecoms regulation 260
telecoms revenue 90,91
telecoms services 94,128
telecoms services industry 94
Telefonica 122
Telefonica International 121
Telekom 2000 147
Telekomunikacja Polska SA 156, 457,462
telephone competition 228
telephone density 121
telephone penetration 122
telephone service margins 54
telepublishing 51
telework 51
TELIA 60,83
TelSource 184
TEN-project 411
tender for GSM concessions 191
thermal production 402
THIEME, H.J. 390,392
third-party access (TPA) 29,307, 321,326,340,342,347,355,360,363
time-oriented pricing 358
TIROLE, J. 29,82,235,251,252, 269,332,333
top down 352
total cost of calls 59
TPSA 130
tradables sector 88
trade in energy 405
trade-offs 327
tradeable frequency rights 68
trading 318
training programme 184,409
trans European networks 340,411, 413,421
transaction costs 223,358,427,433
transeuropean telecom networks 9
transformation 88
transformation program 337
transforming economies 85
transition department 129
transmission 318
transmission companies 319
transmission facilities 309
transmission infrastructure 413
transmission of data 11
transmission system 413
transparency 3,86,303,346
transparency of decision-making 387
transparent criteria 349
trunk lines 244
trunk networks 88
trunk operations 130
trunk services 63
Turkey 92
turnkey projects 134
TYLER, M. 122,140
type approval 42
types of entrants 244

UCPTE 412
UCPTE network 395,397
UCPTE system 396,400,401
Ukraine 92
unbundling 131,247,342,363
unbundling of marketing 248
unbundling regulated bottleneck monopolies 233

UNCE 140,194
under-investment 279
undue benefits 274
UNGERER, H. 30,33,35,39,42, 49,51,83
uniform pricing 98
uniform subscriber pricing 98
United Kingdom 11,91,287,288, 417,434,440,465
United Stated 91,302,307,310,361, 439,440,465
universal service 123,128,218, 249,256
universal service obligations 65
universal service principles 45
universal services access 163
Uniworld 184
unprofitable markets 257
unregulated operations 217
UPS 398
upstream companies 319
Uruguay 92
US Bell Operating Companies (BOCs) 60
US market 3
use of energy 409
USSR 371,396,406,420

value added tax 381
value productivity 130
value-added services 11,49,89,127, 134
VAN AS, H.R. 38,83
VEAG 416
VED 398
VELJANOVSKI, C. 436,447, 448,449,472,474,475
vertical integration 5,112,317,359
vertical separation 325
vertical structures 325
vertically integrated monopolistic structures 376
veto rights 178
VICKERS, J. 332,333,469,471, 472,475
video conferencing 51
videotelephone 37
VINING, A. 462,474
Visegrad countries 3,96,121
VOGELSANG, I. 140
VON WEIZSÄCKER, C.C. 27, 49,83
voucher privatisation 273
voucher privatization 183,193

waiting lines 90
waiting list 131,279
WALKE, B. 39,81
WALTER, H.-P. 143,52
WALTER, J. 470,475
WAS 148,152
WATERBURY, J. 475
WAVERMAN, L. 252,256,269
welfare 98
welfare analysis 104
welfare gains 54,55
WELFENS, M. 422,475
WELFENS, P.J.J 9,12,41,83,85, 87,140,141,394,422,469,475
WELLENIUS, B. 129,141
Western Europe 370
Western Germany 404
Western Siberia 410
WEYMAN-JONES, T.G. 442, 443,470,475
wholesale business 100
wholesale discount 248
wholesale market 132
wholesale supply 318
Wide Area Networks 36
WILLIAMS, J.R. 223,269
WILLIG, R.D. 332,333,475
WILSON, R.W. 27,83
WINSTON, C. 117,141
wireless communications 170
wireless communications 280

wireless local loop 204
wireless transmission 21
wireline local loops 244
WORLD BANK 92,96,120,130,
167,200,332,333
World Bank Group 133
WORLD DEVELOPMENT
REPORT 92
World Trade Organization (WTO)
222
World War II 364

X-inefficiencies 15

Yamal field 408
yardstick competition 131
YARROW, G. 5,307,332,333,448,
469,470,471,472,474,475
YPSILANTI, D. 91,141
Yugoslavia 91,92

MIX
Papier aus verantwortungsvollen Quellen
Paper from responsible sources
FSC® C105338

If you have any concerns about our products,
you can contact us on
ProductSafety@springernature.com

In case Publisher is established outside the EU,
the EU authorized representative is:
Springer Nature Customer Service Center GmbH
Europaplatz 3, 69115 Heidelberg, Germany

Printed by Libri Plureos GmbH
in Hamburg, Germany